Quality Management

Tools and Methods for Improvement

THE IRWIN SERIES IN PRODUCTION OPERATIONS MANAGEMENT

Aquilano and Chase
Fundamentals of Operations Management, 1/e

Chase and Aquilano
Production and Operations Management, 6/e

Berry et al.
ITEK, Inc., 1/e

Hill
Manufacturing Strategy: Text & Cases, 2/e

Klein
Revitalizing Manufacturing: Text & Cases, 1/e

Lambert and Stock
Strategic Logistics Management, 3/e

Leenders, Fearon, and England
Purchasing and Materials Management, 9/e

Lofti and Pegels
Decision Support Systems for Production & Operations Management for Use with IBM PC, 2/e

Nahmias
Production and Operations Analysis, 2/e

Niebel
Motion and Time Study, 8/e

Sasser, Clark, Garvin, Graham, Jaikumar, and Maister
Cases in Operations Management: Analysis & Action, 1/e

Schonberger and Knod
Operations Management: Continuous Improvement, 5/e

Stevenson
Production/Operations Management, 4/e

Vollmann, Berry, and Whybark
Manufacturing Planning & Control Systems, 3/e

Whybark
International Operations Management: A Selection of Imede Cases, 1/e

THE IRWIN SERIES IN STATISTICS

Aczel
Complete Business Statistics, 2/e

Duncan
Quality Control & Industrial Statistics, 5/e

Cooper and Emory
Business Research Methods, 5/e

Gitlow, Oppenheim, and Oppenheim
Quality Management: Tools and Methods for Improvement

Hall
Computerized Business Statistics, 3/e

Hanke and Reitsch
Understanding Business Statistics, 2/e

Lind and Mason
Basic Statistics for Business and Economics, 1/e

Mason and Lind
Statistical Techniques in Business and Economics, 8/e

Neter, Wasserman, and Kutner
Applied Linear Statistical Models, 3/e

Neter, Wasserman, and Kutner
Applied Linear Regression Models, 2/e

Siegel
Practical Business Statistics, 2/e

Webster
Applied Statistics for Business and Economics, 1/e

Wilson and Keating
Business Forecasting, 2/e

THE IRWIN SERIES IN QUANTITATIVE METHODS AND MANAGEMENT SCIENCE

Bierman, Bonini, and Hausman
Quantitative Analysis for Business Decisions, 8/e

Knowles
Management Science: Building and Using Models, 1/e

Lofti and Pegels
Decision Support Systems for Management Science & Operations Research, 2/e

Stevenson
Introduction to Management Science, 2/e

Turban and Meredith
Fundamentals of Management Science, 6/e

Second Edition Quality Management

Tools and Methods for Improvement

Howard Gitlow
University of Miami

Alan Oppenheim
Montclair State University

Rosa Oppenheim
Rutgers The State University of New Jersey

Irwin
McGraw-Hill

Boston, Massachusetts   Burr Ridge, Illinois   Dubuque, Iowa
Madison, Wisconsin   New York, New York   San Francisco, California   St. Louis, Missouri

Irwin/McGraw-Hill

A Division of The **McGraw·Hill** *Companies*

In recognition of the fact that our company is a large end-user of fragile yet replenishable resources, we at IRWIN can assure you that every effort is made to meet or exceed Environmental Protection Agency (EPA) recommendations and requirements for a "greener" workplace.

To preserve these natural assets, a number of environmental policies, both companywide and department-specific, have been implemented. From the use of 50% recycled paper in our textbooks to the printing of promotional materials with recycled stock and soy inks to our office paper recycling program, we are committed to reducing waste and replacing environmentally unsafe products with safer alternatives.

The earlier edition was published under the title: *Tools and Methods for the Improvement of Quality*.

© RICHARD D. IRWIN, INC., 1989 and 1995

Senior sponsoring editor: *Richard T. Hercher, Jr.*
Editorial assistant: *Gail Centner*
Marketing manager: *Brian Kibby*
Project editor: *Denise Santor-Mitzit*
Production supervisor: *Laurie Kersch*
Interior designer: *Mike Warrell*
Cover designer: *PIR Design Group*
Art coordinator: *Mark Malloy*
Compositor: *Bi-Comp, Inc.*
Typeface: *10/12 Times Roman*
Printer: *R. R. Donnelley & Sons Company*

Library of Congress Cataloging-in-Publication Data

Gitlow, Howard S.
   Quality management : tools and methods for improvement / Howard
Gitlow, Alan Oppenheim, Rosa Oppenheim. — 2nd ed.
      p.   cm. — (Irwin series in statistics)
   Rev. ed. of: Tools and methods for the improvement of quality /
Howard Gitlow ... [et al.]. 1989.
   Includes bibliographical references and index.
   ISBN 0-256-10665-7
   1. Quality control—Statistical methods.   I. Oppenheim, Alan
V., 1937- .   II. Oppenheim, Rosa.   III. Title.   IV. Title: Tools
and methods for the improvement of quality.   V. Series.
TS156.T587   1995
658.5′62′015195—dc20   94–15814

Printed in the United States of America
4 5 6 7 8 9 0 DOC/DOC 0 9 8 7 6 5 4 3

Dedicated to our families:

Ali Gitlow
Adam and David Oppenheim

Shelly Gitlow

Beatrice and Abraham Gitlow
Sylvia and Norman Oppenheim
Esther and Aaron Blitzer

Preface

Never-ending quality improvement is now recognized as essential for any organization's survival. Leading corporations have demonstrated that improved quality raises profits, reduces costs, and improves competitive position. Government agencies and other not-for-profit organizations have begun to reap the benefits of continuous quality improvement. The seemingly geometric growth in interest in quality bodes well for the future. Meaningful progress requires knowledge in many areas. We attempt here to present a unique and workable approach to the tools and methods necessary for real quality improvement. Furthermore, we believe that a comprehensive approach to this subject must consider the philosophy of the late W. Edwards Deming and have woven that philosophy into this text.

This book is one of a very select few quality control texts that adopt Shewhart's and Deming's views of control charts. Shewhart approached control charts by creating an acceptable economic rule to distinguish between common and special causes of variation. The control limits he used are created using a process statistic plus and minus three times the standard error of that statistic. He demonstrated that almost all processes will generate statistics within those control limits, and that this will be true for a wide variety of distributions of those statistics. Consequently, Shewhart avoided probability models when constructing control charts. It should be pointed out that the traditional paradigm of control charts (which relies on probability models) often makes overly precise and inaccurate statements about the future behavior of a process.

Deming added that it's impossible to obtain a sampling frame in a study whose aim is to take action on the future behavior of a process. This is because a sampling frame must include all past, present, and future observations. Clearly the future observations don't yet exist and can't be measured. Since the sampling frame is unavailable, we can't quantify the two types of errors that can result from action on a process: (1) decide not to change or adjust the process when it should be modified or (2) decide to modify the process when it should be maintained as it is. Any inference in this type of study is conditional on the environmental state of the process when the sample was selected, and that environmental state will never exist again. Information on such a problem can never be complete. As a result, Deming created an alternative view for analyzing and interpreting control charts. Predictions of a process's future functioning are based jointly on process knowledge and analysis of past process data, not on probability models that assume the existence of a sampling frame. This text adopts the views of Drs. Shewhart and Deming.

Unlike many other currently available publications, this book

1. Focuses on Deming's theory of management and ties all discussions of statistical topics to that theory.
2. Distinguishes between enumerative statistical studies (studies whose purpose is to take action on a population) and analytical statistical studies (studies of processes whose purpose is to take actions that will improve the process's functioning).
3. Focuses attention on modern inspection policies (Deming's kp rule) as opposed to traditional acceptance sampling procedures.
4. Provides detailed quality improvement stories (an effective format for presenting and working on statistical process improvement efforts).
5. Has many examples and mini–case studies to help readers understand and appreciate the topics covered.

This book can be used for three different levels of education. The basic level covers Chapters 1 ("Fundamentals of Quality"), 2 ("Fundamentals of Statistical Studies"), 3 ("Defining and Documenting a Process"), 4 ("Basic Probability and Statistics"), 5 ("Stabilizing and Improving a Process with Control Charts"), 6 ("Attribute Control Charts"), 7 ("Variables Control Charts"), 9 ("Diagnosing a Process"), and 11 ("Process Capability and Improvement Studies"). The intermediate level covers all chapters in the basic level plus Chapters 8 ("Out-of-Control Patterns") and 10 ("Specifications"). The advanced level covers all chapters in the basic and intermediate levels plus Chapters 12 ("Taguchi Methods—Quality Improvement in Product and Process Design"), 13 ("Inspection Policy"), 14 ("Deming's 14 Points and the Reduction of Variation"), and 15 ("Some Current Thinking about Statistical Studies and Practice").

We acknowledge and thank the late Dr. W. Edwards Deming for his philosophy and guidance in our personal studies of quality improvement and statistics. We also thank those individuals who helped by reviewing this second edition: Bruce Christensen (Weber State University, Ogden, Utah), Mark Hanna (Miami University, Oxford, Ohio), Carol Karnes (Clemson University, Clemson, South Carolina), J. Keith Ord (Pennsylvania State University, University Park), and Herbert F. Spirer (University of Connecticut, Stamford). Additionally, we thank those who helped by reviewing the first edition: Robert F. Hart (University of Wisconsin, Oshkosh), Chandra Das (University of Northern Iowa, Cedar Falls), Donald Holmes (Stochos Incorporated), Peter John (University of Texas, Austin), Sudhakar Deshmukh (Kellogg Graduate School, Northwestern University, Evanston, Illinois), Edwin Saniga (University of Delaware, Newark), Theresa Sandifer (Kimberly-Clark Corporation), and Jeffrey Galbraith (Greenfield Community College, Greenfield). Finally we thank Richard Hercher Jr., Gail Centner, Denise Santor-Mitzit, and Rita McMullen of Richard D. Irwin, Inc., for all their help in writing and editing this book. In the final analysis, we, the authors, accept total responsibility for its information.

Most importantly we thank Aaron and Esther Blitzer, Sylvia and the late Norman Oppenheim, Shelly Gitlow, Judy Dimmerman, and Abraham and Beatrice Gitlow for their unwavering support and confidence.

We sincerely hope that you, the reader, find this book to be a valuable aid in your studies of quality improvement and statistics. Best of luck in your studies.

Howard Gitlow
Alan Oppenheim
Rosa Oppenheim

Contents

PART IV

Process Performance in Analytic Studies   337

Chapter 10   Specifications   339

Chapter 11   Process Capability and Improvement Studies   352

PART V

Process/Product Design 407

Chapter 12 Taguchi Methods—Quality Improvement in Product and Process Design 409

Foundations of Quality

This text is based on concepts presented in this section. Chapter 1 discusses the three types of quality and the losses society incurs from the lack of quality in goods and services. Chapter 1 also provides a brief history of quality and focuses on W. Edwards Deming's theory of management, key to quality improvement in any service or manufacturing organization, education, or government. Chapter 2 analyzes the basic distinction between enumerative and analytic statistical studies and discusses the significance of each type of statistical study as regards quality improvement efforts. Chapter 3 details the procedures used to document a process. When a process has been properly documented and defined, and characteristics to be studied have been operationalized, management can begin process improvement efforts using data, not guesswork and opinion.

The material presented in this section is critical to any effort towards quality and process improvement. Any improvement efforts must be preceded by a clear understanding of quality and the statistical studies required to improve quality. The remainder of this book focuses on the tools and methods needed to accomplish these goals.

CHAPTER 1 Fundamentals of Quality

Definition of Quality

Quality is "a predictable degree of uniformity and dependability, at low cost and suited to the market."[1] For example, an individual buying a container of milk expects the milk to remain fresh at least until the expiration date stamped on the container, and wants to purchase it at the lowest possible price. If the milk spoils before the expiration date, the customer's expectation won't have been met, and he'll perceive the milk's quality as poor. Further, if this happens repeatedly, the customer will lose confidence in the milk provider's ability to supply fresh milk; in other words, the customer will feel he can't predict with a high degree of belief that the milk will be uniformly and dependably fresh.

Here's another example of quality: If an assembly line worker receives parts that are predictably dependable and uniform from the worker before her, her needs will be met and she'll perceive the quality of those parts as good. Similarly, if a hotel guest finds a clean, comfortable room containing all of the amenities promised, he'll feel that his expectations were met. But if the room isn't made up properly or lacks soap, the guest will perceive that the quality is poor.

The Quality Environment

Pursuit of quality requires that organizations globally optimize their system of interdependent stakeholders. This system includes employees, customers, investors, suppliers and subcontractors, regulators, and the community. The

FIGURE 1.1   System of Interdependent Stakeholders

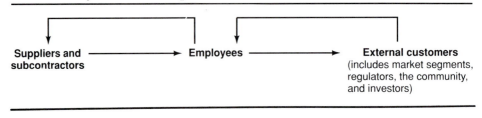

organization, which consists of employees and investors, must work together with suppliers and subcontractors to satisfy the needs of all stakeholders. Figure 1.1 depicts the system of interdependent stakeholders.

At one end of the system of interdependent stakeholders are an organization's external customers (its market segments). Each market segment's needs must be communicated to the organization through an ongoing process that conveys how an organization's products and services are performing in the marketplace and what improvements and innovations would optimize the system of interdependent stakeholders. The concept of customer also includes regulatory agencies, the community, and investors. The concept of customer should be applied to all areas and people within an organization. For example, customers are areas and people down the line.

At the other end of the system of interdependent stakeholders are the organization's suppliers and subcontractors. The organization communicates its customers' needs to its suppliers and subcontractors so that they can aid in the pursuit of quality for all stakeholders.

Employees are the most critical stakeholders of an organization. In the words of quality expert Kaoru Ishikawa,

> In management, the first concern of the company is the happiness of people who are connected with it. If the people do not feel happy and cannot be made happy, that company does not deserve to exist. . . . The first order of business is to let the employees have adequate income. Their humanity must be respected, and they must be given an opportunity to enjoy their work and lead a happy life.[2]

Types of Quality

Three types of quality are critical to the production of products and services with a predictable degree of uniformity and dependability, at low cost, that are suited to the market. They are (1) quality of design or redesign, (2) quality of conformance, and (3) quality of performance.[3]

FIGURE 1.2   Design/Redesign

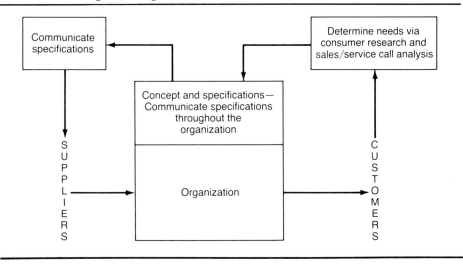

Quality of Design

Quality of design focuses on determining the quality characteristics of products that are suited to the needs of a market, at a given cost; that is, quality of design develops products from a customer orientation. Quality of design studies begin with consumer research, service call analysis, and sales call analysis, and lead to the determination of a product concept that meets the consumer's needs. Next, product specifications are prepared for the product concept (Figure 1.2).

The process of developing a product concept involves establishing and nurturing an effective interface between all areas of an organization—for example, between marketing, service, and design engineering. Design engineering is one of marketing's customers, and vice versa.

Continuous, never-ending improvement and innovation of an organization's product and service concept require that consumer research and sales/service call analysis be an ongoing effort. Consumer research is a collection of procedures whose purpose is to understand the customer's needs, both present and future. Consumer research procedures include both nonscientific and scientific studies. An example of consumer research is a study into the reasons why dog food purchasers buy or don't buy a particular brand of dog food. The investigation's goal is to determine the customer's needs and redesign the dog food product around those needs—for example, redesign the package size, make the package resealable, or alter the dog food's composition. Consumer research should be ongoing so that the firm will always be in touch with changing customer needs.

Consumer research can also be performed internally within an organization. For example, employees are the customers of some management policy decisions. Hence employee surveys are a form of consumer research that could lead to improved management policy.

Sales call analysis involves the systematic collection and evaluation of information concerning present and future customer needs. Information is collected during sales interactions with customers. The analysis helps determine customer needs by examining the questions and concerns people express about products or services at the time of purchase. Sales call analysis is an important window into the customer's needs. An example of sales call analysis is a formal investigation into salesperson–customer interactions at a personal computer distributorship. The investigation's purpose could be to collect information about the questions customers most frequently ask, and to use this data to improve the selling protocol.

Service call analysis is the systematic investigation of the problems customers/users have with the product's performance. Service call analysis provides an opportunity to understand which product features must be changed to surpass the customer's present and future needs. An example of service call analysis is SONY Corporation's formal collection of information from field service technicians of customers' problems with SONY KV-32TW76 TV sets. The basic source document for the service call analysis data is the service ticket, which indicates the problem and the work done to solve it. This information is collected and over time may indicate problems that could require changes to work methods and/or materials—for example, redesigning the TV tuner or reducing the time between a customer's request for service and the completed service call.

Service call analysis can also be performed within an organization. For example, an area supervisor may examine the problems the next operation encounters using the parts/service forms which his area delivers to the next operation. The purpose of the analysis could be to learn what the supervisor must do to pursue process improvement and innovation within his own area.

Quality of Conformance

Quality of conformance is the extent to which a firm and its suppliers can produce products with a predictable degree of uniformity and dependability, at a cost that is in keeping with the quality characteristics determined in a quality-of-design study. As Figure 1.3 shows, once the specifications are determined via a quality-of-design study, the organization must continuously strive to surpass those specifications. The ultimate goal of process improvement and innovation efforts is to create products and services whose quality is so high that consumers (both external and internal) brag about them.

Some readers may question why the preceding paragraph states that specifications should be surpassed, rather than merely met. The rationale for this statement is that there's a loss associated with products that conform to specifications but deviate from the nominal or target value. Figure 1.4 shows the traditional view of losses arising from deviations from nominal: Losses are zero until the lower specification limit (LSL) or upper specification limit (USL) is reached. Then suddenly they become positive and constant, regardless of the size of the deviation from nominal.[4]

Figure 1.5 shows a more realistic loss function. Losses begin to accrue as soon as products deviate from nominal. As we can see, the view represented in

FIGURE 1.3   Conformance

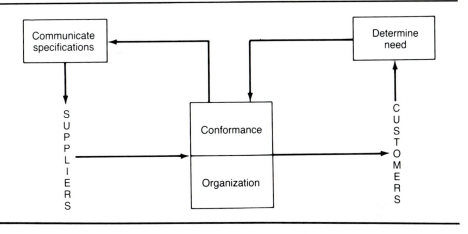

FIGURE 1.4   Traditional View of Losses Arising from Deviations from Nominal

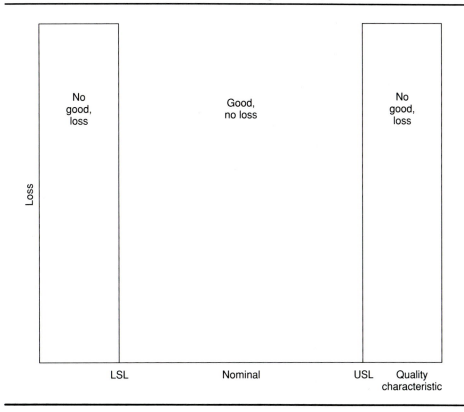

FIGURE 1.5   Realistic View of Losses Arising from Deviations from Nominal

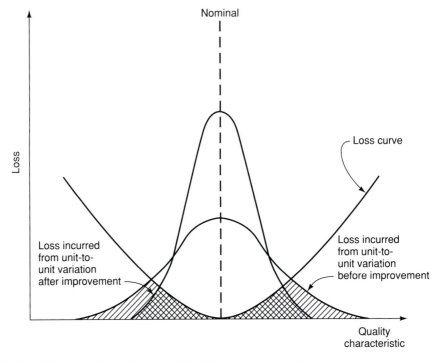

this figure requires the never-ending reduction of process variation around nominal; that is, it requires surpassing specifications.

Quality of Performance

Quality-of-performance studies focus on determining how the quality characteristics determined in quality-of-design studies, and improved and innovated in quality-of-conformance studies, are performing in the marketplace (Figure 1.6). The major tools of quality-of-performance studies are consumer research and sales/service call analysis. These tools are used to study after-sales service, maintenance, reliability, and logistical support, as well as to determine why consumers do not purchase the company's products.

The Relationship between Quality, Market Size, and Market Share

Consumers can be grouped into market segments once the product characteristics (features) they desire are known and operationally defined. A characteristic is operationally defined if it is stated in terms of a criteria, a test, and a decision so

FIGURE 1.6   Performance

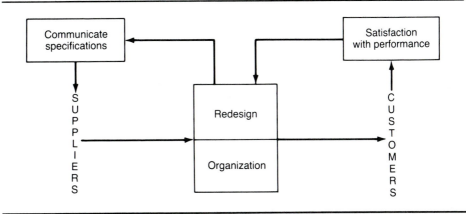

that people can communicate with respect to its value. (Operational definitions are discussed further in Chapter 3.) Features and price determine if a consumer will initially enter a market segment; hence features and price determine market size. After the initial purchase, consumers' decisions to brag about a product or purchase it again are based on their experience with the product—that is, the product's dependable and uniform performance. Dependability and uniformity determine a product's success within a market segment; therefore dependability and uniformity determine market share within a market segment. Ultimately features, dependability, uniformity, and price determine market size and market share.

Features

A loss in quality occurs when a process generates products whose features deviate from the needs of the individual (or group of individuals in a market segment)—that is, when the products and/or price don't suit the market. This type of loss can be remedied by tailoring the product to the consumer's requirements and/or by modifying the product's price. For example, shirt neck sizes may be marketed in tenths of an inch rather than in half inches, or Velcro® may be used in shirt collars instead of buttons. This segmentation strategy minimizes the loss in quality caused when the nominal levels of a product's feature package deviate from the needs of an individual (or group of individuals) in a market segment.

Dependability and Uniformity

A loss in quality also results when a process generates products whose quality characteristics lack a predictable degree of uniformity and dependability (that is, when there's high unit-to-unit variation). Lack of predictable uniformity and dependability in a product's features will cause customers to lose confidence in that product. For example, if shirt neck sizes are manufactured to be 15½ inches and customers notice variation from shirt to shirt, then this shirt-to-shirt variation will cause a loss in quality. This loss in quality can be reduced by understanding and resolving the causes of process variation. The two basic types of process variation—common and special variation—are discussed next.

Common and Special Variation

All systems (processes) vary over time. Consider a system such as your own appetite. Some days you're hungrier than usual, while some days you eat less than usual, and perhaps at different times. Your system varies from day to day to some degree. This is common variation. However, if you go on a diet or become ill, you might drastically alter your eating habits for a time. This would be a special cause of variation because it would have been caused by a change in the system. If you hadn't gone on a diet or become ill, your system would have continued on its former path of common variation.

Understanding the difference between common and special variation is critical to understanding W. Edwards Deming's theory of management. According to his theory, managers must realize that unless a change is made in the system (which only they can make) the system's capability will remain the same. This capability is determined by common variation, which is inherent in any system. Employees can't control a common cause of variation and shouldn't be held accountable for, or penalized for, its outcomes. Common variation can be caused by such things as poor lighting, lack of ongoing job skills training, or poor product design. On the other hand, special variation can be caused by new raw materials, a broken die, or a new operator. Employees should become involved in creating and utilizing statistical methods so that common and special causes of variation can be differentiated, special variation can be resolved, and common variation can be reduced by management action. These actions will result in process improvements. Since unit-to-unit variation decreases the customer's ability to rely on products' dependability and uniformity, managers must understand how to reduce and control variation. Understanding and controlling variation leads to product improvement and innovation.[5]

Managers must balance the cost of having many market segments with the benefits of high consumer satisfaction caused by small deviations between an individual consumer's needs and the product characteristic package for his market segment. Also managers must continually strive to reduce variation in product characteristics for all market segments.

The two sources of loss in quality must be detected in quality-of-performance studies. This information is then fed back into the quality-of-design studies and quality-of-conformance studies.

Sources of Customer Dissatisfaction and Delight

Customer dissatisfaction isn't the opposite of customer satisfaction (or in the extreme, customer delight); that is, if all sources of customer dissatisfaction are eliminated, the customer is neutral—neither satisfied nor delighted. This concept

FIGURE 1.7   Types of Quality

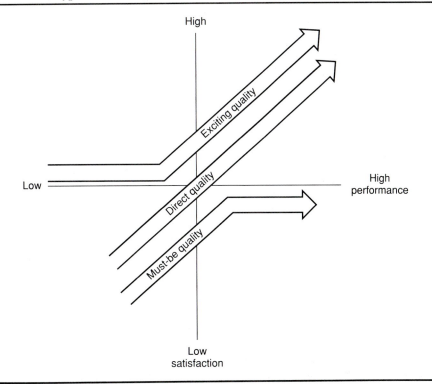

was first explained by Dr. Noriaki Kano of the Science University of Tokyo in his categorization of quality into "must be" quality, "direct" quality, and "exciting" quality (Figure 1.7).[6] These terms explain how both the features and uniformity characteristics of products and processes combine to create performance/nonperformance and satisfaction/dissatisfaction dimensions on the part of the consumer. Traditionally a linear relationship is assumed between performance and satisfaction; that is, the more performance, the more satisfaction. This type of quality is called "direct" quality. However the relationship between performance and satisfaction can be nonlinear in two different ways. First, a quality characteristic can behave in such a way that as nonperformance is decreased (that is, as performance is increased), dissatisfaction is decreased, but satisfaction isn't increased; this leaves the consumer in a neutral position at best. This type of quality characteristic is called a "must be" quality characteristic. Airline safety may be an example of a "must be" quality characteristic because its absence can cause great customer dissatisfaction, but its presence isn't noticed by the customer. Second, a quality characteristic can behave in such a way that nonperformance doesn't create customer dissatisfaction, but performance creates feelings of customer satisfaction and customer delight in the extreme. This type of quality characteristic is called an ''exciting'' quality characteristic. For many, the tastiness of airline meals may be an example of exciting quality because its absence doesn't create customer dissatisfaction (customers have come not to expect it), but its presence may cause customer delight.

Improvement versus Innovation (Quality Creation)

Improvement and innovation are both required if a firm is to be healthy in the future. The purpose of process improvement is to modify current methods to continuously reduce the difference between customer needs and process performance. Tools such as consumer research and service/sales call analysis are helpful in this endeavor. The purpose of innovation is twofold: (1) to create a dramatic breakthrough in decreasing the difference between customer needs and process performance and (2) to discover the customer's future needs. Examples of improvement and innovation can be seen in the design and manufacture of shirts. Improvement in the production of shirts (of a given neck size) could be realized through changing methods to reduce unit-to-unit variation in neck sizes. Innovation in the design of shirts could be realized by using Velcro® strips to close collars, instead of buttons and holes. This method would create a breakthrough in the shirt designer's ability to satisfy customers with changing neck sizes (due to changing body weight) by creating a range of available neck sizes around a nominal neck size. Also, innovation in the design of shirts to meet customers' future needs could be realized by developing a new material that expands and contracts with body weight (e.g., neck size) and always remains 0.25 inches larger than a user's current neck size.

Ideas for innovation with respect to the customer's future needs can't come from direct queries to customers; rather, they must come from the producer. In this regard, consumer research is backward looking. That is, asking customers what they want can only help producers improve existing products or services; it can't help producers anticipate the customer's future needs.[7] Consumers don't know what innovations they'll want in the future. For example, a consumer could not have told you he wanted a facsimile machine or an automatic loading camera before such things existed. These types of breakthroughs must be discovered by the producer studying the problems customers have when using products and services—and not by asking customers what they want in the future.

For example, in 1974 the camera market was saturated with cameras that satisfied customers' current needs; cameras were reliable, were relatively inexpensive to use, and gave good pictures. This created a nightmare for the camera industry. Consequently Konica decided to ask consumers, ''What more would you like in a camera?'' Consumers replied that they were satisfied with their cameras. Unfortunately, asking consumers what more thay would like in a camera didn't yield the information Konica needed to create a breakthrough. In response to this situation, Konica studied negatives at film processing laboratories and discovered that the first few pictures on many rolls were overexposed. Further research indicated that camera operators had a low degree of confidence in their ability to properly load film into their camera. After loading a roll of film and winding the first few pictures, users would open their camera to make sure that the film had been properly loaded, thereby exposing the first few pictures on the roll. This presented an opportunity to innovate camera technology. In response to this analysis, Konica developed the automatic loading camera. The customer couldn't have been expected to think of this innovation. The same type of procedure could have been used to discover consumer needs for a completely new product.[8]

The Relationship between Quality and Productivity

Why should organizations try to improve quality? If a firm wants to increase its profits, why not raise productivity? For years, W. Edwards Deming has worked to change the thinking in organizations that operate with the philosophy that if productivity increases, profits will increase. The following example illustrates the folly of such thinking.

For the past 10 years the Universal Company has produced an average of 100 widgets per hour, 20 percent of which are defective. The board of directors now demands that top management increase productivity by 10 percent. The directive goes out to the employees, who are told that instead of producing 100 widgets per hour, the company must produce 110. Responsibility for producing more widgets falls on the employees, creating stress, frustration, and fear. They try to meet the new demands, but must cut corners to do so. Pressure to raise productivity creates a defect rate of 25 percent and only increases production to 104 units, yielding 78 good widgets, two fewer than the original 80 (Figure 1.8a).[9]

Stressing productivity often has the opposite effect of what management desires. The following example demonstrates a new way of looking at productivity and quality.

The Dynamic Factory produces an average of 100 widgets per hour with 20 percent defective. Top management is continually trying to improve quality, thereby increasing productivity. Top management realizes that Dynamic is making 20 percent defectives, which translates into 20 percent of the total cost of production being spent to make bad units. If Dynamic's managers can improve the process, they can transfer resources from the production of defectives to the manufacture of additional good products. Management can improve the process by making some changes at no additional cost, so that only 10 percent of the output is defective on average. This results in an increase in productivity (Figure 1.8b). Here management's ability to improve the process results in a decrease in defectives, yielding an increase in good units, quality, and productivity.

Benefits of Improving Quality

Deming's approach to the relationship between quality and productivity stresses improving quality to increase productivity. Several benefits result:

FIGURE 1.8 Productivity versus Quality Approach to Improvement

	(a) Universal Company Output	
	Before Demand for 10% Productivity Increase (Defect Rate = 20%)	*After Demand for 10% Productivity Increase (Defect Rate = 25%)*
Widgets produced	100	104*
Widgets defective	20	26
Good widgets	80	78

	(b) Dynamic Factory Output		
	Before Improvement (Defect Rate = 20%)		*After Improvement (Defect Rate = 10%)*
Widgets produced	100	→	100
Widgets defective	20	Process	10
Good widgets	80	improvement	90
		→	

* Only reached 104, not required 110, but defect rate rose 20 percent to 25 percent. More widgets were produced; but more were defective, yielding less productivity.

1. Productivity rises (in our Dynamic Factory example, from an average of 80 good units in 100 produced to an average of 90 good units in 100 produced).
2. Quality improves (from 80 percent good units on average to 90 percent good units on average in the Dynamic Factory example).
3. Cost per good unit is decreased.
4. Price can be cut.
5. Workers' morale goes up because they aren't seen as the problem. This last aspect leads to further benefits:
   a. Less employee absence.
   b. Less burnout.
   c. More interest in the job.
   d. Motivation to improve work.

In sum, stressing productivity means sacrificing quality and possibly decreasing output. Employee morale plunges, costs rise, customers are unhappy, and stockholders become concerned. On the other hand, stressing quality can produce all the desired results: less rework, greater productivity, lower unit cost, price flexibility, improved competitive position, increased demand, larger profits, more jobs, and more secure jobs. Customers get high quality at a low price, vendors get predictable long-term sources of business, and investors get profits. Everybody wins!

The History of Quality

Issues of quality have existed since tribal chiefs, kings, and pharaohs ruled. An example of a quality issue in ancient times is found in the Code of Hammurabi dating from as early as 2000 BC. Item 229 states, ''If a builder has built a house for a man, and his work is not strong, and the house falls in and kills the householder, that builder shall be slain.'' Phoenician inspectors eliminated any repeated violations of quality standards by chopping off the hand of the maker of the defective product. Inspectors accepted or rejected products and enforced government specifications. The emphasis was on equity of trade and complaint handling. In ancient Egypt (approximately 1450 BC), inspectors checked stone blocks' squareness with a string as the stonecutter watched. This method was also used by the Aztecs in Central America.

In 13th-century Europe, apprenticeships and guilds developed. Craftsmen were both trainers and inspectors. They knew their trades, their products, and their customers, and they built quality into their goods. They took pride in their work and in training others to do quality work. The government set and provided standards (e.g., weights and measures) and, in most cases, an individual could inspect all the products and establish a single quality standard. If the world had remained small and localized, this idyllic state of quality could have thrived and lasted. However, as the world became more populated, more products were needed.

During the 19th century the modern industrial system began to emerge. In the United States, Frederick Taylor pioneered scientific management, removing work planning from the purview of workers and foremen and placing it in the hands of industrial engineers. The 20th century ushered in a technical era that enabled the masses to avail themselves of products previously reserved for only the wealthy. Henry Ford introduced the moving assembly line into Ford Motor Company's manufacturing environment. Assembly line production broke down complex operations that could be performed by unskilled labor. This resulted in the manufacture of highly technical products at low cost. As part of this process, an inspection operation was instituted to separate good and bad products. Quality, at this point, remained under the purview of manufacturing.

It soon became apparent that the production manager's priority was meeting manufacturing deadlines—achieving product quality wasn't a priority. Managers knew they'd lose their jobs if they didn't meet production demands, whereas they'd only be reprimanded if quality was poor. Upper management eventually realized that quality was suffering as a result of this system, so a separate position of ''chief inspector'' was created.

Between 1920 and 1940 industrial technology changed rapidly. The Bell System and Western Electric, its manufacturing arm, led the way in quality control by instituting an Inspection Engineering Department to deal with problems created by defects in their products and lack of coordination between their departments. George Edwards and Walter Shewhart, as members of this department, provided leadership in this area. According to George Edwards,

> Quality control exists when successive articles of commerce have their characteristics more nearly like its fellows' and more nearly approximating the designer's intent, than would be the case if the application were not made. To me, any procedure, statistical or otherwise, which has the results I have just mentioned, is quality control, and any procedure which does not have these results is not quality control.[10]

Edwards coined the term *quality assurance* and advocated quality as part of management's responsibility. He said,

> This approach recognizes that good quality is not accidental and that it does not result from mere wishful thinking, that it results rather from the planned and interlocked activities of all the organizational parts of the company, that it enters into design, engineering, technical and quality planning specification, production layouts, standards, both workmanship and personnel, and even into training and fostering the point of view of administrative, supervisory, and production personnel. This approach means placing one of the officers of the company in charge of the quality control program in a position at the same level as the controller or as the other managers in the operation. Its objective would be elimination of the hunch factors that at present so largely determine the product quality of too many companies. It puts a man at the head of the quality control program in a position to establish and make effective a company-wide policy with respect to quality, to direct the actions to be taken where it is necessary and to place responsibility where it belongs in each instance.[11]

In 1924 mathematician Walter Shewhart introduced statistical quality control. This provided a method for economically controlling quality in mass production environments. Shewhart was concerned with many aspects of quality control. In his book of lectures at the graduate school of the U.S. Department of Agriculture, he asked the reader to write several letter A's as carefully as possible. He then suggested that the reader observe them for variations. Clearly, no matter how carefully one formed the letters, variations occurred. This was a simple yet powerful example of variation in a process. Although Shewhart's primary interest was statistical methods, he was very aware of principles of management and behavioral science.

World War II quickened the pace of quality technology. The need to improve the quality of products being manufactured resulted in increased study of quality control technology and more sharing of information. In 1946 the American Society for Quality Control (ASQC) was formed, and George Edwards was elected its president. He stated at the time,

> Quality is going to assume a more and more important place along side competition in cost and sales price, and the company which fails to work out some arrangement for securing effective quality control is bound, ultimately, to find itself faced with a kind of competition it can no longer meet successfully.[12]

In this environment, basic quality concepts expanded rapidly. Many companies implemented vendor certification programs. Quality assurance professionals developed failure analysis techniques to problem-solve, quality engineers became involved in early product design stages, and environmental performance testing of products was initiated. But, as World War II ended, progress in quality control began to wane. Many companies saw it as a wartime effort and felt that it was no longer needed in the booming postwar market.

In 1950 W. Edwards Deming, a statistician who had worked at the Bell System with George Edwards and Walter Shewhart, was invited by JUSE (the Union of Japanese Scientists and Engineers) to speak to Japan's leading industrialists. They were concerned with rebuilding Japan after the war, breaking into foreign markets, and improving Japan's reputation for producing poor-quality goods. Deming convinced them, despite their reservations, that by instituting his methods, Japanese quality could become the best in the world. The industrialists took Deming's teaching to heart. Over the following years, Japanese quality, productivity, and competitive position were improved and strengthened tremendously. Dr. Deming was awarded the Second Order Medal of the Sacred Treasure by Emperor Hirohito for his contribution to Japan's economy. The coveted Deming Prizes are awarded each year in Japan to the company that has achieved the greatest gain in quality and to an individual for developments in statistical theory.[13] Prize-winning Japanese companies include Nissan, Toyota, Hitachi, and Nippon Steel. In 1989 Florida Power & Light Company became the first non-Japanese company to receive the Deming prize.

Deming's ideas have spread in the United States and the rest of the world. Companies such as Nashua Corporation, Ford Motor Company, and General Motors have listened to Dr. Deming and are seeking the benefits of continuous process improvement. Deming's clients have included railways, telephone companies, consumer researchers, hospitals, law firms, government agencies, and university research organizations. While a professor at the New York University Graduate School of Business Administration and at Columbia University, he wrote extensively on statistics and management.

In 1951 Armand V. Feigenbaum published his book *Total Quality Control*, which advanced the concept of quality control in all areas of business, from design to sales. Up until then, quality efforts had been primarily directed toward corrective activities, not prevention.

The Korean War sparked increased emphasis on reliability and end-product testing. However, all of the additional testing did not enable firms to meet their quality and reliability objectives, so quality awareness and quality improvement programs began to emerge in manufacturing and engineering areas. Service Industry Quality Assurance (SQA) also began to focus on the use of quality methods in hotels, banks, government, and other service systems. By the end of the 1960s quality programs had spread throughout most of America's major corporations. But American industry was still enjoying the top position in world markets as Europe and Japan continued to rebuild.

Foreign competition began to threaten U.S. companies in the 1970s. The quality of Japanese products such as cars and TVs began to surpass American-made goods. Consumers began to consider the long-term life of a product in purchase decisions. Foreign competition and consumers' increased interest in quality forced American management to become more concerned with quality. The late 1970s, 1980s, and 1990s have been marked by striving for quality in all aspects of businesses and service organizations including finance, sales, personnel, maintenance, management, manufacturing and service. The focus is on the entire system, not just the manufacturing line. Reduced productivity, high costs, strikes, and high unemployment have caused management to turn to quality improvement as the means to organizational survival. Some of the quality leaders in the United States have been W. Edwards Deming, Joseph Juran, and Armand Feigenbaum.[14] In this book we focus largely on the ideas of W. Edwards Deming.

The Purpose of Deming's Ideas on Quality

Deming's ideas on quality require that managers transform themselves so that they will

1. Improve and innovate the system of interdependent stakeholders of an organization over the long term to allow all people to experience joy in their work and pride in the outcome.

2. Optimize the system of interdependent stakeholders of an organization over the long term so that everybody wins; and not optimize one stakeholder group's welfare at the expense of another stakeholder group's welfare. Stakeholders include employees, customers, suppliers and subcontractors, regulators, investors, the community, and competitors.[15]

3. Improve and innovate the condition of society. Society includes local, regional, national, and international systems—for example, the entire educational system (public and private primary and secondary schools and universities), the environment, public health, and the economic and social well-being of communities and countries.

In Dr. Deming's words,

The aim [purpose of the transformation] will be to unleash the power of human resource contained in intrinsic motivation. Intrinsic motivation is the motivation an individual experiences from the sheer joy of an endeavor. In place of competition for high rating, high grades, to be Number 1, there will be cooperation between people, divisions, companies, governments, countries. The result will in time be greater innovation, science, applied science, technology, expanded market, greater service, greater material reward for everyone. There will be joy in work, joy in learning. Anyone that enjoys his work is a pleasure to work with. Everyone will win; no loser.[16]

W. Edwards Deming's Theory of Management

Dr. Deming has developed his ideas on quality into a theory of management that helps individuals learn through the acquisition of process knowledge gained from experience coordinated by theory. He calls this theory "a system of profound knowledge.'' An explanation of the system of profound knowledge can be found in his last book, *The New Economics for Industry, Government, Education.*[17] Dr. Deming discussed the system of profound knowledge extensively in his famous four-day seminars on management.

The system of profound knowledge is an appropriate theory for leadership in any culture or society. However, applying this theory in a particular society or culture requires a focus on issues that are unique to that society or culture. For example, in the Western world, managers frequently operate using the following paradigms (a paradigm being a filter through which an individual or group interprets data about conditions and circumstances—often without realizing it):

1. Reward and punishment are the most important motivators for people and organizations.
2. Winners and losers are necessary in most interactions between people and between organizations.
3. Results are achieved by focusing on productivity (as opposed to quality).
4. Quality is inversely related to quantity.
5. Rational decisions can be made based on guesswork and opinion, using only visible figures.
6. Construction, execution, and control of plans is solely the function of management.
7. Organizations can be improved in the long term by fighting fires.
8. Superiors are your most important customers.
9. Competition is a necessary aspect of personal and organizational life.

Western leaders who manage in the context of the preceding paradigm are lost in the new economic age. They have no idea of how to manage their organizations because they don't know the new paradigms required for success in today's marketplace. Such leaders need a perspective from which they can understand the new paradigms of Total Quality Management (TQM). That perspective is called the ''14 Points for Management.''

The 14 Points for Management

The system of profound knowledge generates an interrelated set of 14 points for leadership in the Western world. These 14 points provide guidelines for the shifts in thinking required for organizational success in the 21st century. They form a highly interactive system of management; no one point should be studied in isolation.

Point 1: Create constancy of purpose toward improvement of product and service, with the aim to become competitive and to stay in business and to provide jobs.

Leaders must state their organization's values and beliefs. They must create statements of vision and mission for their organizations based on these values and beliefs. Values and beliefs are the fundamental operating principles that provide guidelines for organizational behavior and decision making. A vision statement is a wish or dream, for all to see, that seeks to communicate the desired future state of the organization to the stakeholders. A mission statement serves to inform stakeholders of the current reason for the existence of the organization. The values and beliefs plus the vision and mission statements provide a frame of reference for focused, consistent behavior and decision making by all stakeholders of an organization. This framework permits stakeholders to feel more secure because they understand where they fit into the organization.

An organization's stakeholders will change over time, both by stakeholder group and by individuals within a stakeholder group. The process of understanding the needs and wants of the ever-shifting set of stakeholders is known collectively as the "voice of the stakeholder." Its components include the "voice of the customer" representing external customers and regulatory agencies that watch over issues important to external customers; the "voice of the business" representing employees; and the "voice of the shareholder" representing shareholders. The voice of the stakeholder consists of a series of techniques that can be used to collect data concerning the needs and wants of relevant stakeholder groups. These techniques range in diversity from a simple analysis of complaint data to sophisticated market research surveys. Voice of the stakeholder data is used to help define long-term strategies and related short-term plans including methods, indicators, and rational targets to allocate resources that optimize the system of interdependent stakeholders.

As an organization's stakeholders change over time, it may become necessary to refocus an organization's market segments or strategic plan for a given market segment. The transition out of a market segment is just as important as the transition into a market segment because these transitions aren't viewed as independent events by the individuals in these market segments; today's young person is tomorrow's senior citizen. These changes will be based on modification of the organization's mission.

Point 2: Adopt the new philosophy. We are in a new economic age. Western management must awaken to the challenge, must learn their responsibilities, and take on leadership for change.

Point 2 encompasses the thinking changes that Western leaders must accept as a consequence of Dr. Deming's system of profound knowledge. There are four new paradigms.

Paradigm 1. People are best inspired by intrinsic motivation, not extrinsic motivation. Intrinsic motivation comes from the sheer joy of performing an act. It releases human and systems energy that can be focused into improvement and innovation of the system of interdependent stakeholders via the mission statement and strategic plans. It is management's responsibility to create an atmosphere that fosters intrinsic motivation; this atmosphere is a basic element of Deming's theory of management.

Extrinsic motivation comes from the desire for reward or the fear of punishment. It restricts the release of energy from intrinsic motivation by judging, policing, and destroying the process and the individual. Management based on extrinsic motivation will "squeeze out from an individual, over his lifetime, his innate intrinsic motivation, self-esteem, dignity, and build into him fear, self-defense."[18]

Paradigm 2. Manage using both a process and results orientation, not only a results orientation. Management's job is to improve and innovate the processes that create results, not just to manage results. This thinking shift allows management to define the capabilities of its processes and, consequently, to predict and plan the future of the system of interdependent stakeholders to achieve organizational optimization. This type of optimization requires that managers make decisions based on facts, not on guesswork and opinion. Managers must consider both visible and invisible figures as well as unknown and unknowable figures (for example, the cost of an unhappy customer or the benefit of a prideful employee). Finally managers must understand that quality and quantity are directly related (as in the case when management controls the quality of outgoing products by improving and innovating the process that makes the products).

Paradigm 3. Management's function is to optimize the system so that everyone wins, not to maximize stockholder's wealth in the short term. Managers shouldn't view their roles in such a narrow sense that they attempt only to construct, execute, and control plans to achieve profit in the short term. The latter view is myopic in its focus on only the needs of stockholders, and it is hedonistic in its focus on stockholders' short-term gratification. Managers must understand that individuals, organizations, and systems of organizations are interdependent. Optimization of one component may cause suboptimization of another component. Management's job is to optimize the global system.

Paradigm 4. Cooperation works better than competition. In a cooperative environment, everybody wins. Customers win products and services they can brag about. The firm wins returns for investors and secure jobs for employees. Suppliers win long-term customers for their products. The community wins an excellent corporate citizen.

In a competitive environment, most people lose. The huge but unknown and unknowable costs resulting from competition include the costs of rework, waste, and redundancy as well as the costs for warranty, retesting, reinspection, customer dissatisfaction, schedule disruptions, and destruction of the individual's joy in work and pride in the outcome. Individuals and organizations can't reap the benefits of a win-win point of view when they're forced to compete. Competition causes individuals or departments to optimize their own efforts at the expense of other stakeholders. This form of optimization seriously erodes the performance of the system of interdependent stakeholders.

If Western leaders adopt and live all of the preceding paradigms, they will reap enormous benefits.

Point 3: Cease dependence on inspection to achieve quality. Eliminate the need for inspection on a mass basis by building quality into the product in the first place.
There is a hierarchy of views on how to pursue predictable dependability and

uniformity at low cost: (1) defect detection, (2) defect prevention, and (3) never-ending improvement.

1. Defect detection involves dependence upon mass inspection to sort conforming material from defective material. Mass inspection doesn't make a clean separation of good from bad. It involves checking products with no consideration of how to make them better. Management must eliminate the need for inspection on a mass basis and build quality into the processes that generate goods and services. Mass inspection does nothing to decrease the variability of the quality characteristics of products and services. Dependence on mass inspection to achieve quality forces quality to become a separate subsystem (called *quality assurance*) whose aim is to police defects without the authority to eliminate the defects. As the Quality Assurance Department optimizes its efforts, it causes other departments to view quality as someone else's responsibility.

2. Defect prevention involves improving processes so that all output is predictably within specification limits; this is often referred to as *zero defects*. Defect prevention leaves employees with the impression that their job (with respect to reducing variation) is accomplished if they achieve zero defects. Unfortunately zero defects will be eroded by a force similar to the concept of entropy in thermodynamics, or the natural tendency of a system to move toward disorder or chaos. This force makes a stable and capable process eventually stray out of specification limits. Further, when people are rewarded for zero defects, they may attempt to widen specification limits, rather than improve the process's ability to predictably create output within specification limits.

3. Never-ending improvement is the continuous reduction of process (unit-to-unit) variation, even within specification limits. Both Deming and Taguchi contend that products, services, and processes must be improved in a relentless and never-ending manner. Figure 1.9 shows this reasoning. Distribution A depicts unit-to-unit variation before improvement. The costs (loss) incurred using the system that produced distribution A are represented by the lightly shaded area under the total cost curve, $L(y)$. Distribution B depicts unit-to-unit variation after improvement. Costs (loss) incurred using the improved system that produced distribution B are represented by the densely shaded area under the total cost curve, $L(y)$. The loss incurred under the system with lower unit-to-unit performance variation, system B, is clearly lower than the loss incurred with system A. Following the preceding logic, it is always economical to reduce unit-to-unit variation around nominal, even when a process is producing output within specification limits.

kp Rule.   Deming advocates a plan that minimizes the total cost of incoming materials and final product. Simply stated, the rule is an inspect all-or-none rule. Its logical foundation has statistical evidence of quality as its base. The rule for minimizing the total cost of incoming materials and final product is referred to as the kp rule. (Chapter 13 discusses Deming's kp rule.) The kp rule specifies when all items should be inspected and when none should be inspected. This method facilitates the collection of process or product data such that variation can be

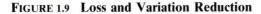

FIGURE 1.9   Loss and Variation Reduction

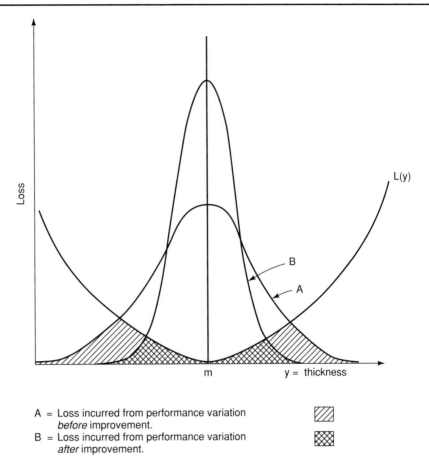

A = Loss incurred from performance variation *before* improvement.

B = Loss incurred from performance variation *after* improvement.

continually reduced; this means moving from defect detection to never-ending improvement.

Point 4: End the practice of awarding business on the basis of price tag. Instead, minimize total cost. Move toward a single supplier for any one item on a long-term relationship of loyalty and trust.

Buyers and vendors form a system. If each individual player in this system attempts to optimize her own position, the system between buyers and vendors will be suboptimized. With this point we can understand how to purchase materials, products, services, and processes to optimize the system of interdependent stakeholders in the long term (i.e., minimize total cost). Optimization requires that policy makers understand the different scenarios in which purchasing can take place. Deming has defined three scenarios that cover most purchasing situations. He calls these three scenarios World 1, World 2, and World 3.

World 1 describes a purchasing situation in which the customer knows what she wants and can convey this information to a supplier. In this scenario, purchase price is the total cost of buying and using the product (for example, no supplier provides better service than any other supplier), several suppliers can precisely meet the customer's requirements, and the only difference between suppliers is the price. In this world, purchasing on lowest price is the most rational decision.

In World 2 the customer knows what she wants and can convey this information to a supplier. The purchase price isn't the total cost of buying and using the product (for example, one supplier may provide better service than any other supplier), several suppliers can precisely meet the customer's requirements, and all suppliers quote identical prices. In this world, purchasing based on best service is the most rational decision. World 2 frequently includes the purchasing of commodities.

In World 3 the customer thinks she knows what she wants and can convey this information to a supplier. However, she'll listen to advice from the supplier and make changes based on that advice. Purchase price isn't the total cost of buying and using the product. (For example, there's also a cost to use the purchased goods.) Several suppliers tender their proposals (all of which are different in many ways) and all suppliers quote different prices. In this world selecting a supplier will be difficult.

In World 3, after careful and extensive research, it makes sense for customers and suppliers to enter into long-term relationships based on trust (i.e., relationships without the fear caused by threat of alternative sources of supply) and statistical evidence of quality. Such long-term relationships promote continuous improvement in the predictability of uniformity and reliability of products and services, and, hence, lower costs. The ultimate extension of reducing the supply base is moving to a single supplier and purchasing agent for a given item.[19] Additionally, single supplier relationships should include contingencies on the part of the supplier and customer for disasters.

The concept of single supplier extends far beyond the purchasing function. For example, employees should focus on improvement of existing internal and external information channels to get data, rather than create additional information channels to get data when the main channel doesn't yield the desired information.[20]

The importance of Point 4 is that people become aware of the differences between purchasing in Worlds 1, 2, and 3. Some purchasing agents buy as if all purchases were World 1 scenarios; that is, they purchase solely on the basis of price, without adequate measures of quality and service.

Point 5: Improve constantly and forever the system of production and service to improve quality and productivity, and thus constantly decrease costs.
Point 5 explains management's responsibility for the system of production and service. Improvement and innovation of the system of production and service require statistical and behavioral methods that should be used by everyone in the organization.

Management's Responsibility.   Management should understand the difference between special and common causes of variation and the appropriate type of managerial response to each within their system. This will stop managerial tampering with the system. Improvement of a system comes from rectification of special causes of variation and the reduction of common causes of variation. Management must understand the capability of a system. They must realize that only when a system is stable (that is, when it exhibits only common causes of variation) can management use process knowledge to predict the future condition of the system. This allows management to plan the future state of the system. Further, management of a system requires knowledge of the interrelationships between all functions and activities in the system; this includes the interactions between people and the system, as well as between other people.

Operational Definitions.   There's no such thing as a fact concerning an empirical observation. Any two people may have different ideas about what constitutes knowledge of an event. This leads to the need for people to agree on the definitions of characteristics that are important about a system. Operational definitions increase communication between people and help to optimize a system; they require statistical and process knowledge. Operational definitions are fully discussed in Chapter 3.

SDSA Cycle.   The SDSA cycle is a tool that helps employees document a process by creating an identity for the process. It includes four steps: (1) Standardize: Employees study the process and develop "best practice" methods using tools such as flowcharts. Do: (2) Employees use the best practice methods on a trial basis. (3) Study: Employees research the effectiveness of the best practice methods. (4) Act: Managers establish standardized best practice methods and formalize them through training.

Individual workers must be educated to understand that increased variability in output will result if each worker follows his own best practice method. They must be educated about the need to reach consensus on one best practice method. Management should understand the differences between workers and channel these differences into the development of the best practice method in a constructive (team-building) manner.

The best practice method will consist of generalized procedures and individualized procedures. Generalized procedures are standardized procedures that all workers must follow. The generalized procedures can be improved or innovated through *quality control circle* (QCC) activities. Individualized procedures are procedures that afford each worker the opportunity to utilize his individual differences by creating his own standardized procedure. However, the outputs of individualized procedures must be standardized across individuals. The individualized procedures can be improved through individual efforts. In the beginning of a quality improvement effort, management may not have the knowledge to allow for individualized procedures.

FIGURE 1.10   The Deming (PDSA) Cycle

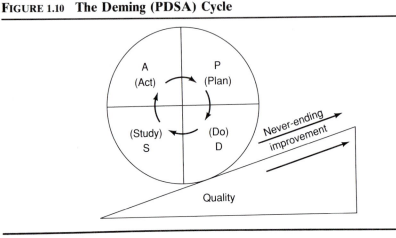

Deming (PDSA) Cycle.   The Deming cycle (Figure 1.10) can aid management in improving and innovating processes (that is, in helping to reduce the difference between customers' needs and process performance).[21] The Deming cycle consists of four basic stages: a "plan" stage, a "do" stage, a "study" stage, and an "act" stage. Hence the Deming cycle is sometimes referred to as the PDSA cycle (plan-do-study-act cycle). Initially a plan is developed to improve or innovate a standardized best practice method. This may involve using tools such as a flowchart. The plan is then tested on a small scale or trial basis (Do Stage), the effects of the plan are studied (Study Stage), and appropriate corrective actions are taken (Act Stage). These corrective actions can lead to a new or modified plan, and are formalized through training. The PDSA cycle continues forever in an uphill progression of never-ending improvement.

The PDSA cycle requires a theory (a method as visualized by an improved flowchart) in the plan stage to initiate improvement. The success of another person or organization isn't a rational basis for another iteration of the PDSA cycle. For example, isolating one component of System A and expecting it to work within the context of System B isn't rational. Reasons behind the success in System A may not be present in System B. Hence copying (benchmarking), without a true understanding of the conditions surrounding the copied (benchmarked) system, can lead to misapplication of the component and, hence, misuse of the PDSA cycle.

The plan stage of the PDSA cycle requires making predictions about future events; these are the assumptions of the plan. Realization of these assumptions can only happen if the process for which the plan is developed is stable with a low degree of variation; consequently there's need for the SDSA cycle and earlier iterations of the PDSA cycle.

Empowerment.   *Empowerment* is a term commonly used by managers in today's organizational environment.[22] However, empowerment hasn't been operationally defined and its definition varies from application to application. Currently the prevailing definition of empowerment relies loosely on the notion of dropping decision making down to the lowest appropriate level in an organization. Empowerment's basic premise is that if people are given the authority to make decisions, they'll take pride in their work, be willing to take risks, and work harder to make things happen. This sounds great—but frequently employees are empowered until they make a mistake, and then the hatchet falls. Most employees know this and treat the popular definition of empowerment with the respect it deserves—not much. Consequently empowerment in its current form is destructive to Total Quality Management.

Empowerment in a TQM sense has a dramatically different aim and definition. The aim of empowerment in TQM is to increase pride in work and joy in the outcome for all employees. Empowerment can be defined so as to translate the preceding aim into a realistic objective. Empowerment is a process that provides an individual or employees with (1) opportunity to define and document their key systems, (2) opportunity to learn about systems through training and development, (3) opportunity to improve and innovate the best practice methods that make up systems, (4) latitude to use their own judgment to make decisions within the context of best known methods, and (5) an environment of trust in which superiors won't react negatively to the latitude taken by people in decision making within the context of a best practice method.

Empowerment starts with leadership, but requires the commitment of all employees. Leaders need to provide employees with all five of the preceding conditions. Item 5 requires that the negative results emanating from employees using their judgment within the context of a best practice method lead to improvement or innovation of best practice methods, not to judgment and punishment of employees. Employees need to accept responsibility for (1) increasing their training and knowledge of the system, (2) participating in the development, standardization, improvement, and innovation of best known methods that make up the system, and (3) increasing their latitude in decision making within the context of best known methods.

We should point out that latitude to make decisions within the context of a best known method refers to the options an employee has in resolving problems within the confines of a best known method, not to modification of the best known method. Differentiating between the need to change the best known methods and latitude within the context of the best known methods must take place at the operational level.

Teams must work to improve or innovate best known methods. Individuals can also work to improve or innovate best known methods; however, the efforts of individuals must be shared with and approved by the team. Empowerment can only exist in an environment of trust that supports planned experimentation concerning ideas to improve and innovate best known methods. Ideas for improvement and innovation can come from individuals or from the team, but tests of ideas' worthiness must be conducted through planned experiments under the auspices of the team. Anything else will result in chaos because everybody will do her own thing.

Empowerment is operationalized at two levels. First, employees are empowered to develop and document best known methods using the SDSA cycle. Second, employees are empowered to improve or innovate best known methods through application of the PDSA cycle.

Point 6: Institute training on the job.

Employees are an organization's most important asset. Organizations must make long-term commitments to employees that include the opportunity to take joy in their work and pride in the outcome. Point 6 requires that training in job skills becomes an integral part of that long-term commitment.

Management of Training Efforts.   Job skills training is a system whose aim must be optimization of the system of interdependent stakeholders. If the aim of the system of interdependent stakeholders changes, the job skills training required for optimization changes.

Effective training changes the skill distribution for each job skill, as Figure 1.11 shows. Management must understand the capability of the training process and the current distribution of trainee job skills to improve the future distribution of trainee job skills.

Training, like quality and safety, is a line function, not a staff function. Data, (not guesswork or opinion) should be used to guide the training plans for employees. The training needs of employees must be prioritized to effectively allocate resources and to optimize the system of interdependent stakeholders.

Training Methods.   Training is a part of everyone's job and should include formal class work, experiential work, and instructional materials. Training courseware must take into consideration how the trainee learns and the speed at which she learns, and it must incorporate these factors into training courseware.

FIGURE 1.11   Distribution of Job Skills

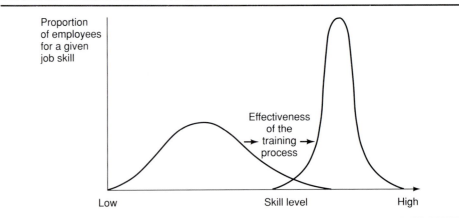

Training must include theory; it can't just be a presentation of other people's experiences. Experiences, and the observations caused by them, depend upon the condition of the observer. Hence different observers experience the same event differently. Theory creates a common view by which to understand an experience. Theory also improves our ability to understand and predict future experiences. Experience can be used to improve theory.

Employees should be trained in their job skills using statistical methods demonstrating that their training is complete by indicating when they reach a state of statistical control. If an employee isn't in statistical control with respect to a job characteristic, then more training of the type he's receiving will be beneficial. However if an employee is in a state of statistical control with respect to a job characteristic, then more training of that type won't be beneficial; the employee has learned all that's possible from the training program.

Point 7: Institute leadership. The aim of leadership should be to help people and machines and gadgets to do a better job. Leadership of management is in need of overhaul, as well as leadership of production workers.

A leader must see the organization as a system of interrelated components, each with an aim, but all focused collectively to support the aim of the system of interdependent stakeholders. This type of focus may require suboptimization of some system components.

A leader uses plots of points and statistical calculations, with knowledge of variation, to try to understand both his performance and that of his people. Leaders know when their people are experiencing problems that make their performance fall outside of the system, and leaders treat the problems as special causes of variation. These problems could be common causes to the individual (e.g., long-term alcoholism), but special causes to the system (an alcoholic works differently from his peers).

A leader must understand that experience without theory doesn't facilitate prediction of future events. For example, a leader can't predict how a person will do in a new job based solely on experience in the old job. A leader must have a theory to predict how an individual will perform in a new job.

A leader must be able to predict the future to plan the actions necessary to pursue the organization's aim. Prediction of future events requires that the leader continuously work to create stable processes with low variation to facilitate rational prediction.

Point 8: Drive out fear so that everyone may work effectively for the company.

Fear and Anxiety.   There are two kinds of negative reaction behaviors: fear and anxiety. Fear is a reaction to a situation in which the person experiencing the fear can identify its source. Anxiety is a reaction to a situation in which the person experiencing the anxiety can't identify its source. We can remove the source of fear because it's known, which is not the case with anxiety. Consequently Point 8 focuses on driving out fear.

Impact of Fear.   Fear in organizations has a profound impact on those working in the organization and on the functioning of the organization. On an individual level, fear can cause physical and physiological disorders such as a rise in blood pressure or an increase in heart rate. Behavioral changes, emotional problems, and physical ailments often result from fear and stress generated in work situations, as do drug and alcohol abuse, absenteeism, and burnout. These maladies impact heavily on any organization. An employee subjected to a climate dominated by fear experiences poor morale, poor productivity, stifling of creativity, reluctance to take risks, poor interpersonal relationships, and reduced motivation to optimize the system of interdependent stakeholders. The economic loss to an organization from fear is immeasurable, but huge.

A statistically based system of management will not work in a fear-filled environment. This is because people in the system will view statistics as a vehicle for policing and judging, rather than a method that provides opportunities for improvement.

Sources of Fear.   Fear emanates from lack of job security, possibility of physical harm, ignorance of company goals, shortcomings in hiring and training, poor supervision, lack of operational definitions, failure to meet quotas, blame for the problems of the system (fear of being below average and being punished), and faulty inspection procedures, to name a few causes. Management is responsible for changing the organization to eliminate the causes of fear.

Point 9: Break down barriers between departments. People in research, design, sales, and production must work as a team to foresee problems of production and in use that may be encountered with the product or service.
Management's job is to optimize the system of interdependent stakeholders. This may require suboptimization of some or all parts of the system. An example of suboptimization of a part, which leads to optimization of the whole, is a supermarket's ''loss leader'' product (a product carrying an extremely low price). The aim of a loss leader is to entice buyers into a store. Once in the store, buyers purchase other products, thereby creating a greater profit for the store. Profit from the loss leader is suboptimized to optimize store profit. Managers must remove incentives for suboptimization if they want to optimize the organization. For example, rating departments or divisions with respect to profit alone will foster suboptimization.

Barriers between areas cause suboptimization of the system of interdependent stakeholders by decreasing communication and thwarting cooperation. The greater the interdependence between the components of a system, the greater the need for communication and cooperation between them.

One last thought on barriers: Paradigms, such as Dr. Deming's theory of management, are constructive because they organize and focus their holder's thoughts and perceptions. However, paradigms are destructive because they can prevent their holders from seeing data that exist in the real world. What may be obvious to people with one paradigm may be invisible to people with another paradigm. Thus, breaking down barriers will often involve helping people see and understand the paradigms of other people and new paradigms.

Point 10: Eliminate slogans, exhortations, and targets for the work force that ask for zero defects and new levels of productivity.

Slogans, exhortations, and targets don't help to form a plan (method) to improve or innovate a process, product, or service. They don't operationally define process variables in need of improvement or innovation. They don't motivate individuals or clarify expectations. Slogans, exhortations, and targets are meaningless without methods to achieve them. Generally targets are set arbitrarily by someone for someone else. If a target doesn't provide a method to achieve it, it's a meaningless plea. Examples of slogans, exhortations, and targets that don't help anyone do a better job are

> Do it right the first time.
>
> Safety is job number 1.
>
> Increase return on net assets 3 percent next year.
>
> Decrease costs 10 percent next year.

These kinds of statement don't represent action items for employees; rather, they show management's wishes for a desired result. How, for example, can an employee "do it right the first time" without a method? People's motivation can be destroyed by slogans.

Slogans, exhortations, and targets shift responsibility for improvement and innovation of the system from management to the worker. The worker is powerless to make improvements to the system. This causes resentment, mistrust, and other negative emotions.

Point 11a: Eliminate work standards (quotas) on the factory floor. Substitute leadership.

Work standards, measured day work, and piecework are names given to a practice in American industry that has devastating effects on quality and productivity. A work standard is a specified level of performance determined by someone other than the worker who's actually performing the task.

The effects of work standards are, in general, negative. They don't provide a road map for improvement, and they prohibit good supervision and training. In a system of work standards, workers are blamed for problems beyond their control. In some cases, work standards actually encourage workers to produce defectives to meet a production quota. This robs workers of their pride and denies them the opportunity to produce high-quality goods and thus to contribute to the stability of their employment.

When work standards are set too high or too low, there are additional devastating effects. Setting work standards too high increases pressure on workers and results in the production of more defectives. Worker morale and motivation are diminished because the system encourages the making of defectives. Setting work standards too low also has negative effects. Workers who've met their quota spend the end of the day doing nothing; their morale is also destroyed.

Work standards are negotiated values that have no bearing on a process or its capability. Changes in the process's capability aren't considered. Consequently the work standards don't reflect the potential of the new system. Further, work standards create pressure to work at a given speed. This pressure inhibits employees' desire to improve beyond the work standard because they believe management will raise their quota.

Work standards are frequently used for budgeting, planning, and scheduling, and provide management with invalid information on which to base decisions. Planning, budgeting, and scheduling would improve greatly if they were based on process capability studies as determined by statistical methods. These will be discussed extensively in Chapter 11.

Point 11b: Eliminate management by objective. Eliminate management by numbers and numerical goals. Substitute leadership.

The Old Way

Setting Arbitrary Goals and Targets Is Dysfunctional.   Numerical goals are frequently set without understanding a system's capability. In a stable system, the proportion of the time an individual is above or below a specified quota/goal is a random lottery. This causes people below the quota to copy the actions of those above the quota even though they're both part of the same common cause system. This increases the variability of the entire system.

Numerical goals don't consider why a system is in its current state or how to improve it. Numerical goals are devoid of methods. They don't facilitate learning about the system.

Deploying Arbitrary Goals and Targets.   Managers use management by objectives to systematically break down a "strategic plan" into smaller and smaller subsections. Next managers assign the subsections to individuals or groups who are accountable for achieving results. This is considered fair because subsection goals emerge out of a negotiation between supervisor and supervisee. For example, an employee may negotiate a 3 percent increase in output instead of a 3.5 percent increase as long as the subsection's goals yield the goals of the strategic plan. It's important to note that employees aren't being given any new tools, resources, or methods to achieve the 3 percent increase. Consequently they must abuse the existing system to meet the goal. This type of behavior may allow an employee to meet a goal, but the system will fail somewhere else due to a lack of resources.

Evaluating People with Respect to Dysfunctional Goals and Targets.   Management by objectives, management by numbers, and management by numerical goals are just sophisticated methods to legitimize the managerial use of arbitrary numerical goals to hold people accountable for the problems of the system and consequently to steal their pride of workmanship. Management by objectives typifies the evils of management by numbers and management by numerical goals.

FIGURE 1.12   Methods/Result Matrix

		Result	
		Good	*Bad*
Method	Good	1	2
	Bad	3	4

SOURCE: This matrix was developed by Dr. Brian Joiner on April 1, 1990, in Atlanta, Georgia.

Numerical goals, without methods, that are motivated by extrinsic rewards often force individuals to choose between what's best for them and what's best for the system. This form of management rewards employees for meeting numerical goals regardless of the long-term impact to the system and fails to reward (sometimes punishes) employees for improving the system at the expense of short-term numerical goals. This is an example of a lose-lose situation; that is, the organization loses optimization and the individual loses joy in work.

The New Way

Relationship between the Aim of a System, Methods, and Goals (Targets).   A group of components come together to form a system with an aim. The aim requires that the components organize in such a way that they create subsystems. The subsystems are complex combinations of the components. The subsystems require certain methods to accomplish the aim. Resources are allocated between the methods by setting goals (targets) that may be numerical and that optimize the overall system, not the subsystems, with respect to the aim. For example, a group of individuals form a team with an aim. The individuals must combine their efforts to form subsystems. These combinations may require complex interactions between the individuals. The subsystems require methods, and the methods require resources. Resources are allocated between the methods and ultimately the subsystems and individuals by setting goals (targets) that optimize the team's aim. The aim, methods, and goals (targets) are all part of the same system; they can't be broken into three separate entities. Separation of the aim, methods, and goals (targets) destroys them because they are defined by their interactions.

Methods and Results.   Variation can cause a good method to yield undesirable results. Therefore one should not overreact (tamper) and change methods by considering negative results in the absence of theory. Figure 1.12 illustrates this point. Only cell 1 leads to positive results based on methods. All other cells represent either bad results (cells 2 and 4) or bad methods (cells 3 and 4); these cells are very misleading to managers who don't operate from a base of theory. Consequently managers should study the system of profound knowledge to understand theory's role in decision making.

Point 12: Remove barriers that rob the hourly worker of his right to pride of workmanship. The responsibility of supervisors must be changed from stressing sheer numbers to quality. Remove barriers that rob people in management and engineering of their right to pride of workmanship. This means, *inter alia,* abolishment of the annual merit rating and of management by objective.

Joy versus Pride.   Dr. Deming and others have replaced the concept of "pride in workmanship" (workmanship being the continuous improvement of art, skill, or technique applied to a particular task) with the concept of "joy in ownership through joy of workmanship."[23] In the system of interdependent stakeholders, sources of joy in ownership for different stakeholders include joy in workmanship for employees, joy in products and services for investors, joy in long-term, healthy relationships with customers for suppliers, and joy in economic and social health for the community. This expansive concept of joy defines a win-win scenario.

Joy in Workmanship.   People are born with the right to find joy in their workmanship. Joy in workmanship provides the impetus to perform better and to improve quality for the worker's self-esteem, for the company, and ultimately for the customer. People enjoy taking pride in their work, but very few are able to do so because of poor management. Management must remove the barriers that prevent employees from finding joy in their workmanship.

Barriers to Joy in Workmanship.   In the current system of Western management, there are many barriers to joy in workmanship. Examples include (1) employees not understanding their company's mission and what's expected of them with respect to the mission, (2) employees being forced to act as automatons who aren't allowed to think or use their skills, (3) employees being blamed for problems of the system, (4) hastily designed products and inadequately tested prototypes, (5) inadequate supervision and training, (6) faulty equipment, materials, and methods, and (7) management systems that focus only on results, such as daily production reports. The system of interdependent stakeholders will reap tremendous benefits when management removes barriers to joy in workmanship.

One of the great barriers to joy of workmanship is the traditional *performance appraisal system*. The annual or merit rating (performance appraisal) system robs people in management and engineering of their right to joy of workmanship. It does this in several ways:

1. It destroys teamwork by encouraging every person and every department to focus on individual goals, rather than on optimization of the interdependent system of stakeholders, to obtain a positive rating in the annual review. This leads to serious suboptimization with respect to the organizational aim. Simultaneous seeking of different, and possibly conflicting, goals creates variability in management's behavior, which creates confusion and fear as to exactly what everybody's job is. This destroys employees' ability to take pride in their workmanship.

2. The annual review reduces initiative or risk taking because once an objective is reached, effort stops. This is counter to the notion of continuous, never-ending improvement.

3. The annual review can cause an employee to lower her planned performance level to increase the chances of meeting her objectives.

4. The annual review can foster banking of performance (e.g., not contributing all ideas for improvements this year so that some can be submitted next year) to create a cushion for the next annual review cycle.

5. The annual review assumes that people are directly and solely responsible for their output. This assumption fails to consider the distinction between the individual's effect and the system's effect on output.

6. The annual review can increase variability in employee performance by rewarding everyone who's above average and penalizing everyone who's below average. In this type of situation, below-average employees try to emulate above-average employees. However, since the employees who are above average and those who are below average are part of the same system (only common variation is present), those who are below average are adjusting their behavior based on common variation; they are tampering.

7. The annual review focuses on the short term. A short-term focus promotes feelings of frustration and fear for employees because it decreases their ability to work toward long-term improvement and innovation.

In conclusion, the annual review is used for (1) judging individual performance, (2) rewarding and punishing individuals through promotion or demotion, and/or changes in salary and bonuses, and (3) providing direction and input for improvement of job skills and other functions. The annual review is woefully inadequate with respect to all the preceding functions. It was never designed to carry such a heavy burden.

An Idea that Will Promote Joy in Workmanship.   A view of performance improvement that's consistent with Dr. Deming's theory of management states that an employee can only exist in one of three categories: in the system, out of the system on the negative side, and out of the system on the positive side.[24] The latter two categories signify that the employee needs special attention. For example, if an employee shows improving or deteriorating performance through a lack of statistical control, that can be used as evidence that the employee needs special attention; this illustrates the institution of modern methods of supervision.

If all employees are within the system, then to reward or punish them on the basis of performance (or merit) would be destructive. This is because the punished employees will try to emulate the rewarded employees, when in fact there's no difference between the two types of employees; they are both in the system and any difference is only due to common variation, which is out of the individual's control. This type of reward/punish management will increase the variation between employees. Alternatively, if all employees are within the system, improvement results from application of the PDSA cycle to best practice methods; this will benefit all employees. If one or more employees aren't within the system (special employees), then management must (1) determine if the special employees form their own special system and (2) determine if the rest of the employees are within the old revised system. Next management must determine the sources of variation for the special employees and use this information to reduce variation, and create consistency and improvement.

An improved method for conducting the annual review is called "unbundling." [25] Unbundling calls for developing separate but interdependent systems for giving feedback to employees, providing an alternative basis for salary and bonuses, giving direction to employees, providing an occasion for communication, identifying candidates for promotion, and assessing training needs of employees.

Point 13: Encourage education and self-improvement for everyone.

Education and self-improvement are important vehicles for continuously improving employees, both professionally and personally. Leaders are obligated to educate and improve themselves and their people to optimize the system of interdependent stakeholders. Educators for leaders will probably have to come from outside the system.

A leader is educated by learning theory. A theory can be improved by continually studying its application in the real world. But an application can't be improved upon without the aid of a theory.

Point 14: Take action to accomplish the transformation.

The transformation of a system won't occur without the expenditure of energy by its stakeholders. Top management will expend this energy due to a variety of causes—for example, if they're confronted with a crisis or if they have a vision (aim) that they want to pursue. Other stakeholders will expend this energy if stimulated by top management. The transformation can't take place without a critical mass of stakeholders. The critical mass must include some policy makers.

Individuals have different reasons for wanting to, or not wanting to, accomplish the transformation. Individuals will have different interpretations of what's involved in the transformation. To be able to plan, control, and improve the transformation, a leader must know (1) each of her people's reasons for wanting (or not wanting) the transformation and (2) how each of those different reasons interact with each other and with the aim of the transformation.

Transformation of Management

Figure 1.13 presents issues involved in understanding the transformation of people and organizations from Western management's prevailing style to its new style. The figure displays (1) the prevailing paradigm of Western leadership and

FIGURE 1.13 Issues Involved in Transformation

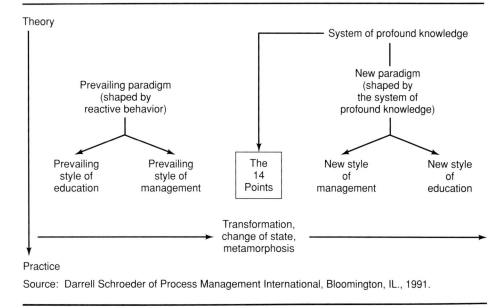

Source: Darrell Schroeder of Process Management International, Bloomington, IL., 1991.

the business and education systems it creates, (2) the new paradigm of Western leadership, the theory on which it is based, and the business and education systems it creates, and (3) the 14 points' role in the transformation process from the prevailing style of Western management to the new style of management.

The Prevailing Paradigm of Western Leadership

In Deming's words, "The prevailing style of management was not born with evil intent. It grew up little by little by reactive behavior, unsuited to any world, and especially unsuited to the new kind of world of dependence and interdependence that we are in now."[26] The prevailing paradigm of Western management, shown on the left side of Figure 1.13, isn't based on any holistic or comprehensive theory; it's just the cumulative result of assorted theories and experiences.

The New Paradigm of Leadership

The new style of management allows leadership to change and to develop a new basis for understanding the interrelationships between themselves and their environment. The environment includes people, systems, and organizations.

Transformation

It's not easy to move from the prevailing style of Western leadership to the new style of leadership. The 14 points provide a framework that helps explain the relationship between the prevailing style and the new style of management. They provide a window for managers operating under the prevailing techniques to compare and contrast their business practices with business practices in the new style of management. The real work of transformation comes from understanding the system of profound knowledge and the 14 points by building a new paradigm with which to operate in the 21st century.

"Transformation of American style of management is not a job of reconstruction, nor is it revision. It requires a whole new structure, from foundation upward,"[27] writes Deming.

Managers in one organization shouldn't use experiences of managers in another organization to focus their transformation efforts. This is because organizations are unique, having their own idiosyncrasies and nuances. Conditions that led to the experiences of managers in one organization may not exist for managers of the other organization. But this isn't to say that managers' experiences in one organization can't stimulate development of theories for improvement and innovation on the part of another organization's managers.

FIGURE 1.14   Service Applications in the Ford Motor Company

Organization	Application
Central laboratory	Time to process request of customer
	Errors in laboratory (based on audits)
Power train and chassis engineering	Time for supplier to notify company of failure
	Number of failures per month
Ford parts and service division	Errors in filling orders for dealer
Accounting	Time to process travel expense
Ford tractor operations engineering	Time to process engineering changes
Manufacturing staff, manufacturing engineering and systems	Time to review report on productivity sent from various Ford locations
Computer graphics	Variation in time of usage of disks
Product engineering office	Number of computer sign-on calls that gave busy signal
	Number of times that files (cabinets) were used for information
Product development	Run chart on number of revisions in word processing
	Wasted labor-hours resulting from late start of meetings
Comptroller's office	Errors in accounts payable resulting in late payment of invoices to suppliers
Saline plant	Costs from errors in scheduling
Purchasing staff, transportation, and traffic office	Transit time by rail from manufacture of part to assembly plants
Transmission and chassis division	Errors in shipments of components to assembly plants (wrong quantity or parts)

Quality in Service

The U.S. Census shows that 86 percent of Americans work in service organizations or perform service functions in manufacturing organizations. Since so many people work in service in the United States, improvement in our standard of living is highly dependent on better quality and productivity in the service sector.[28] Service organizations and service functions in manufacturing organizations require never-ending improvement of their system of interdependent stakeholders. Inefficient service increases price to the customer and lowers quality.[29]

A denominator common to manufacturing and any service organization is that mistakes and defects are costly. The further a mistake goes without correction, the greater the cost to correct it. A defect that reaches the consumer or recipient may be costliest of all.[30] Figure 1.14 lists some active administrative (service) applications of analytic studies in the Ford Motor Company.[31]

The principles and methods for process improvement are the same in service and manufacturing. The system of profound knowledge and 14 points apply equally to both sectors of the economy.

Summary

Quality is predictable uniformity and dependability at low cost, and is suited to the needs of the market. Three types of quality are integral to improving the system of interdependent stakeholders: quality of design or redesign, quality of conformance, and quality of performance.

Issues of quality have existed since tribal chiefs, kings, and pharaohs ruled. The modern history of quality is marked by great advances between 1920 and the 1950s by George Edwards, Walter Shewhart, W. Edwards Deming, Armand Feigenbaum, and Joseph Juran. The 1970s, 1980s, and 1990s have been characterized by foreign competition threatening American companies. A renewed emphasis on quality control has been the response. Deming, Juran, and Feigenbaum are among the leaders in this area. The book focuses on Deming's philosophy and methods.

Total Quality Management (TQM) is the management theory that Western leaders must adopt to (1) improve and innovate the system of interdependent stakeholders of an organization over the long term to allow all people to take joy in their work and pride in the outcome, (2) optimize the system of interdependent stakeholders of an organization over the long term so that everybody wins, and (3) improve and innovate the condition of society. The theoretical base of TQM is the "system of profound knowledge." This system helps people learn through the acquisition of process knowledge gained from experience and coordinated by theory.

The 14 points for management follow naturally from the system of profound knowledge. They are a vehicle to understand the relationship between the prevailing style and the new style of Western management. They provide a window for managers operating in the prevailing paradigm to compare and contrast their business practices with the business practices in the new style of management.

Acknowledgments

The authors would like to thank Darrell Schroeder and Louis Schultz of Process Management International, Bloomington, Minnesota, for their significant assistance in developing this chapter.

Exercises

1.1 Discuss the concept of a system of interdependent stakeholders.

1.2 Define quality.

1.3 Define quality of design, quality of conformance, and quality of performance.

1.4 Products have been changed over time because customers have been dissatisfied with certain product characteristics such as the smell of canned dog food. Describe another such product change motivated by customer dissatisfaction.

1.5 Service call analysis has resulted in changes to consumer products (such as the change from mechanical to electronic tuners on TVs). Describe one such case.

1.6 Discuss the two sources of loss in quality.

1.7 Describe the relationship between quality and productivity. Do efforts at productivity improvement help or hurt quality? Why? Do quality improvement efforts help or hurt productivity? Why? Construct examples for all of the preceding cases.

1.8 List the benefits of process improvement.

1.9 List and briefly describe Dr. Deming's 14 points.

1.10 Describe the Deming cycle and discuss its importance to never-ending improvement.

1.11 List some service applications for never-ending improvement in your organization.

1.12 Discuss the similarity in process improvement efforts in service and manufacturing.

1.13 Discuss the difference between product features and product quality. Select a particular consumer product for the focus of your discussion.

1.14 Discuss the benefits to quality that can result from cooperation as opposed to competition between the departments of an organization.

Endnotes

1. Paraphrased from W. Edwards Deming, *Quality, Productivity, and Competitive Position* (Cambridge, Mass.: Massachusetts Institute of Technology, 1982), p. 229.

2. Kaoru Ishikawa and David Lu, *What Is Total Quality Control? The Japanese Way* (Englewood Cliffs, N.J.: Prentice-Hall, 1985), p. 97.

3. The source document for this material is J. Juran, *Quality Control Handbook,* 3d ed. (New York: McGraw-Hill, 1979), pp. 2-4 through 2-9.

4. G. Taguchi and Y. Wu, *Introduction to Off-Line Quality Control* (Nagoya, Japan: Central Japan Quality Control Association, 1980), pp. 7–9.

5. H. Gitlow and S. Gitlow, *The Deming Guide to Quality and Competitive Position* (Englewood Cliffs, N.J.: Prentice-Hall, 1987), pp. 9–10.

6. Kano, Seraku, Takahashi, and Tsuji, "Attractive Quality versus Must Be Quality," *Hinshitsu* (Quality), vol. 14, no. 2 (1984), pp. 39–48.

7. Paraphrased from a lecture by Dr. W. E. Deming concerning managerial education in the United States at Fordham University, November 6, 1989.

8. The Cannon camera example was presented as part of a lecture by Dr. Noriaki Kano of the Science University of Tokyo, at the University of Miami, Coral Gables, Florida, on September 27, 1988.

9. Gitlow and Gitlow, pp. 29–31.

10. H. J. Harrington, "Quality's Footprints in Time," IBM Technical Report, TR02.1064, September 20, 1983, p. 7.

11. Ibid., p. 8.

12. Ibid.

13. Gitlow and Gitlow, p. 7.

14. Harrington, pp. 29–31.

15. Dr. Deming has discussed the benefits of cooperation between competitors. For example, all railroads agree on using the same gauge track. In another example, two competing gas stations, neither of which could afford its own tow truck share one tow truck to mutual advantage.

16. Quotation from W. E. Deming, paper delivered at a meeting of The Institute of Management Sciences in Osaka, July 24, 1989.

17. W. E. Deming, *The New Economics for Industry, Government, Education* (Cambridge, Mass.: Massachusetts Institute of Technology, 1993), pp. 94–118.

18. W. E. Deming, "Foundation for Management of Quality in the Western World," revised April 1, 1990, delivered at a meeting of The Institute of Management Sciences in Osaka, July 24, 1989.

19. William Scherkenbach, *The Deming Route to Quality and Productivity: Road Maps and Roadblocks* (Washington, D.C.: CeePress, 1986), pp. 131–136.

20. Ibid.

21. W. E. Deming, *Out of the Crisis* (Cambridge, Mass.: Massachusetts Institute of Technology, Center for Advanced Engineering Study, 1986), pp. 86–89.

22. Information in this section was developed by David Pietenpol (Employers Health Insurance, Green Bay, Wis.) and Howard Gitlow.

23. William Scherkenbach, *Deming's Road to Continual Improvement* (Knoxville, Tenn.: SPC Press, 1991), p. 4.

24. William Scherkenbach, *The Deming Route to Quality and Productivity: Road Maps and Road Blocks* (Washington, D.C.: CeePress, 1986), p. 68.

25. Peter Scholtes, *The Deming Users Manual—Parts 1 and 2* (Joiner Associates, 1988).

26. W. E. Deming, ''Foreword by W. Edwards Deming,'' in Henry Neave, *The Deming Dimension* (Knoxville, Tenn.: SPC Press, 1990), p. vii.

27. Deming (1986), flyleaf cover.

28. Ibid., pp. 184–85.

29. Ibid., p. 183.

30. Ibid., p. 190.

31. Ibid., p. 183.

CHAPTER 2 Fundamentals of Statistical Studies

Purpose and Definition of Statistics

The purpose of statistics is to study and understand process and product variation, interactions among product and process variables, operational definitions (definitions of process and product variables that promote effective communication between people), and ultimately to take action to reduce variation[1] in a process or population. Hence statistics can be broadly defined as the study of data to provide a basis for action on a population or process.[2]

Types of Statistical Studies

Statistical studies can be either one of two types: enumerative or analytic. *Enumerative studies* are statistical investigations that lead to action on a static population (that is, a group of items, people, etc. that exist in a given time period and/or at a given location).

An example of an enumerative problem is the estimation of the number of residents in South Florida left homeless by Hurricane Andrew in 1992 and the number of tents required to temporarily house them. This estimation might be performed to determine how much food must be shipped in to feed the residents. This is an enumerative study because it investigates the number of people in need at a specific point in time. Dynamic questions such as why the people are where they are or why they need the supplies that they need aren't considered.[3]

Other examples of enumerative studies are determining market share for product X in households with a certain demographic profile; calculating statistics

43

on births, deaths, income, education, and occupation, by area; estimating the frequency of an illness in a given city; and assaying samples from a barge of coal to determine an appropriate price. All of these examples are time specific and static; there's no reference to the past or the future.

Analytic studies are statistical investigations that lead to action on a dynamic process. They focus on the causes of patterns and variations that take place from year to year, from area to area, from class to class, or from one treatment to another.[4] An example is determining why grain production in an area is low and how it can be increased in the future.

Other examples of analytic studies are testing varieties of wheat to determine the optimal type for a particular area's future production; comparing the output of two machine types over time to determine if one is more productive; comparing ways of advertising a product or service to increase market share; and measuring the effects of an action (such as change in speed, temperature, or ingredients) on an industrial process output.[5] All of these examples focus on the future, not on the present. The information gathered in the preceding examples is used to make dynamic decisions about a process: What type of wheat should be grown to achieve an optimal yield? Should we replace machine A with machine B to raise production? Is TV advertising more effective than print media for increasing market share? If we change the ingredients in our product, can we improve its quality and lower its cost?

Both types of statistical studies focus on quality-related issues in subsequent chapters. However our primary focus is on analytic studies.

Enumerative Studies

Basic Concepts.   As stated before, enumerative studies are statistical investigations that lead to action on static populations. A *population* (or universe) is the total number of units, items, or people of interest that exist in a given time period and/or given location and so on. If a population is to be studied, it must be operationally defined by listing all of its units, items, people, and so on. This list is called a *frame*. It's assumed that the population and frame are identical. If they are not, then bias and error will occur in any study results. There are many reasons that a frame may differ from the population it's supposed to define: omissions, errors, or double counting, for example. Regardless of the type of error, the difference between the frame and the population is called the *gap*.

A *sample* is a portion of the population under investigation, and is selected so that information can be drawn from it about the population. For example, 100 randomly selected accounts receivable drawn from a list, or frame, of 10,000 accounts receivable constitute a sample. There are two basic types of samples: nonrandom and random.

Nonrandom samples are selected on the basis of convenience (convenience sample), the opinion of an expert (judgment sample), or a quota to ensure proportional representation of certain classes of items, units, or people in the sample (quota sample). All nonrandom samples have the same shortcoming: They're subject to an unknown degree of bias in their results. This bias is caused by the absence of a frame. The sampled items aren't selected from an operationally defined population via a frame; hence classes of items or people in the population may be systematically denied representation in the sample. This may cause bias in the results. Consequently nonrandom samples should be used only when better information is too costly to obtain.

To demonstrate the bias of a nonrandom sample, let's look at a case in which we wish to estimate the average salary in a firm's manufacturing area. The list of names is organized by work crews. Every fifth name is selected as part of the sample. But since the list is organized by work crews, it first lists the foreman and then 4 workers. Thus the sample will contain either all foremen or all workers, depending on the starting point. This creates bias in the results—the average salary won't reflect the true average because of the method of selecting the sample.

Random samples are selected so that every element in the frame has a known probability of selection. Types of random samples include simple, stratified, and cluster. These types of samples are discussed in many of the sampling and basic statistics references at the end of this book. All random samples have the same strengths. They allow generalized statements to be made about the frame from the sample. These generalized statements form the basis for action on the population under study.

A random sample is selected by operationally defining a procedure that utilizes random numbers in the selection of the sampled items from the frame to eliminate bias and hold uncertainty within known limits. Seven steps are involved in selecting a simple random sample:

Step 1. Count the number of elements in the frame. Let N be this number.

Step 2. Number the elements in the frame from 1 through N. If N is 25, then the elements in the frame should be numbered from 01 through 25, as in Figure 2.1. All elements must receive an identification number with the same number of digits.

FIGURE 2.1   Identification Frame

Item	Identification Number
A	01
B	02
C	03
D	04
E	05
F	06
.	.
.	.
.	.
Y	25

FIGURE 2.2

(a) Random Numbers from Page 495 of Table 2 in the Appendix

19	30	40
09	28	78
31	13	98
67	60	69
61	13	39
04	34	62
05	28	56
73	59	90
54	87	09
42	29	34
27	62	12
49	38	69
29	40	93

(b) Random Sample Size $n = 6$ from Frame of Size $N = 25$

Sample Number	Identification Number	Items
1	19	S
2	09	I
3	04	D
4	05	E
5	13	M
6	12	L

Step 3. Select a page in Table 2, a table of random numbers. For example, select page 495.

Step 4. On the selected page of random numbers, randomly select a column of numbers, select a random starting point in that column, and use as many digits as there are digits in N (two digits in the case of $N = 25$). For example, beginning with the first column in Table 2 on page 495, selecting the seventh line of that column as the starting point, and using the first two digits of each number in that column, the first random number is 19. Figure 2.2a shows a list of 39 random numbers obtained in this way.

Step 5. Determine the necessary sample size. This calculation is discussed in many textbooks. Let n be this number. Assume $n = 6$ in this example.

Step 6. From the chosen column on the selected page, select the first six two-digit numbers between 01 and 25, inclusive. If a number is encountered that's smaller than 01 (e.g., 00) or larger than 25 (e.g., 31), ignore the number and continue down the column. If an acceptable number appears more than once, ignore every repetition and continue moving down the column until six unique numbers have been selected. If the bottom of the page is reached before six unique random numbers are obtained, go to the top of the page and move down the next two-digit column. Figure 2.2b shows the six random numbers selected and the corresponding items comprising the sample.

Step 7. Finally, analyze the information as a basis for action.

Two important points to remember are (1) different samples of size six will yield different results and (2) different methods of measurement will also yield different results. There's no true value possible, only one with which we may do business. Random samples, however, don't have bias, and the sampling error can be held to any given limits by increasing the sample size. These are the advantages of random sampling over nonrandom sampling. Because of their inherent bias, nonrandom samples are rarely used in enumerative studies, even though they're easier to implement.

Conducting an Enumerative Study.   The following 13 steps present a guide for conducting an enumerative study. All steps are the same whether the study is based on a complete count or a sample.[6]

Step 1. Specify the reason(s) you want to conduct the study (for example, to estimate the average number of sick days per employee in the XYZ Company in 1994). If this average is greater than 8.0 days, then a preventive health care plan will be instituted. If it's less than or equal to 8.0 days, the current plan will be maintained.

Step 2. Specify the population to be studied. In our example, the population would be all full-time employees in the XYZ Company in 1994. An employee is considered full-time in 1994 if he had full-time status designation at any time during the year.

Step 3. Construct the frame (a list of all full-time employees). Everyone who'll use the study's results as a basis for action must agree that the frame represents the population upon which they want to take action.

Step 4. Perform secondary research (such as the examination of prepublished data) to determine how much information is already available about the problem under investigation. For example, check the Personnel Department's records.

Step 5. Determine the type of study to be conducted (for example, mail survey, personal interviews, chemical analysis of units). In this example, we would survey employee absentee cards for 1994.

Step 6. Make it possible for respondents to give clear, understandable information, and/or for the researcher to elicit clear, understandable information. For example, the method for analyzing absentee cards should be clear and straightforward. Consider the problem of nonresponse. Refusal to answer, no one at home, and missing items to be studied are all possible causes of nonresponse. How significant is the nonresponse? Will it impair the study results? What can be done to reduce it? Establish a procedure for dealing with and reducing nonresponse problems. In this example, be sure no absentee cards are missing, and make sure the data gatherers know how to interpret the absentee cards.

Step 7. Establish the sampling plan to be used, determine the amount of allowable error in the results, and calculate the cost of the sampling plan. At this stage, Steps 1 and 2 may need revisions due to cost considerations. For example, we may decide to draw a simple random sample of employee absentee cards using random numbers, at a cost of $1 per card, assuming an allowable error of one quarter of a day in the estimate.

Step 8. Establish procedures to deal with nonresponse problems and differences between interviews, testers, inspectors, and so on. For example, be able to assess differences in collected data due to differences in data gatherers' abilities.

Step 9. Prepare unambiguous instructions for the data gatherers that cover all phases of data collection. Supervisors may require special additional instructions. Train all gata gatherers and supervisors. Use statistical methods to determine when their training is complete, as discussed in Point 6 of Chapter 1.

Step 10. Establish plans for data handling including format of tables, headings, and number of classes.

Step 11. Pretest the data gathering instrument and data gathering instructions. If pretests show a high refusal rate or generate unsatisfactory-quality data, the study may be modified or abandoned. Conduct a dry run with the data gatherers to ensure that they understand and adhere to established procedures. Revise the data gathering instrument and instructions based on the information collected during the pretest. Finalize all aspects of the study's procedure.

Step 12. Conduct the study and the tabulations. It's critical that the study be carried out according to plans. From the gathered data, calculate the sampling errors of interest. This is important so that the study's users can understand the degree of uncertainty present in the study results. In our example, we would calculate the standard error for the average number of days absent per employee.

Step 13. Interpret and publish the results so that decision makers can take appropriate action. For example, if the average number of days absent per employee is greater than eight, then establish the preventative health care program.

In conducting an enumerative study, you must realize that Step 7 requires random sampling, not nonrandom sampling, so that the information can serve as a basis for action with a known degree of uncertainty. The result of a nonrandom sample in an enumerative study is worth no more than the reputation of the person who signs the report.[7] This is because the margin of uncertainty in the estimate reported depends entirely on his knowledge and judgment, rather than on objective, quantifiable methods.

Analytic Studies

Basic Concepts.   Analytic studies are statistical investigations that lead to actions on the cause-and-effect systems of a process (that is, the systems creating the past, present, and future output of a process).[8] An analytic study's aim is prediction of a process's future state so that it can be improved and/or innovated over time.

As stated earlier, in an enumerative study, a *population* is the total number of units, items, or people of interest that exist in a given time period and/or location. The concept of a population doesn't exist for an analytic study because future process output, which doesn't yet exist, can't be part of the population. Consequently, a frame can't exist without a population. Lack of a population and frame makes it impossible to draw a random sample to study the cause-and-effect systems that dictate a process's behavior.

The inability to study a process in an enumerative sense seemingly creates difficulty in understanding its underlying cause-and-effect systems. However, models can be used to study these cause-and-effect systems. These models include simulations of the process, prototypes of a product, flowcharts of a process, operational definitions of process or product quality characteristics, and cause-and-effect diagrams describing the factors and conditions that influence quality characteristics.[9] Unfortunately it's impossible to consider all of the factors that will influence the process's output. Conditions will change. There will always be some uncertainty. In an analytic study (unlike an enumerative study), there is no theory that enables quantification of this uncertainty.

Expert opinion is invaluable in understanding the magnitude of the uncertainty caused by changes to a process (changes from, for example, new equipment, new workers, new tools, new methods, or different operating conditions). Generally this uncertainty is best explained by an expert who's involved with the process under study. In general, the individual(s) directly involved with a process know more about it than anyone else.

Conducting an Analytic Study.   Activities aimed at improvement or innovation of a process are structures using the plan-do-study-act (PDSA) cycle. The PDSA cycle is used to study, improve, and innovate a process. Improvements and innovations are aimed at narrowing the difference between process performance and customer (either internal or external) needs and wants.

In the **plan** stage of the PDSA cycle, the aim is to determine an improvement or innovation to a process that will narrow the difference between process performance and customer needs and wants. In the plan stage, where we study and understand a process, the many tools and methods used include (but aren't limited to) brainstorming, cause and effect diagrams, linear programming, simulation models, control charts, and Taguchi methods. Several of these methods are discussed later in this text.

In analytic studies, judgment samples can be drawn from a stable process to study its output. Judgment samples are samples selected based on an expert's opinion. Generally, if a process is stable, any slice of process output (a set of judgment samples) will be very revealing about the process's future behavior. The use of judgment samples to study a process's output will be discussed in great detail in Chapters 5 through 7.

The do stage of the PDSA cycle requires that planned experiments be conducted to determine how best to implement the plan established in the previous stage. Experiments should be conducted on a small scale. Experiments may be conducted using a laboratory, an office or plant site, or even a customer's location (with the consent of that customer). The aim here is to place the plan into action in a realistic setting.

The results of the work in the do stage are examined in the study stage. The aim of the study stage is to determine if the plan has been effective in decreasing the difference between process performance and customer needs and wants. If the plan hasn't been effective, return to the plan stage to attempt to devise a new course of action. However, if the plan has been effective, the process owner should move on to the act stage.

In the act stage, the plan should be integrated into the process. This may involve anything from minor alterations to the process to a major overhaul of an entire operating procedure. Regardless, the act stage includes training all relevant personnel with respect to the details of the revised process. Keep in mind that actions to improve a process's future functioning are rational only if the process is stable. Process stability means that there are not special sources of variation present for the process. Methods for discovering special sources of variation will be discussed in Chapter 8. Having a stable process allows a process expert to predict a process's future performance with a high degree of belief.

Distinguishing between Enumerative and Analytic Studies

A simple rule exists for deciding whether a study should be enumerative or analytic. If a 100 percent sample of the frame answers the question under investigation, the study is enumerative; if not, the study is analytic.[10] A 100 percent sample can be obtained in an enumerative problem because items are drawn from the frame, which is composed of all the elements in the population. A 100 percent sample can't be obtained in an analytic problem because there's no frame. We're merely studying the output of a dynamic process with an eye to taking action on the process that created the output data.

Statistical Studies and Quality

Statistical studies are critical components of quality improvement efforts. When conducting research for the improvement of quality of design, quality of conformance, and quality of performance, you must understand the different types of statistical studies so that the information generated by the study can be used rationally as a basis for quality improvement action.

Summary

The goal of a statistical study is to provide a basis for action on a population or process. There are two types of statistical studies: enumerative (which lead to action on a static population) and analytic (which lead to action on dynamic process). Chapter 2 focused on tasks necessary to undertake both kinds of studies.

There are two basic types of sampling plans: random and judgment. Random samples are required to provide useful information from an enumerative study. Judgment samples serve the same function for analytic studies.

We must be able to determine whether a study should be enumerative or analytic because the two types of study require different sampling and computational techniques. Here's a simple rule: If a 100 percent sample of the frame answers the question under investigation, the study is enumerative; if not, the study is analytic. We must also understand the different types of statistical studies so that the results can be used for quality improvement actions.

Exercises

2.1 List the two types of statistical studies. What's the purpose of any statistical study? Describe each type of statistical study. List three examples of each type of study in quality improvement efforts.

2.2 Define the following terms: *population, frame, gap, nonrandom sample, judgment sample, random sample*.

2.3 List the steps required to draw a simple random sample.

2.4 List the steps required to perform an enumerative study.

2.5 Discuss why judgment samples can't be used as a rational basis for action in enumerative studies.

2.6 Discuss why judgment samples can be used as a rational basis for action in analytic studies.

2.7 Specify a rule to distinguish between enumerative and analytic studies.

Endnotes

1. Stated by Dr. W. Edwards Deming, Seminar on Improved Practice of Statistics, March 13–15, 1989, New York University.
2. W. Edwards Deming, *Some Theory of Sampling* (New York: John Wiley and Sons, 1950), p. 247.
3. Ibid., pp. 247–48.
4. Ibid., p. 249.
5. W. Edwards Deming, ''On the Use of Judgment-Samples,'' *Reports of Statistical Applications,* vol. 23 (March 1976), p. 26.
6. Deming (1950), pp. 4–9.
7. Deming (1976), p. 29.
8. Idea taken from R. Moen, T. Nolan, and L. Provost, *Improving Quality through Experimentation* (New York: McGraw-Hill, 1991), pp. 54–55.
9. Paraphrased from ibid., p. 55.
10. Deming (1976), p. 26.

3 Defining and Documenting a Process

This chapter explains process basics and questions to consider when documenting and defining a process using empowerment and the SDSA cycle in the context of analytic studies.

A Brief Review

Empowerment Revisited

We learned in Chapter 1 that empowerment provides an employee with (1) the opportunity to define and document processes, (2) the opportunity to learn about processes through training and development, (3) the opportunity to improve and innovate best practice methods that make up processes, (4) the latitude to use her own judgment to make decisions within the context of best practice methods, and (5) an environment of trust in which superiors won't react negatively to the latitude taken by people making decisions within the context of best practice methods.

Empowerment is operationalized at two levels. First, employees are empowered to develop and document best practice methods using the SDSA cycle. Second, employees are empowered to improve or innovate best practice methods through application of the PDSA cycle.

SDSA Cycle Revisited

Recall from Chapter 1 how the SDSA cycle is a tool that helps employees define and document a process; it helps to create an identity for a process. It includes the following four steps:

Standardize: Employees study the process and develop best practice methods using tools such as flowcharts.

Do: Employees conduct experiments using the proposed best practice methods on a trial basis.

Study: Employees research the effectiveness of the best practice methods.

Act: Managers establish standardized best practice methods and formalize them through training.

Analytic Studies Revisited

Analytic studies are statistical investigations that lead to action on a dynamic process. Consequently analytic studies are the vehicle for using the SDSA cycle in the context of the first level of empowerment to define and document a process.

Process Basics

What exactly is a process? We've used the term previously in our discussion of analytic studies. Now let's be more specific about its definition.

A process is the transformation of inputs (manpower/services, equipment, materials/goods, methods, and environment) into outputs (manpower/services, equipment, materials/goods, methods, and environment). The transformation involves the addition or creation of value in one of three aspects: time, place, or form:

Time value: Something is available *when* it is needed. For example, you have food when you're hungry, or material inputs are ready on time.

Place value: Something is available *where* it's needed. For example, gas is in your tank (not in an oil field), or wood chips are in a paper mill.

Form value: Something is available *how* it's needed. For example, bread is sliced so it can fit in a toaster, or paper has three holes so it can be placed in a binder.

Figure 3.1 shows a basic process.

FIGURE 3.1   Basic Process

Inputs	Process	Outputs
Manpower/Services Equipment Materials/Goods ⟶ Methods Environment	Transformation of inputs, value (time, place, form) is added or created ⟶	Manpower/Services Equipment Materials/Goods Methods Environment

FIGURE 3.2   Feedback Loop

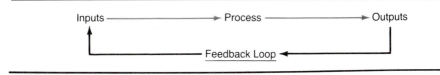

Processes exist in all aspects of organizations, and our understanding of them is crucial. Many people mistakenly think of only production processes. However administration, sales, service, human resources, training, maintenance, paper flows, interdepartmental communication, and vendor relations are all processes. These processes can be studied, documented, defined, improved, and innovated. We must study processes in every area of an organization.

An organization is a multiplicity of micro subprocesses, all synergistically building to the macro process of that firm. We must realize that all processes have customers and suppliers. These customers and suppliers can be internal or external to the organization. A customer can be an end user or the next operation down the line. The customer doesn't even have to be a human; it could be a machine. A vendor could be another firm supplying subassemblies or services, or the prior operation up the line. Similarly, the vendor doesn't even have to be human.

The Feedback Loop

An important aspect of a process is a feedback loop. A feedback loop relates information about outputs back to the input stage so that an analysis of the transformation process can be made. Figure 3.2 depicts the feedback loop in relation to a basic process. The tools and methods discussed in this book provide vehicles for relating information about outputs back to the input stage. Decision making about processes is aided by the transmission of this information. A major purpose of analytic studies is to provide the information (flowing through a feedback loop) needed to take action with respect to a process.

Examples of Processes

An example of a process that most people are familiar with is the hiring process (Figure 3.3). The figure shows the inputs, process, and outputs of hiring an employee. The inputs (which include the candidate's resume, the information gathered from the interview with the candidate, and other data from references, former employers, and schools attended) are transformed into the output (which is an employee to fill a vacant position). The process involved in this transformation of inputs into output includes synthesizing and evaluating the information, making a decision, and hiring the candidate. An important aspect of this process is the feedback loop that enables the new employee's supervisor to report back to the

FIGURE 3.3   Hiring Process

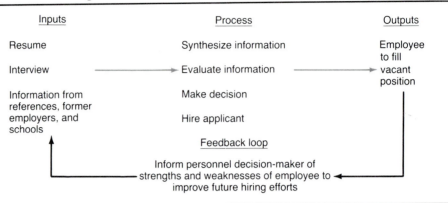

FIGURE 3.4   Memo-Sending Process

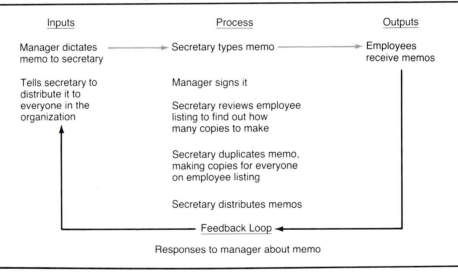

personnel decision maker as to the employee's appropriateness. The personnel decision maker (supplier) and supervisor (customer) can use this information to work together to improve the hiring process.

Clerical functions are also processes. Figure 3.4 depicts the process of sending a memo in an organization. First, the manager inputs the information for the memo and her instructions regarding its distribution. Then the secretary trans-

FIGURE 3.5   Production Process

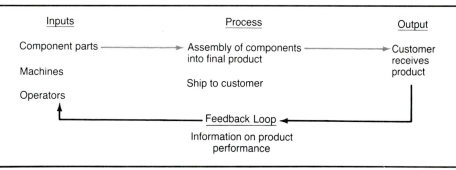

forms the input, by typing it and distributing it, into the output (communication to the employees). The feedback loop is important because the manager and the employees can work together to improve the communication process.

Figure 3.5 shows an example of a generic production process. The inputs (component parts, machines, and operators) are transformed in the process of making the final product and shipping it to the customer. The output is the customer receiving the product. Again, the feedback loop (which in this case is the customer's reporting back to the supplier on the product's performance) is vital. Communicating this way promotes cooperation on process improvement and innovation.

Defining and Documenting a Process

Defining and documenting a process is an important step toward improvement and/or innovation of the process. The following example demonstrates this point.

In a study of an industrial laundry, an analyst began to diagram the flow of paperwork using a flow chart. While walking the green copy of an invoice through each step of its life cycle, he came upon a secretary feverishly transcribing information from the green invoice copy into large black loose-leaf books. The analyst asked, "What are you doing?" so that he could record it on his flow diagram. She responded, "I'm recording the numbers from the green papers into the black books." He asked, "What are the black books?" She said, "I don't know." He asked, "What are the green papers?" She said, "I don't know." He asked, "Who looks at the black books?" She said, "Easy. I know the answer to that question. Nobody." He asked, "How long have you been going this job?" She said, "Seven and one half years." He asked, "What percentage of your time do you spend working on the black books?" She said, "About 60 percent."

Next, the analyst did two things. First, he examined the black books. Second, he asked the secretary how long ago the person who hired and trained her had left the company. She said that the person who hired her had left the company seven years ago. At this point, the consultant realized he had solved a problem the company didn't know it had. From examining the black books he realized that they were sales analysis books. Every day all sales, by item, were entered into the appropriate page in the black book. At the end of each month, page totals were calculated, yielding monthly sales by items. Nobody was looking at the books because nobody knew what the secretary was doing. The current manager assumed the secretary was doing something important. After all, she seemed so busy.

This problem is an example of a failure to define and document a process to make sure that it's logical, complete, and efficient. As an epilogue, the secretary was reassigned to other needed duties because the sales analysis had been computerized five years earlier.

Questions to Consider

Considering the following questions will help you define and document a process.

1. Who owns the process? Who's responsible for the process's improvement?
2. What are the boundaries of the process?
3. What is the flow of the process?
4. What are the process's objectives? What measurements are being taken on the process with respect to its objectives?
5. Are process data valid?

Now let's discuss the preceding questions.

Who Owns the Process?   Every process must have an owner (an individual who's responsible for the process). Process owners may have responsibilities extending beyond their departments. So they must be high enough in the organization to influence the resources necesssary to take action on the process.

In some cases, a process owner is the focal point for the functions that operate within her process. However each function is still controlled by the line management within that function. The process owner will require representation from each function that comprises her process. These individuals are assigned by the line managers of the subprocesses, provide functional expertise to the process owner, and are the promoters of change within their functions. A process owner is the coach and counsel of her process in an organization.

What Are the Boundaries of the Process?   Boundaries must be established for processes. These boundaries make it easier to establish process ownership and highlight the process's key interfaces with other (customer/vendor) processes. Process interfaces frequently are the source of process problems. These problems are often caused by a failure to understand downstream requirements. Process interfaces must be carefully managed to prevent communication problems. Construction of operational definitions for critical process characteristics that are agreed upon by process owners on both sides of a process boundary will go a long

FIGURE 3.6 **Quality-of-Design Study**

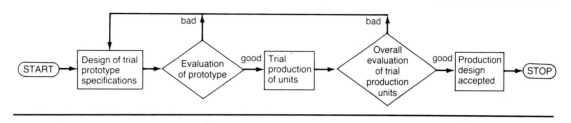

way toward eliminating process interface problems. Operational definitions are discussed in depth later in this chapter.

What Is the Flow of the Process? A flowchart is a pictorial summary of the flows and decisions that comprise a process. It's used for defining and documenting the process. Figure 3.6 shows an example of a flowchart for one type of quality-of-design study (a production design study). In the design phase, design engineers determine detailed specifications for the product. They also prepare a full-scale prototype of the product. Then the design is tested and evaluated. It can be judged as good or bad. If it's bad, it returns to the design stage for redesign. If it's good, it moves to the next phase, trial production. After this stage, the product is again evaluated. If it's judged bad, it returns to the design stage. If it's deemed good, the production design is accepted and full-scale production begins. During appropriate stages of the process, the results are fed back to the designers. Also overall evaluations determine how the product compares with competitors' products and whether it meets consumers' expectations.

Once a flowchart is constructed, it can provide important information about a process. This information can help a manager, designer, analyst, or anyone else seeking to define, document, study, improve, or innovate the process.

Flowchart Symbols. The American National Standards Institute, Inc. (ANSI) has approved a standard set of flowchart symbols that are used for defining and documenting a process. The shape of the symbol and the information written within the symbol provide information about that particular step of the process. Figure 3.7 shows the basic symbols for flowcharting.[1] Using these symbols standardizes process definition and documentation.

Types of Flowcharts. This section, while not exhaustive,[2] describes several types of flowcharts used in defining and documenting a process.

Systems Flowchart. A systems flowchart is a pictorial summary of the sequence of operations that make up a process. It shows what's being done in a process. Figure 3.6's flowchart is an example of a systems flowchart. It represents each phase or stage in the process of a quality-of-design study using standard flowcharting symbols.

FIGURE 3.7   **Flowchart Symbols**

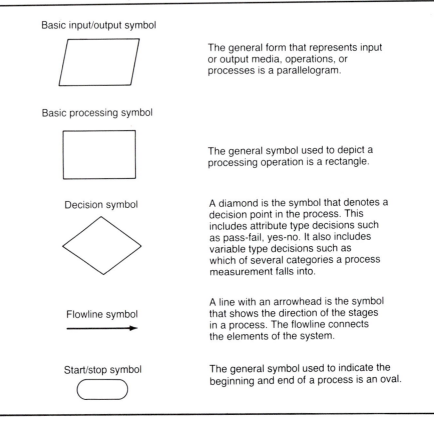

Basic input/output symbol

The general form that represents input or output media, operations, or processes is a parallelogram.

Basic processing symbol

The general symbol used to depict a processing operation is a rectangle.

Decision symbol

A diamond is the symbol that denotes a decision point in the process. This includes attribute type decisions such as pass-fail, yes-no. It also includes variable type decisions such as which of several categories a process measurement falls into.

Flowline symbol

A line with an arrowhead is the symbol that shows the direction of the stages in a process. The flowline connects the elements of the system.

Start/stop symbol

The general symbol used to indicate the beginning and end of a process is an oval.

Layout Flowchart.   A layout flowchart depicts the floor plan of an area, usually including the flow of paperwork or goods and the location of equipment, file cabinets, storage areas, and so on. These flowcharts are especially helpful in improving the layout to more efficiently utilize a space. Figure 3.8 shows a layout flowchart before and after flowcharting analysis and flow improvement. The existing system is shown in Figure 3.8a; the new proposed system appears in Figure 3.8b. This flowchart includes the flow of work along with the floor plan and location of desks and files. Comparing the existing and proposed systems is simple when the process's flow is documented this way.

Advantages of a Flowchart.   Flowcharting a process, as opposed to using written or verbal descriptions, has several advantages.

   • *It functions as a communications tool.* A flowchart provides an easy way to convey ideas between engineers, managers, hourly personnel, vendors, and others in the extended process. It's a concrete, visual way of representing complex systems.

FIGURE 3.8   Layout Flow Chart

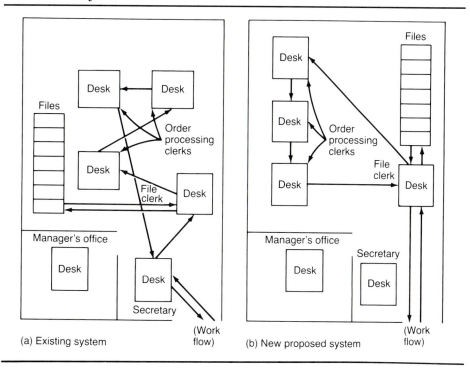

(a) Existing system   (b) New proposed system

• *It functions as a planning tool.* Designers of processes are greatly aided by flowcharts. Flowcharts enable them to visualize the elements of new or modified processes and their interactions while still in the planning stages.

• *It provides an overview of the system.* The critical elements and steps of the process are easily viewed in the context of a flowchart. It removes unnecessary details and breaks down the system so designers and others get a clear, unencumbered look at what they're creating.

• *It defines roles.* A flowchart demonstrates the roles of personnel, workstations, and subprocesses in a system. It also shows the personnel, operations, and locations involved in the process.

• *It demonstrates interrelationships.* Flowcharts show how the elements of a process relate to each other.

• *It promotes logical accuracy.* Flowcharts enable the viewer to easily spot errors in logic. Planning is facilitated because designers have to clearly break down all of the elements and operations of the process.

• *It facilitates trouble-shooting.* Flowcharts are an excellent diagnostic tool. Problems in the process, failures in the system, and barriers to communication can be detected by using a flowchart.

• *It documents a system.* The record of a system that a flowchart provides enables anyone to easily examine and understand the system. Flowcharts facilitate changing a system because the documentation of what exists is available.

A widely utilized tool, flowcharts can be applied in any type of organization to aid in defining and documenting a process and ultimately improving and innovating a process.

Guidelines for Systems-Type Flowcharts.   Flowcharts are simple to use, providing that the following guidelines are followed, in keeping with standard practices.[3]

1. Flowcharts are drawn from the top of a page to the bottom and from left to right.
2. The activity being flowcharted should be carefully defined and this definition should be made clear to the reader.
3. Where the activity starts and where it ends should be determined.
4. Each step of the activity should be described using ''one-verb'' descriptions (e.g., ''prepare statement'' or ''design prototype'').
5. Each step of the activity should be kept in its proper sequence.
6. The scope or range of the activity being flowcharted should be carefully observed. Any branches that leave the activity being charted shouldn't be drawn on that flowchart.
7. Use the standard flowcharting symbols.

Constructive Opportunities to Change a Process.   When using a flowchart, changing a process is facilitated by (1) finding the sections of the process that are weak (for example, parts of the process that generate a high defect rate), (2) determining the parts of the process that are within the process owner's control, and (3) isolating the elements in the process that affect customers. If these three conditions exist simultaneously, an excellent opportunity to constructively modify a process has been found.

What Are the Process's Objectives?   A key responsibility of a process owner is to clearly state process objectives that are consistent with organizational objectives. An example of an organizational objective is ''Strive to continually provide our customers with higher-quality products/services at an attractive price that will meet their needs.''

Each process owner can find meaning and a starting point for stating process objectives in the adaptation of this organizational objective to his process's objectives. For example, a process owner in the Purchasing Department could translate the preceding organizational objective into the following subset of objectives:

1. Continuously monitor Purchasing's customers (e.g., Maintenance Department, Administration Department) to determine their needs with respect to:
   a. Number of days from purchase request to item/service delivery.
   b. Ease of filling out purchasing forms.
   c. Satisfaction with purchased material.

2. Continuously train and develop Purchasing personnel with respect to job requirements. For example, take measurements on the following as a basis for process improvement:

   a. Number of errors per purchase order.

   b. Number of minutes to complete a purchase order.

Whatever the objectives of a process are, all persons involved with that process must understand its objectives and devote their efforts toward those objectives, not toward some other objectives. A major benefit of clearly stating a process's objectives is that it gets everybody involved with the process working toward the same aim.

Are Process Data Valid?   Management's (process owners') attempts to define and document a process must include precise definitions of process objectives and subobjectives, specifications, products and services, and jobs.[4] Such definitions are a prerequisite for understanding between process members. Operational definitions used to collect data must have the same meaning to everyone so that the data can be used as a basis for action.

> It is useful to illustrate the confusion which can be caused by the absence of operational definitions. The label on a shirt reads "75% cotton." What does this mean? Three quarters cotton, on the average, over this shirt, or three quarter cotton over a month's production? What is three quarters cotton? Three quarters by weight? If so, at what humidity? By what method of chemical analysis? How many analyses? Does 75 percent cotton mean that there must be some cotton in any random cross-section the size of a silver dollar? If so, how many cuts should be tested? How do you select them? What criterion must the average satisfy? And how much variation between cuts is permissible? Obviously, the meaning of 75% cotton must be stated in operational terms, otherwise confusion results.[5]

As another example, one operation in a production process is a deburring operation. Hence it's reasonable to ask for the definition of a burr. The supervisor in charge of the deburring operation was asked for a definition and stated that a burr is a bump or protrusion on a surface. (A surface is a face.) Also, he said, "Deburring's five inspectors all have at least 15 years experience and certainly know a burr when they see one."

A test was conducted to determine if the definition of a burr was consistent among all five inspectors. Ten parts were drawn from the production line and placed into a tray so that each part could be identified by a number; each of the inspectors was shown the tray and asked to determine which parts had burrs. Figure 3.9 shows the results.

Although inspectors B and C always agree, they always disagree with inspector E. Inspector A agrees with inspectors B and C 40 percent of the time, with inspector E 60 percent of the time, and with inspector D 50 percent of the time. Inspector D also agrees with inspectors B, C, and E 50 percent of the time. The preceding doesn't paint a pretty picture. Absence of an operational definition of a burr creates mayhem. Deburring's manager (process owner) and inspectors have no consistent concept of their jobs. This creates fear (Deming's 8th point) and steals their pride of workmanship (Deming's 12th point).

FIGURE 3.9 Identification of Burrs on 10 Parts

| | Inspector | | | | |
Part Number	A	B	C	D	E
1	0	1	1	1	0
2	0	1	1	0	0
3	1	1	1	1	0
4	1	1	1	0	0
5	0	1	1	1	0
6	0	1	1	0	0
7	1	1	1	1	0
8	1	1	1	0	0
9	0	1	1	1	0
10	0	1	1	0	0

Legend: 0 = No burr on part.
       1 = Burr on part.

Operational Definitions

Major problems can arise when process measurement definitions are inconsistent over time, or when their applications and/or interpretations are different over time. Employees may be confused about what constitutes their job. Also major problems between customers and suppliers may result from the absence of agreed-upon operational definitions. Endless bickering and ill will are inevitable results.

Operational definitions establish a language for process improvement and innovation. An operational definition puts communicable meaning into a process, product, service, job, or specification. Specifications like defective, safe, round, 5 inches long, reliable, and tired have no communicable meaning until they are operationally defined. Everyone concerned must agree on an operational definition before action can be taken on a process. A given operational definition is neither right nor wrong. The importance of an operational definition is that there's agreement between the stakeholders involved with the definition. As conditions change, the operational definition may change to meet new needs.

Establishing Operational Definitions

An operational definition consists of (1) a criterion to be applied to an object or to a group, (2) a test of the object or group; and (3) a decision as to whether the object or group did or did not meet the criterion.[6] The three components of an operational definition are best understood through some examples.

A firm produces washers. One of the critial quality characteristics is roundness. The following procedure is one possible procedure to use to arrive at an operational definition of roundness as long as a buyer and seller agree on it.

Step 1: Criterion for roundness.

Buyer: "Use calipers that are in reasonably good order." (You perceive at once the need to question every word.)

Seller: "What is 'reasonably good order'?" (We settle the question by letting you use your calipers.)

Seller: "But how should I use them?"

Buyer: "We'll be satisfied if you just use them in the regular way."

Seller: "At what temperature?"

Buyer: "The temperature of this room."

Buyer: "Take six measures of the diameter about 30 degrees apart. Record the results."

Seller: "But what is 'about 30 degrees apart'? Don't you mean exactly 30 degrees?"

Buyer: "No, there's no such thing as exactly 30 degrees in the physical world. So try for 30 degrees. We'll be satisfied."

Buyer: "If the range between the six diameters doesn't exceed .007 centimeters, we'll declare the washer to be round." (They've determined the criterion for roundness.)

Step 2: Test of roundness.

a. Select a particular washer.

b. Take the six measurements and record the results in centimeters: 3.365, 3.363, 3.368, 3.366, 3.366, and 3.369.

c. The range is 3.369 to 3.363, or a 0.006 difference. They test for conformance by comparing the range of 0.006 with the criterion range of less than or equal to 0.007 (Step 1).

Step 3: Decision on roundness.

Because the washer passed the prescribed test for roundness, they declare it to be round.

If a seller has employees who understand what round means and a buyer who agrees, many of the problems the company may have had satisfying the customer will disappear.[7]

Here's another illustration: A salesperson is told that her performance will be judged with respect to the percent change in this year's sales over last year's sales. What does this mean? Average percent change each month? Each week? Each day? For each product? Percent change between December 31, 1992, and December 31, 1993, sales?

How are we measuring sales? Gross, net, gross profit, net profit? Is the percent change in constant or inflated dollars? If it's in constant dollars, is it at last year's prices or this year's prices? Under what economic conditions?

A loose definition of percent change can only lead to confusion, frustration, and ill will between management and the salesforce—which is hardly the way to improve productivity. How should management operationally define a percent change in sales?

Step 1: Criterion for percent change in sales. A percent change in sales is the difference between 1993 (January 1, 1993, to December 31, 1993) sales and 1992

(January 1, 1992, to December 31, 1992) sales divided by 1992 sales:

$$\text{Percentage change (92, 93)} = (S93 - S92)/S92$$

where:

S93 = dollar sales volume for the period January 1, 1993, through December 31, 1993.

S92 = Dollar sales volume for the period January 1, 1992, through December 31, 1992.

S92 is measured in constant dollars, with 1991 as the base year, using June 15, 1991, and June 15, 1992, prices to derive the constant dollar prices, and total unit sales less returns (due to any cause) as of December 31, 1992, for each product.

$$S92 = \sum_{i=1}^{m} [(P_i 91/P_i 92)(TS_i 92 - R_i 92)]$$

where:

m = The number of products in the product line.

$P_i 91$ = The price of product i as of June 15, 1991.

$P_i 92$ = The price of product i as of June 15, 1992.

$TS_i 92$ = The total unit sales for product i between January 1, 1992, and December 31, 1992.

$R_i 92$ = The total of unit returns (for any reason) for product i between January 1, 1992, and December 31, 1992.

$R_i 92$ recognizes that products sold late in 1992 that may be returned will be reflected in $R_i 93$, next year's return for product i. S93 is measured in constant dollars, with 1991 as the base year, using June 15, 1991, and June 15, 1993, prices to derive the constant dollar prices, and total unit sales less returns (for any reason), as of December 31, 1993, for each product. ($P_i 91$ remains the same for all products.)

$$S93 = \sum_{i=1}^{m} [(P_i 91/P_i 93)(TS_i 93 - R_i 93)]$$

where all items are defined as in $S_i 92$ with the appropriate shift in time frame. This procedure for computing the percentage change in sales between 1992 and 1993 will be in effect, regardless of the economic conditions. Further, management may revise the definition of a percent change in sales after the 1996 sales evaluation, but not before, unless the salesforce and sales management agree.

Steps 2 and 3: Test and decision on percent change in sales.

a. The salesperson and her sales records are the object under study.

b. The sales manager will use all 1992 and 1993 invoices and sales return slips to compute the net number of units sold for each product in 1992 and 1993. The sales manager will record the computations and results.

The prior definition of sales might not suit another manager and salesforce. However if the sales manager adopts it and the salesforce understands it, it's an operational definition.

Operational definitions aren't trivial; if management doesn't operationally define control items and control points so that employees and customers agree, serious problems will follow. Statistical methods become useless tools in the absence of operational definitions.

Summary

This chapter discussed defining and documenting a process. A process is defined as the transformation of inputs into outputs, and it involves the addition of value in time, place, or form. Processes exist in all aspects of all organizations, including administration, sales, training, vendor relations, and production. Two important aspects of processes are (1) they all have customers and suppliers and (2) they all need feedback loops for improvement. Before a process can be improved or innovated, it must be defined and documented. Flowcharts provide a means to define and document a process.

Operational definitions bring about increased communication between the stakeholders of a process because they provide precise meanings for specifications, products and services, and jobs; that is, they establish a language for improvement and innovation of a process. If operational definitions aren't used or aren't agreed upon by all concerned, serious problems can occur.

Exercises

3.1 What is a process boundary? Why should process boundaries be established?

3.2 What is a flowchart? List and describe the different types of flowcharts.

3.3 Construct a flowchart for a simple process requiring at least three flowcharting symbols.

3.4 State several examples of process objectives.

3.5 List and explain the three parts of an operational definition. Why are operational definitions important?

3.6 Construct an operational definition. Is your operational definition right? What makes one operational definition right and another operational definition wrong?

Endnotes

1. G. A. Silver and J. B. Silver, *Introduction to Systems Analysis* (Englewood Cliffs, N.J.: Prentice-Hall, 1976), pp. 142–47.

2. J. M. Fitzgerald and A. F. Fitzgerald, *Fundamentals of Systems Analysis* (New York: John Wiley and Sons, 1973), pp. 230–37.
3. Ibid., pp. 227–84.
4. W. Edwards Deming, *Quality, Productivity, and Competitive Position* (Cambridge, Mass.: Massachusetts Institute of Technology, Center for Advanced Engineering Study, 1982), pp. 323, 334.
5. W. Edwards Deming, *Out of the Crisis* (Cambridge, Mass.: Massachusetts Institute of Technology, Center for Advanced Engineering Study, 1986), pp. 287–89.
6. Adapted from Ibid., p. 277.
7. H. Gitlow and P. Hertz, "Product Defects and Productivity," *Harvard Business Review,* September-October 1983, pp. 138–39.

PART II Foundations for Analytic Studies

This section presents the underlying foundations for analytic studies required for managers to think statistically. Thinking statistically is key to developing the ability to manage by data rather than guesswork and opinion.

Chapter 4 outlines the concepts and axioms of probability, summarizes types and characterization of data, and presents the basic notions of descriptive statistics. Finally, an empirical rule for analytic studies is described, which provides the foundation for the construction of control charts.

4 Basic Probability and Statistics

Introduction

Almost everyone makes statements such as "It will probably rain tomorrow" or "Chances are that Bert will outperform Ernie at this new task." These are simple, intuitive statements of likelihood that usually don't require any further explanation. However, there are situations for which we may make more precise likelihood statements—for example, "the probability that a coin lands heads side up when tossed is ½" or "the probability of this same coin landing heads side up twice in a row when tossed is ¼." Understanding some simple probability concepts can help us conceptualize process performance.

In this chapter we look at the issues of quantifying probabilities and examining characteristics of data taken from a population or process. We discuss the types of data that may be encountered and how they're classified; methods for visually displaying data; and the calculation of numerical measures to describe the data.

Probability Defined

There are two popular definitions of probability: the classical definition and the relative frequency definition.

The Classical Definition of Probability

The *classical* approach to probability states that if an experiment has N equally likely and mutually exclusive outcomes (that is, the occurrence of one event precludes the occurrence of another event), and if n of those outcomes correspond to the occurrence of event A, then the probability of event A is

$$P(A) = \frac{n}{N} \qquad (4.1)$$

where

      n = The number of experimental outcomes that correspond to the occurrence of event A and

      N = The total number of experimental outcomes.

For example, tossing a fair die results in one of six equally likely and mutually exclusive outcomes occurring (N = 6). If we're interested in the probability of tossing an even number (e.g., 2, 4, or 6, so n = 3), then the probability of tossing an even number on a die is

$$P(\text{Even}) = \frac{3}{6} = 0.50000$$

The Relative Frequency Definition of Probability

The *relative frequency* approach to probability states that if an experiment is conducted a large number of times, then the probability of event A occurring is

$$P(A) = \frac{k}{M} \qquad (4.2)$$

where

      M = The maximum number of times that event A could have occurred during these experiments and

      k = The number of times A occurred during these experiments.

For example, suppose in the die-tossing example that a fair die was tossed 100,000 times and that a 2, 4, or 6 appeared in 50,097 throws. Consequently, the relative frequency probability of tossing an even number on a fair die is

$$P(\text{Even}) = \frac{50,097}{100,000} = 0.50097$$

The classical definition of the probability of an even toss and the relative frequency definition of an even toss differ as a result of the methods used in their respective calculations.

Analytic studies are conducted to determine process characteristics. But process characteristics have a past and present and will have a future; hence there's no frame from which classical probabilities can be computed. Probabilities concerning process characteristics must be obtained empirically through experimentation and must therefore be relative frequency probabilities.

For example, if we were to say that a certain process has a very high probability of producing a camshaft with a case hardness depth within specification limits, no frame of events exists from which to make probabilistic statements or quantify the probability. This is because we're dealing with a process that has a past and present and will have a future. Hence we must rely on the relative frequency approach to probability to describe and predict the process's behavior.

As another illustration, a newly hired worker is supposed to perform an administrative operation, entering data from sales slips into a computer terminal. Is it possible to predict the percentage of sales slips per hour so entered that will be in error? Unfortunately not; the best we can do is to train workers properly, and then observe them performing their jobs over a long period of time. If a worker's efforts represent a stable system (more on this in Chapters 5 through 7), we can compute the relative frequency of sales slips entered per hour with errors as an estimate of the probability of a sales slip's being entered with errors. This empirical probability can be used to predict the percentage of sales slips per hour entered with errors.

Calculating Classical and Relative Frequency Probabilities

A bin contains 4,000 screws; 3,000 are good, and 1,000 are defective. We'll use this bin of screws to compare the two definitions of probability.

Calculating the Classical Probability.   The classical definition of the probability of drawing a defective screw is (from Equation 4.1) $1,000/(1,000 + 3,000) = 0.250$. This is a reasonable conclusion if the screws are selected through a random sampling plan; that is, using a table of random numbers. This would involve labeling each screw, a clearly impractical procedure. Since in the real world screws aren't selected through a random sampling plan, the classical probability of a defective screw is of little value for process improvement work.

Calculating the Relative Frequency Probability.   The relative frequency definition of the probability of drawing a defective screw is very different. Suppose we place the screws in the bin, mix them thoroughly, and draw one screw at random. We record the status of the screw (0 if good and 1 if defective) and replace the screw in the bin. We again thoroughly mix the screws, draw another screw, and record the number 0 or 1. Suppose we repeat this process 1,250 times. Assuming that the screws are thoroughly mixed between each drawing, we can say that the 1,250 numbers were drawn in a random manner. Further, suppose we put the numbers into groups of 50 and record the number of defective screws in each group, as shown in the first four columns of Figure 4.1.

If we examine more samples of size 50, we'll note that the fluctuations in the cumulative fraction defective diminish because the estimate of the cumulative fraction of defective screws is based on a larger sample. The last three columns of Figure 4.1 and the whole of Figure 4.2 illustrate this. From Figure 4.2, we see that after accumulating 23 subgroups the fraction defective of screws appears to have stabilized around 0.208. This number is the relative frequency estimate of the probability of drawing a defective screw. We use the term *stabilized* to be synonymous with the concept of a constant system of variation. In other words, the subgroup fraction defective is within stable and predictable limits. Chapters 5 through 7 extensively discuss the concept of stability.

FIGURE 4.1   **Cumulative Samples of 50 Screws**

1	*2*	*3*	*4*	*5*	*6*	*7*
Subgroup Number	Subgroup Size	Number of Defective Screws in Subgroup	Fraction of Defective Screws in Subgroup	Cumulative Number of Defective Screws	Cumulative Number of Screws	Cumulative Fraction of Defective Screws
1	50	10	.20	10	50	.200
2	50	9	.18	19	100	.190
3	50	15	.30	34	150	.227
4	50	7	.14	41	200	.205
5	50	9	.18	50	250	.200
6	50	17	.34	67	300	.223
7	50	13	.26	80	350	.229
8	50	8	.16	88	400	.220
9	50	10	.20	98	450	.218
10	50	7	.14	105	500	.210
11	50	7	.14	112	550	.204
12	50	11	.22	123	600	.205
13	50	9	.18	132	650	.203
14	50	9	.18	141	700	.201
15	50	8	.16	149	750	.199
16	50	9	.18	158	800	.198
17	50	12	.24	170	850	.200
18	50	17	.34	187	900	.208
19	50	8	.16	195	950	.205
20	50	7	.14	202	1000	.202
21	50	17	.34	219	1050	.209
22	50	9	.18	228	1100	.207
23	50	11	.22	239	1150	.208
24	50	11	.22	250	1200	.208
25	50	10	.20	260	1250	.208
		Total: 260				

Comparing the Probabilities.   Figure 4.3 compares the two definitions of probability. We've seen that the classical probability of selecting a defective screw from the bin is 0.250. The relative frequency probability of selecting a defective screw—or the long-run cumulative fraction defective—is 0.208. In the absence of a complete frame, and hence in the absence of the classical fraction defective in the bin, the best we can do is to use the most recent cumulative fraction defective.

It's important to consider why the relative frequency approach didn't yield a 0.250 probability of selecting a defective screw. The difference between the two probabilities indicates that the mechanical procedure of mixing and selecting the screws under the relative frequency definition must have been biased against defective screws; it wasn't equivalent to random sampling. A possible explanation of this bias is that defective screws are broken and consequently smaller. As a result, they either fall to the bottom of the bin during mixing or present a smaller

FIGURE 4.2   Cumulative Fraction of Defective Screws

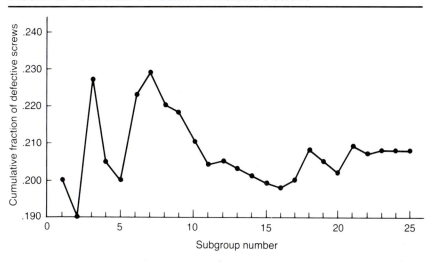

FIGURE 4.3   Comparison of the Two Definitions of the Probability of a Defective Screw

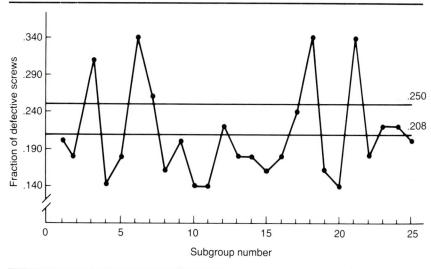

target for selection. If the mixing and selecting procedure used under the relative frequency approach were equivalent to a random sampling procedure, the two probabilities would be the same.

Types of Data

Information collected about a product, service, process, person, or machine is called *data*. In the analytic study of a process, data are often used as a basis for predictions on which to take actions. For example, if we collect data about the weight of a particular grade of paper, we can monitor and understand the production process and take actions that will allow us to make paper with a weight consistently around some desired value. As another example, if we collect data about the number of payroll entry errors by pay period, we can understand when and why errors are made so that we can take appropriate actions to reduce the error rate.

We classify data into two types: attribute data and variables data.

Attribute Data

Attribute data arise (1) from the classification of items, such as products or services, into categories, (2) from counts of the number of items in a given category or the proportion in a given category, and (3) from counts of the number of occurrences per unit.

Classification of Items into Categories.   Often data are classified into two categories or grades, representing conformity or nonconformity with some quality characteristic. Whether a purchase order is filled out properly; whether a bearing in a piece of equipment has excessive vibration; whether the seal on a jar of instant coffee is broken; whether a nine-volt battery is operating; or whether a car radio is defective are all examples of attributes.

Items may be similarly classified into three or more categories or grades. Classifying the grade of a stick of butter as Grade AA, Grade A, or Grade B; classifying employees by department; or classifying the type of defect in a roll of paper as dished, crushed core, or wrinkled are all examples of attributes.

As an illustration, union grievances are classified as first level, second level, third level, or fourth level; a first-level grievance can be resolved by a foreman, whereas a fourth-level grievance must be handled by a special arbitration board. That is, the higher the level of the union grievance, the more serious the grievance. Figure 4.4 lists 11 union grievances and their classifications for 1994.

Counts of the Number of Items in a Given Category, or Proportion in a Given Category.   The number of items in a given category also represents attribute data. For example, the second column of Figure 4.5 shows the number of union grievances in each of the four categories discussed in the preceding example.

FIGURE 4.4   Attribute Data: Classification of 1994 Union Grievances into Four Categories

Grievance Number	Grievance Level
1	1
2	1
3	1
4	2
5	4
6	1
7	1
8	1
9	3
10	1
11	1

FIGURE 4.5   Attribute Data: Number of 1994 Union Grievances per Level

Category	Number of Grievances	Proportion of Grievances
First level	8	8/11 = .73
Second level	1	1/11 = .09
Third level	1	1/11 = .09
Fourth level	1	1/11 = .09

The *proportion* of items in a given category is also an attribute. From Figure 4.5, we see that the proportion of second-level grievances can be calculated as

$$\text{Proportion} = \frac{\text{Number of second-level union grievances}}{\text{Number of union grievances}}$$

$$= 1/11 = 0.09$$

The last column of Figure 4.5 shows the proportion of grievances in each category.

Counts of the Number of Occurrences per Unit.   An attribute of interest in some quality studies is the number of defects per unit, where a unit can be a single item, a batch of the item under question, or a unit of time or space. The number of defects per reel of paper, the number of errors per typed page, the number of accidents per month, the number of union grievances per week, the number of blemishes per square yard of fabric, or the number of defective car radios in a lot of 1,000 such radios are all examples of attribute data.

As an illustration, in the production of stainless steel washers, the manufacturing process occasionally results in a defective product. Four types of defects are observed: cracks, dents, scratches, and discoloration. An individual washer may have some, all, or none of these defects. Every hour, 10 washers are examined for defects. Figure 4.6 shows the number of defects for each washer examined in an 8-hour day.

Variables Data

Variables data arise (1) from the measurement of a characteristic of a product, service, or process and (2) from the computation of a numerical value from two or more measurements of variables data.

Measurement of a Characteristic.   When we make precise measurements of quality characteristics—such as the outside diameter of stainless steel washers (in

---

FIGURE 4.6   **Attribute Data: Number of Defects per Washer in Eight Hourly Samples of 10 Stainless Steel Washers**

Hour #	Item #	Number of Defects	Hour #	Item #	Number of Defects	Hour #	Item #	Number of Defects
1	1	1		8	0		5	1
	2	0		9	3		6	0
	3	0		10	0		7	0
	4	2	4	1	1		8	0
	5	1		2	1		9	3
	6	0		3	0		10	2
	7	0		4	4	7	1	0
	8	0		5	0		2	1
	9	4		6	1		3	0
	10	0		7	0		4	0
2	1	1		8	0		5	1
	2	0		9	3		6	0
	3	0		10	1		7	0
	4	3	5	1	0		8	2
	5	1		2	2		9	0
	6	0		3	0		10	0
	7	2		4	0	8	1	1
	8	0		5	0		2	0
	9	0		6	3		3	0
	10	0		7	0		4	0
3	1	1		8	0		5	1
	2	0		9	1		6	0
	3	0		10	2		7	0
	4	2	6	1	1		8	0
	5	4		2	1		9	2
	6	0		3	0		10	0
	7	1		4	0			

FIGURE 4.7   Variables Data: Length of Logs in a Sample of 95 Trees

Day #	Log #	Length (cm)	Day #	Log #	Length (cm)	Day #	Log #	Length (cm)
1	1	783.2		3	1393.9		5	1719.7
	2	1322.8		4	1968.6	14	1	638.8
	3	1012.5		5	1561.2		2	1834.0
	4	1408.6	8	1	1908.7		3	730.0
	5	1589.6		2	1612.2		4	1338.5
2	1	400.2		3	1585.6		5	1621.4
	2	1456.1		4	1510.4	15	1	1192.1
	3	1126.7		5	898.7		2	1542.6
	4	1563.8	9	1	1087.8		3	1886.1
	5	1362.5		2	938.1		4	1688.2
3	1	1537.8		3	1472.3		5	1069.7
	2	505.4		4	1072.0	16	1	1835.3
	3	1686.7		5	489.7		2	1485.2
	4	1508.0	10	1	1208.6		3	1521.4
	5	1235.8		2	1868.5		4	2104.4
4	1	1827.2		3	1321.7		5	1225.2
	2	2148.3		4	2067.7	17	1	929.4
	3	1175.2		5	835.9		2	1732.1
	4	1657.5	11	1	1830.6		3	1340.2
	5	410.3		2	1227.1		4	1445.5
5	1	1577.7		3	1423.3		5	460.7
	2	827.6		4	1022.2	18	1	1021.8
	3	1353.1		5	1690.0		2	1515.2
	4	1410.2	12	1	625.6		3	753.6
	5	1050.2		2	1131.1		4	1117.3
6	1	1468.7		3	1193.9		5	1490.0
	2	1229.2		4	1252.8	19	1	572.5
	3	1346.3		5	1551.8		2	1436.8
	4	1725.6	13	1	1481.4		3	1356.6
	5	693.1		2	1763.6		4	1112.6
7	1	1639.9		3	1392.2		5	1512.7
	2	1786.3		4	1498.9			

centimeters), the weights of boxes of detergent (in ounces), or the lifetimes of incandescent bulbs (in hours)—we are recording the values of a *variable*.

As an illustration, in a logging operation, trees are cut at age 20 years. Figure 4.7 records the length, in centimeters, of each such log as measured in 19 daily samples of five trees each.

Computation of a Numerical Value from Two or More Measurements of Variables Data.   Figure 4.8 shows the recorded values of two variables for each of six trucks: the number of miles driven since the last refueling, and the number of gallons of fuel consumed since the last refueling.

FIGURE 4.8   **Variables Data: Calculation of the Variable Miles per Gallon**

Truck	Miles since Refueling	Gallons of Fuel Consumed	Miles per Gallon
1	308	17.3	17.8
2	256	15.3	16.7
3	274	16.5	16.6
4	310	16.9	18.3
5	302	17.1	17.7
6	296	17.3	17.1

From these measurements, we may calculate another variable, miles per gallon for each truck, by calculating the ratio of the number of miles driven since the last refueling divided by the number of gallons of fuel consumed since the last refueling. This computed ratio, shown in the last column of Figure 4.8, is an illustration of variables data computed from two other measurements of variables data.

As another illustration, the computation of the volume of a rectangular container from the product of the measurements of its length, width, and height results in a new computed variable value.

In any quality study we usually want to rearrange and/or manipulate data to gain some insight into the characteristics of the process under study. The remainder of this chapter discusses ways of interpreting and appropriately describing data, both visually and numerically.

Characterizing Data

Enumerative Studies

A complete census of the frame in an enumerative study provides all the information needed to take action on the frame. For example, suppose a lumber supplier must determine how much to charge for a particular large shipment to a lumberyard. The total charge is based on the lengths of the logs in the shipment (the frame). Thus, if we knew the length of each log in the shipment (that is, 100 percent sampling in an enumerative study), we would know exactly how much to charge for the shipment, assuming accuracy in the measurement process.

As discussed in Chapter 2, we often sample rather than perform a complete census; we then use that information to develop visual and/or numerical measures that enable us to take action on the frame. As we're examining only a portion of the frame, the possibility of taking an incorrect action does exist. In our example, suppose the lumber supplier bases the charge on the average of the 95 log lengths shown in Figure 4.7. Two types of error are possible in this procedure: (a) the charge is higher by an amount C than what it would be if she had measured every log (often called the *consumer's risk*) or (b) the charge is lower by an amount C' than what it would be if she had measured every log (the *producer's risk*).

As another illustration, consider the sample of stainless steel washers examined for defects in Figure 4.6. If all the washers produced in this 8-hour day make up a single lot to be shipped (the frame), in an enumerative study the manufacturer may wish to determine the number of defects in the entire lot to decide whether or not to ship the lot. In using the sample data as a basis for action, again there are two possible errors: (a) he decides to ship a lot that, in fact, has an unacceptably large number of defects (the consumer's risk) or (b) he decides to reject a lot that, in fact, has an acceptably small number of defects (the producer's risk).

If the information used as the basis for action constitutes a *random sample* from the frame, these errors can be quantified, and valid statistical inferences can be made on the frame in question. Chapter 2 discussed the nature of and procedures for selecting a random sample. This chapter discusses how to describe and characterize such data to ultimately make appropriate inferences.

Analytic Studies

A complete census of the frame is impossible in an analytic study, for a frame consists of all past, present, and future observations, and the latter can't be measured. As we're dealing with an ongoing process, we wish to characterize the data to take action on that process for the future. For example, in the manufacture of washers, suppose we aren't interested, as before, in one particular day's output but instead are interested in the process itself. That is, we wish to determine the necessity for action: whether the process should be left unchanged for the future or modified in some way. A sample selected from the process, such as that in Figure 4.6, may lead to two types of error: (a) decide to retain the process, even though it should be modified or (b) decide to modify the process although it's actually unnecessary to do so.

However, unlike the enumerative study, as the frame is unknown (the future can't be measured) it's not possible to quantify these errors. Any inferences we make in an analytic study are conditional on the environmental state when the sample was selected; that environmental state will never again exist. Information on such a problem can never be complete. If, however, a knowledge of the process and the environment and an analysis of the data indicate that the process is *stable* and predictable, and will remain so in the near future (as discussed in Chapters 5 through 7), the visual and numerical characterizations discussed in this chapter can be used to make inferences and take action in the near future.

Visually Describing Data

Tabular Displays

Frequency Distributions.   A *frequency distribution* shows us, in tabular form, the number of times, or the frequency with which, a given value or group of values occurs. For example, when analyzing variables type data, we generally group the data into class intervals and count the number of items whose values fall within each interval. Thus Figure 4.9 shows a frequency distribution for the data of Figure 4.7, where the class intervals are 400 but less than 700 cm, 700 but less than 1,000 cm, 1,000 but less than 1,300 cm, 1,300 but less than 1,600 cm, 1,600 but less than 1,900 cm, and 1,900 but less than 2,200 cm. The *class limits* are the endpoints of each interval; in our example, the lower class limits are the lower limits of each interval: 400, 700, 1,000, 1,300, 1,600, and 1,900. The upper class limits are 700, 1,000, 1,300, 1,600, 1,900, and 2,200. The *midpoint* of each class is the value halfway between the class limits.

Figure 4.9 indicates *relative frequencies* (the percentage of the total number of observations in each class) as well as *absolute frequencies* (the actual number of observations in each class). Especially in situations where we want to compare two frequency distributions with unequal total numbers, we often use relative frequencies—or the class frequencies divided by the total number of observations—expressed as a percentage.

For attribute type data, we may construct frequency distributions by counting the number of items possessing a particular attribute. Thus Figure 4.10 shows a frequency distribution of the number of defects in 80 stainless steel washers.

FIGURE 4.9   Frequency Distribution of Lengths of Logs in a Sample of 95 Trees

Length (cm)	Class Midpoint	Absolute Frequency	Relative Frequency (percent)
400 but less than 700	550	9	9.5%
700 but less than 1,000	850	8	8.4
1,000 but less than 1,300	1,150	20	21.1
1,300 but less than 1,600	1,450	35	36.8
1,600 but less than 1,900	1,750	18	18.9
1,900 but less than 2,200	2,050	5	5.3

FIGURE 4.10   Frequency Distribution of Number of Defects in a Sample of 80 Stainless Steel Washers

Number of Defects	Absolute Frequency	Relative Frequency (percent)
0	46	57.50%
1	18	22.50
2	8	10.00
3	5	6.25
4	3	3.75

As a general rule of thumb, we select anywhere from 5 to 20 classes when constructing a frequency distribution for variables type data; too few or too many classes may fail to reveal patterns of interest. For example, taking the data of Figure 4.7 and constructing a frequency distribution consisting of two classes (400 but less than 1,300, with 37 observations, and 1,300 but less than 2,200, with 58 observations) offers little insight in comparison with the distribution of Figure 4.9. Where possible, we make the class widths the same for all intervals. This allows us to logically compare frequencies in different classes. For example, if we're measuring the lifetime of 100 special-purpose incandescent bulbs and count 3 bulbs in a class interval of 1 to 5 hours and 35 bulbs in a class interval of 5.1 to 20 hours, we couldn't easily determine whether the large difference in frequency results from actual variations in bulb lifetime or simply from the huge difference in class widths. Sometimes, however, we use an open-ended interval to include a small number of highly dispersed values at the top and/or bottom of the distribution. Using the same example, if 1 or 2 of the 100 bulbs burned for 90 hours, while all the rest burned for 50 hours or less, it might make sense to specify the top interval to be "more than 50 hours." The number of classes in a frequency distribution for attribute type data is determined by the number of values the attribute can assume; for example, an attribute frequency distribution depicting the gender of employees would have two classes.

Cumulative Frequency Distributions.   For many processes, we're interested in the frequency of items with a value less than some measurement. For example, how many of the 100 incandescent bulbs burned for less than 30 hours, how many burned for less than 40 hours, and so on. This information is presented in a cumulative frequency distribution. Figure 4.11(a) shows the cumulative frequencies for the data of Figure 4.7, calculated by adding the successive frequencies in the distribution of Figure 4.9. For example, the number of logs with a length of less than 1,600 cm is the sum of the frequencies for all the class intervals up to and including "1,300 but less than 1,600" or $9 + 8 + 20 + 35 = 72$. Relative cumulative frequency, or the percentage of the items with a value less than the upper limit of each class interval, is calculated as the absolute cumulative frequency divided by the sample size. Thus, for example, 75.8 percent of the logs in the sample have lengths of less than 1,600 cm. Figure 4.11(b) shows the cumulative frequency distribution for the number of defects in the 80 washers.

Limitations of Frequency Displays.   It's important to note that the frequency displays we've discussed don't include information on the time-ordering of data. Recall that the diameters of our 80 stainless steel washers were actually measured by examining 10 washers every hour over a period of eight consecutive hours. (See Figure 4.6.) But the frequency distribution in Figure 4.10 doesn't indicate that the observations were made in any particular sequence. In an analytic study, where we examine the sample to take action on the process, a frequency display would fail to show trends that may be occurring over time. This loss of information is critical.

FIGURE 4.11   Cumulative Frequency Distributions

(a)   Cumulative Frequency Distribution of Lengths of Logs in a Sample of 95 Trees

Log Length (cm) *Less than or Equal to*	*Cumulative Frequency*	*Relative Cumulative Frequency (percent)*
700	9	9.5%
1,000	17	17.9
1,300	37	38.9
1,600	72	75.8
1,900	90	94.7
2,200	95	100.0

(b)   Cumulative Frequency Distribution of Number of Defects in a Sample of 80 Stainless Steel Washers

Washers with Number of Defects Less than or Equal to	*Cumulative Frequency*	*Relative Cumulative Frequency (percent)*
0	46	57.50%
1	64	80.00
2	72	90.00
3	77	96.25
4	80	100.00

Consider the production of brass nuts. The process, as a result of equipment wear, is producing nuts with an inside diameter actually increasing over time. Every hour, for 20 consecutive hours, a nut is selected randomly and its inside diameter recorded. The observations are shown in Figure 4.12(a). A frequency distribution is shown in Figure 4.12(b).

If the nuts manufactured during this 20-hour period constitute a population, the frequency distribution indicates all pertinent information about the lot. However, if we're conducting a study of a manufacturing process for the future, we must detect the trend over time and look for the cause; the frequency distribution in this case gives us an incomplete picture.

Graphical Displays

Data are often represented in graphical form. Such displays offer a composite picture of the relationships and/or patterns at a glance.

Frequency distributions of variables data are commonly presented in frequency polygons or histograms. Frequency distributions of attribute data are commonly presented in bar charts. In all these displays, the class intervals are drawn along the horizontal axis, and the absolute or relative frequencies along the vertical axis.

FIGURE 4.12   Inside Diameters of Brass Nuts

(a)   Hourly Observations

Hour #	Inside Diameter (in)	Hour #	Inside Diameter (in)	Hour #	Inside Diameter (in)
1	1.00	8	1.11	15	1.25
2	1.01	9	1.12	16	1.28
3	1.03	10	1.14	17	1.32
4	1.05	11	1.16	18	1.36
5	1.07	12	1.18	19	1.41
6	1.08	13	1.20	20	1.46
7	1.09	14	1.22		

(b)   Frequency Distribution

Inside Diameter (in)	Absolute Frequency
1.00 but less than 1.10	7
1.10 but less than 1.20	5
1.20 but less than 1.30	4
1.30 but less than 1.40	2
1.40 but less than 1.50	2

Frequency Polygons.   In a *frequency polygon,* we plot the midpoint of each class interval on the horizontal axis against the frequency of that class on the vertical axis. We then connect these points with a series of line segments. To complete the polygon, we add two more classes to the distribution—one class preceding the smallest class and one class succeeding the largest class. Both of these are assigned a "0" frequency at their midpoints. In Figure 4.13(a) the frequency distribution of the data in Figure 4.9 is presented as a frequency polygon.

In the particular case where the lower class limit of the smallest class is 0, adding a class preceding the smallest class results in negative values for that class (and its midpoint) on the horizontal axis. To indicate that such values couldn't really occur, we use a dashed line for the portion of the polygon corresponding to negative observations. Figure 4.13(b) illustrates this procedure for a frequency distribution of 10 product weights, in which three observations are between 0 and 10 pounds, five observations are between 10 and 20 pounds, and two observations are between 20 and 30 pounds. Often the dashed line isn't shown on the final frequency polygon.

Histograms.   In a *histogram,* vertical bars are drawn, each with a width corresponding to the width of the class interval and a height corresponding to the frequency of that interval. The bars share common sides, with no space between them. Figure 4.13(c) shows a histogram of the frequency distribution of the data in Figure 4.9.

FIGURE 4.13   **Graphical Displays of Data**

(a) Frequency Polygon for Lengths of Logs in a Sample of 95 Trees

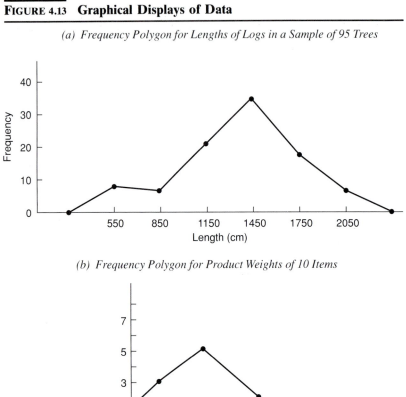

(b) Frequency Polygon for Product Weights of 10 Items

(c) Histogram for Lengths of Logs in a Sample of 95 Trees

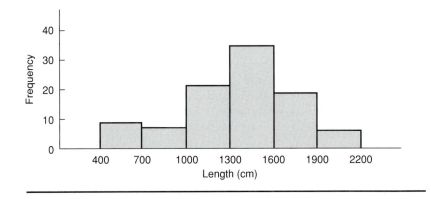

FIGURE 4.14   Bar Chart for Number of Defects in a Sample of 80 Stainless Steel Washers

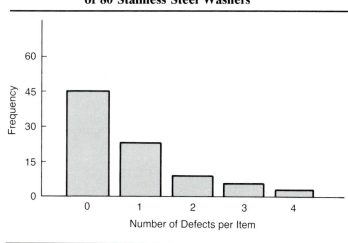

Bar Charts.   For attribute data, a *bar chart* is used to graphically display the data. It's constructed in exactly the same way as a histogram, except that instead of using a vertical bar spanning the entire class interval, we use a line or bar centered on each attribute category. Figure 4.14 shows a bar chart for the frequency distribution in Figure 4.10.

Ogive.   The graph of a cumulative frequency distribution, called an *ogive* (pronounced "oh jive"), is constructed by plotting the upper limit of each class interval on the horizontal axis against the cumulative percentage for that class on the vertical axis. Figure 4.15 shows the ogive for the cumulative frequency distribution of Figure 4.11(a). Note that the lower limit of the first class interval has a cumulative percentage of 0 as there are no observations below that value.

Run Charts: Importance of Time-Ordering in Analytic Studies.   As our discussion of the limitations of tabular frequency displays showed, in analytic studies we want to be able to detect trends or other patterns over time to take action on a process in the near future. The graphical displays we've examined don't incorporate the time-ordering of the data; hence they're of limited value in such studies. In a *run chart* (also called a *tier chart*), this information is preserved by plotting the observed values on the vertical axis and the times they were observed on the horizontal axis.

For example, suppose we are studying a process in which chocolate rectangles are cut from larger blocks of chocolate and then packaged as six-ounce bars. Every 15 minutes, three chocolate bars are weighed, prior to packaging. Figure 4.16 shows the weights for each bar examined in a seven-hour day.

We can construct a frequency distribution of this data using, as lower class limits, 5.98, 6.00, 6.02, 6.04, and 6.06. Figure 4.17(a) is a histogram of this distri-

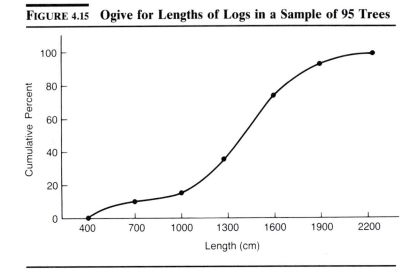

FIGURE 4.15   Ogive for Lengths of Logs in a Sample of 95 Trees

bution indicating, at a glance, that most of the chocolate bars weigh slightly more than the labeled weight of 6 ounces. In an enumerative study, in which we're interested only in the batch of chocolate bars produced during this seven-hour day, such information may be sufficient for taking action on that batch. But in an analytic study, where we seek to take action on the process itself for the future, we'll see that the information gained from Figure 4.17(a) is incomplete.

Figure 4.17(b) is a run chart that shows the observed weights plotted over time. The graph clearly shows an upward drift in the weights of chocolate bars throughout the day and it indicates the need for action on the process.

Numerically Describing Data

Measures of Central Tendency

While tabular and graphical representations give a broad overview, we often need quantitative measures that summarize important characteristics of a population or a process. One such property is *central tendency,* or the behavior of the typical value of a characteristic of the population or process data.

The Mean.   In trying to convey the underlying character of variables data by somehow representing the typical value of the data, the most common numerical representation is the arithmetic average or *mean.* The mean is simply the sum of the numerical values of the measurement divided by the number of items examined. In an enumerative study, if the items constitute a frame, the average is called the *population mean* and is usually denoted by the Greek letter μ (pronounced ''mew''). When the items constitute a sample drawn from a frame, we call the

FIGURE 4.16   Weights of Chocolate Bars Examined at 15-Minute Intervals

Time	Observation #	Weight (oz)	Time	Observation #	Weight (oz)
9:15	1	6.01	12:45	1	6.03
	2	5.99		2	6.02
	3	6.02		3	6.03
9:30	1	5.98	1:00	1	6.03
	2	5.99		2	6.00
	3	6.01		3	6.01
9:45	1	6.03	1:15	1	6.04
	2	6.02		2	6.02
	3	6.02		3	6.03
10:00	1	6.02	1:30	1	6.05
	2	6.03		2	6.02
	3	6.02		3	6.04
10:15	1	6.00	1:45	1	6.03
	2	5.99		2	6.04
	3	6.01		3	6.01
10:30	1	5.99	2:00	1	6.02
	2	6.00		2	6.02
	3	6.00		3	6.02
10:45	1	6.02	2:15	1	6.04
	2	6.01		2	6.05
	3	6.00		3	6.03
11:00	1	6.01	2:30	1	6.06
	2	6.03		2	6.03
	3	6.01		3	6.04
11:15	1	6.01	2:45	1	6.05
	2	6.02		2	6.04
	3	6.00		3	6.02
11:30	1	6.00	3:00	1	6.05
	2	6.02		2	6.04
	3	6.01		3	6.03
11:45	1	6.04	3:15	1	6.04
	2	6.02		2	6.06
	3	6.03		3	6.05
12:00	1	6.02	3:30	1	6.05
	2	6.01		2	6.03
	3	6.00		3	6.04
12:15	1	6.03	3:45	1	6.03
	2	6.02		2	6.04
	3	6.04		3	6.03
12:30	1	6.02	4:00	1	6.06
	2	6.02		2	6.06
	3	6.03		3	6.05

average a *sample mean* and denote it as $\bar{x}$ ("x bar"). Thus, in an enumerative study, we might make reference to either μ or $\bar{x}$. In an analytic study, there is no population, and hence no need to be able to describe a population mean. The mean of a sampled subgroup (e.g., four items from a day's production) is denoted as $\bar{x}$, and when we average the $\bar{x}$ values to calculate a process mean, we denote that value as $\bar{\bar{x}}$ ("x bar-bar").

FIGURE 4.17   Weights of Chocolate Bars Examined at 15-Minute Intervals

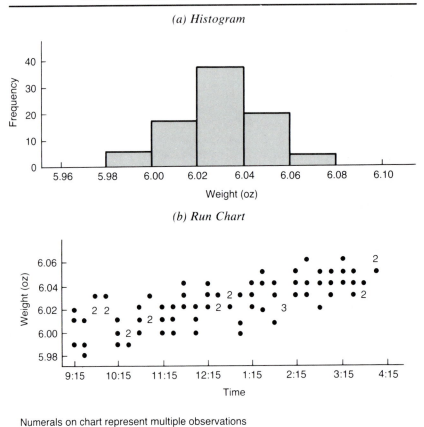

(a) Histogram

(b) Run Chart

Numerals on chart represent multiple observations

To communicate this type of information in compact, precise form, we identify each numerical value as an "x" measurement; the sum of all the measurements we call Σx ("the sum of x") where Σ is the capital Greek letter *sigma* and represents, mathematically, "summation."

The sample mean, x̄, can be calculated as

$$\bar{x} = \frac{\Sigma x}{n} \qquad (4.3)$$

where n represents the size of the sample or the number of items included in the determination of the sample mean.

If we calculate sample means for subgroups of size n in the past and present, we can calculate the process mean by the average of these subgroup means:

$$\bar{\bar{x}} = \frac{\Sigma \bar{x}}{\text{Number of subgroups}} \qquad (4.4)$$

FIGURE 4.18 **Weights of Chocolate Bars Examined at 15-Minute Intervals: Sample Means of Subgroups of Size 3**

Time	$\bar{x}$	Time	$\bar{x}$	Time	$\bar{x}$	Time	$\bar{x}$
9:15	6.01	11:00	6.02	12:45	6.03	2:30	6.04
9:30	5.99	11:15	6.01	1:00	6.01	2:45	6.04
9:45	6.02	11:30	6.01	1:15	6.03	3:00	6.04
10:00	6.02	11:45	6.03	1:30	6.04	3:15	6.05
10:15	6.00	12:00	6.01	1:45	6.03	3:30	6.04
10:30	6.00	12:15	6.03	2:00	6.02	3:45	6.03
10:45	6.01	12:30	6.02	2:15	6.04	4:00	6.06

To illustrate, we may calculate sample means, $\bar{x}$, for each subgroup of three bars observed every 15 minutes, as shown in Figure 4.16. These subgroup sample means are shown in Figure 4.18. From these, we may estimate the mean weight of all chocolate bars produced by this process from Equation 4.4:

$$\bar{\bar{x}} = \frac{\Sigma \bar{x}}{\text{Number of subgroups}} = \frac{168.68}{28} = 6.02$$

The Median. Another measure of the central tendency is given by the *median,* or middle value when the data are arranged in ascending order. When there are an even number of observations, the median value is taken to be the arithmetic average of the middle two values.

Symbolically, if n is the number of data points and the data points are arranged in ascending order, the median can be calculated as

$$M_e = [(n + 1)/2]^{\text{th}} \text{ data point if n is odd or}$$

$$= \text{Average of } (n/2)^{\text{th}} \text{ data point and } (n/2 + 1)^{\text{th}} \text{ data point if n is even}$$

(4.5)

Given an ogive, or the graph of the cumulative frequencies for a given set of data points, the median may also be approximated by reading the data value corresponding to a cumulative frequency of 50 percent. That is, as the median is the middle value, 50 percent of the data points must have values less than the median.

The central tendency of some data is not adequately represented by an arithmetic average. Consider the measurement of burning times (in minutes) of small birthday candles from a production lot. Seven candles are randomly selected from the lot. Burning times are found to be 3 minutes, 2 minutes, 4 minutes, 6 minutes, 19 minutes, 5 minutes, and 3 minutes. As these data points are a sample from the population of all such candles in this production lot, and we're only interested in making some inference on the burning characteristics of this lot, this is an enumerative problem. We may calculate the sample mean from Equation 4.3:

$$\bar{\text{X}} = \frac{3 + 2 + 4 + 6 + 19 + 5 + 3}{7} = \frac{42}{7} = 6 \text{ minutes}$$

Note, however, that in six of the seven data points, the burning time was six minutes or less, so that the average value as a measure of central tendency is somewhat misleading. If we look at a frequency polygon for the data points (Figure 4.19), we see that it's asymmetrical, or *skewed* to the right; only a small proportion of the data has high values.

The median burning time is the middle data point in the sequence of burning times written in ascending order (from lowest value to highest value): 2, 3, 3, 4, 5, 6, 19, or 4 minutes. Note that half the data points have a value lower than the median, and half have a value higher than the median. The median isn't as influenced by the magnitude of the extreme items as is the mean. That is, if the largest data point had been 500 minutes instead of 19 minutes, the median would remain 4 minutes, although the mean would increase by a great deal. Hence the median is particularly appropriate in an enumerative example such as ours, where a single extreme value appears atypical.

The interpretation of the median in an analytic study, however, can be misleading. Where extreme data points are observed in a sample, we've seen that they don't affect the computation of the median. In an enumerative study in which such data points are deemed atypical, the median is an appropriate measure of central tendency. However, in an analytic study, where we're concerned with the process itself, the existence of extreme data points is a critical factor in our analysis. That is, extreme data points may indicate process disturbances and instability, and the need for corrective action on the process. A measure like the median, which is insensitive to extreme data points, must be used with caution.

The Mode. The *mode* of a distribution is the value that occurs most frequently, or the value corresponding to the high point on a frequency polygon or histogram. Like the median, and unlike the mean, it's not affected by extreme data points. A

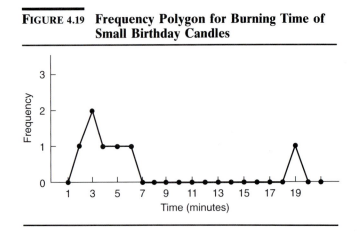

FIGURE 4.19   Frequency Polygon for Burning Time of Small Birthday Candles

frequency distribution with one such high point is called *unimodal;* distributions with more than one high point of concentration are called *multimodal.*

For the data of Figure 4.6, an examination of the frequency distribution in Figure 4.10 shows that the mode, or modal number of defects, is 0, for that value has the largest frequency. This is a unimodal distribution. The frequency polygon in Figure 4.19, with two high points of concentration, indicates a *bimodal* distribution.

In the sample of the burning times of seven randomly selected birthday candles discussed earlier, we found that $\bar{x}$ = 6 minutes and the median was 4 minutes. The mode in this example is the most frequently occurring value (3 minutes). As in the case of the median, had the longest observation been 500 minutes instead of 19 minutes, the mode would remain 3 minutes, although the mean would increase.

As with the median, an important characteristic of the mode is that it's not affected by extreme data points. In analytic studies, where extreme data points may reveal a great deal about the process under investigation, the mode must be used with caution.

The Proportion.   Often data are classified into two nonnumerical attributes, such as broken–not broken, operating–not operating, or defective–conforming. The proportion or fraction of the data possessing one of two such attributes is then a meaningful measure of central tendency.

The data of Figure 4.20 represent the classification of 38 radios as defective or nondefective in an analytic study of a production process. As in the computation of a sample mean in an analytic study, if the process isn't a stable one, the

FIGURE 4.20   Defective Units in a Sample of 38 Radios

Item #	Condition	Item #	Condition
1	Nondefective	20	Defective
2	Defective	21	Nondefective
3	Nondefective	22	Nondefective
4	Nondefective	23	Nondefective
5	Nondefective	24	Nondefective
6	Nondefective	25	Nondefective
7	Nondefective	26	Defective
8	Defective	27	Nondefective
9	Nondefective	28	Nondefective
10	Nondefective	29	Nondefective
11	Nondefective	30	Nondefective
12	Nondefective	31	Nondefective
13	Nondefective	32	Nondefective
14	Defective	33	Defective
15	Nondefective	34	Defective
16	Nondefective	35	Nondefective
17	Nondefective	36	Nondefective
18	Nondefective	37	Defective
19	Nondefective	38	Nondefective

proportion defective computed from this particular sample may be misleading because of process variability. Although we won't discuss process predictability and stability until Chapter 5, under the assumption that the process is stable, the *sample proportion*, or *sample fraction defective*, can be calculated as

$$p = \frac{x}{n} \qquad (4.6)$$

where x is the number of defective items and n is the total number of items in the sample. Thus

$$p = \frac{8}{38} = 0.21$$

Often, in analytic studies of processes that operate over long periods of time, data on defectives are taken on a continuing basis to provide information on the necessity for action on the process. Figure 4.21 shows such data for the manufacture of radios over a seven-day period, where the data of Figure 4.20 were taken on the first day.

If the process is a stable one, the average fraction defective can be obtained by treating the data as a single sample, so that

$$p = \frac{\Sigma x}{\Sigma n} \qquad (4.7)$$

$$= \frac{32}{266} = 0.12$$

Measures of Variability

All populations and processes have some degree of variability, given appropriate sensitivity of the measuring instrument. Input items to a population or process aren't identical. Thus we must be able to quantify not only the central tendency but also the degree of variability in a set of data. To illustrate the need for some

FIGURE 4.21   Defective Units in Daily Samples of 38 Radios

Day #	# Radios Inspected	# Defective Units
1	38	8
2	38	6
3	38	0
4	38	2
5	38	5
6	38	8
7	38	3
	Total: 266	32

FIGURE 4.22   Weights of Ball Bearings from Three Manufacturing Processes

---

			Weight (g)		
Item #	*Process A*	*Item #*	*Process B*	*Item #*	*Process C*
1	5.0	1	5.0	1	7.6
2	5.3	2	7.8	2	7.8
3	8.0	3	7.9	3	8.0
4	9.2	4	8.0	4	8.1
5	10.0	5	8.8	5	8.1
6	10.5	6	10.5	6	8.4

---

numerical measure, consider the three sets of data in Figure 4.22, which represent the output from three different processes. For each process, the mean weight is 8.0 grams. If we limited ourselves to reporting this measure of central tendency only, we would actually fail to adequately characterize the three sets of measurements. While the three groups have the same mean, they differ with respect to how the data are spread around that mean, which we call the *variability* or *dispersion*. The outputs from processes A and B weigh between 5 and 10.5 grams, while those from process C weigh between 7.6 and 8.4 grams. Further, the weights from processes B and C are clustered around the mean, while those from process A are more widely dispersed away from the mean. The two commonly used quantitative measures of such variability are the range and the standard deviation.

The Range.   The *range* is the simplest measure of dispersion; for raw data from an enumerative or an analytic study, it's defined as the difference between the largest data point and the smallest data point in a set of data:

$$R = x_{max} - x_{min} \tag{4.8}$$

Thus, for process A the range is $10.5 - 5.0 = 5.5$ g; for process B the range is $10.5 - 5.0 = 5.5$ g; and for process C it's $8.4 - 7.6 = 0.8$ g. The larger the range, the more dispersed the data. In our illustrations, the outputs from processes A and B have more variability than the output from process C.

The Standard Deviation.   The *standard deviation* as a measure of dispersion takes into account each of the data points and their distances from the mean. The more dispersed the data points, the larger the standard deviation will be; the closer the data points to the mean, the smaller the standard deviation will be.

In an enumerative study, the population standard deviation is computed as the square root of the average squared deviations from the mean and is denoted by the lowercase Greek letter σ (sigma).

For both enumerative and analytic studies, we calculate the standard deviation of a sample (or subgroup) of n observations, called the *sample standard deviation,* s:

$$s = \sqrt{\frac{\Sigma(x - \bar{x})^2}{n - 1}} \tag{4.9a}$$

or the computationally simpler formula

$$s = \sqrt{\frac{\Sigma x^2 - \frac{(\Sigma x)^2}{n}}{n - 1}} \tag{4.9b}$$

The square of the standard deviation is called the *variance, σ^2*. The square of the sample standard deviation is the *sample variance, s^2*.

For Figure 4.22's data, we can calculate the sample standard deviation for process A:

x	x^2
5.0	25.00
5.3	28.09
8.0	64.00
9.2	84.64
10.0	100.00
10.5	110.25
$\Sigma x = 48.0$	$\Sigma x^2 = 411.98$

Using Equation 4.9b,

$$s = \sqrt{\frac{411.98 - \frac{(48)^2}{6}}{5}} = \sqrt{5.596} = 2.37 \text{ grams}$$

Similarly the sample standard deviation from process B is 1.79 grams, and from process C is 0.28 grams. The standard deviation, then, clearly shows not only that the output from process C is less variable than that from the other two processes but that the overall variability for process B is smaller than that for process A, a distinction that the range did not exhibit.

Measures of Shape

In addition to the measures of central tendency and variability that we've discussed, a population or the output of a process can also be characterized by its shape.

Skewness.   We've already mentioned one property of shape in our discussion of the median, the *skewness,* or lack of symmetry of a set of data. Many variables are naturally skewed, such as surface areas, volumes, warpage, and incomes. The frequency distributions in Figure 4.23 illustrate different degrees of skewness.

FIGURE 4.23   Skewness of Frequency Distributions

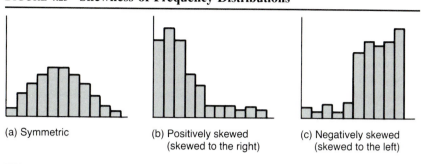

(a) Symmetric   (b) Positively skewed   (c) Negatively skewed
                                 (skewed to the right)   (skewed to the left)

A numerical measure of skewness, *Pearson's coefficient of skewness,* is defined as

$$\text{Skewness}_P = \frac{3(\overline{x} - M_e)}{s} \tag{4.10}$$

Note that for a symmetric distribution, as in Figure 4.23(a), where the mean, the median, and the mode are the same, the coefficient will be 0; for a *positively skewed* distribution, as in Figure 4.23(b), where the mean is larger than the median, which is larger than the mode, the coefficient will be positive; and for a *negatively skewed* distribution, as in Figure 4.23(c), the mode is larger than the median, which is larger than the mean, and the coefficient will be negative. For our 95 log lengths, examination of the frequency displays in Figure 4.13(a) and (c) indicates approximate symmetry. The mean length, from Equation 4.3, is $\overline{x} = 1336.2$ cm. The median, from Equation 4.5, is 1408.6 cm, and the standard deviation, from Equation 4.9b, is 402.8 cm. Using these statistics, we can calculate the Pearson coefficient of skewness to be

$$\text{Skewness}_P = \frac{3(1336.2 - 1408.6)}{402.8} = -0.539$$

Another way to view skewness is as a measure of the relative sizes of the tails of the distribution. Symmetric distributions, where the two tails are the same, thus have a 0 coefficient of skewness; where the difference between the frequencies in the two tails is great, the magnitude of the coefficient of skewness will be large.

Kurtosis.   Another characteristic of the shape of a frequency distribution is its peakedness or *kurtosis.* A distribution with a relatively high concentration of data in the middle and at the tails, but low concentration in the shoulders, has a large kurtosis; one that's relatively flat in the middle, with fat shoulders and thin tails, has little kurtosis. Figure 4.24 illustrates three curves with the same mean and standard deviation, 0 skewness, and different degrees of kurtosis.

FIGURE 4.24   Kurtosis

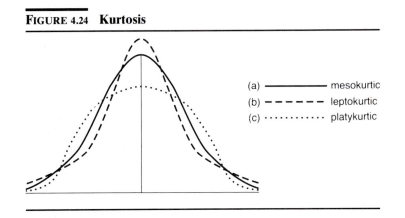

(a)	———————	mesokurtic
(b)	– – – – –	leptokurtic
(c)	··············	platykurtic

A numerical measure of kurtosis is given by[1]

$$\text{Kurtosis} = \frac{\Sigma (x - \bar{x})^4}{ns^4} - 3 \qquad (4.11)$$

The bell-shaped frequency distribution (or normal curve)—labeled curve a in Figure 4.24 has a kurtosis of 0 and is called *mesokurtic*. A more peaked curve (curve b in the figure) is called *leptokurtic* and has a positive numerical kurtosis; the flatter curve c is called *platykurtic* and has a negative kurtosis.

Necessary Sample Sizes for Estimating Measures of Shape.   The measures of shape we've discussed depend largely upon the tails of the distribution. As the central portion of most distributions usually contains a sizable fraction of the observed data points, measures of skewness and kurtosis are often based on the relatively small fraction of the data in the tails. Hence some statisticians suggest large sample sizes for estimating skewness and kurtosis for populations or stable processes. Neither of these statistics can be interpreted for unstable processes.

Interpreting Variability

In the physical world in which we live no two things are exactly alike. There will always be some variation, and, as we've seen, we can measure that variation using one of several methods to examine our data's degree of scatter. Our ability to use the information provided by this measure plays an important role in both enumerative and analytic statistical studies.

Chebychev's Inequality.   Calculating the standard deviation for a data set tells us a lot about the location of most of the points making up those values. The *Chebychev Inequality* provides that if we construct an interval that extends k standard deviations on either side of the mean, at least $1 - 1/k^2$ of the individual points will be within that interval, provided k is greater than 1.0.

For the data in Figure 4.22, the average for process B is 8.00 grams and the standard deviation 2.37 grams. Choosing a value for k of 2, our interval will extend 2(2.37) = 4.74 on either side of the mean. The interval thus extends from 8 − 4.74

to 8 + 4.74, or from 3.26 to 12.74. According to Chebychev's Inequality, at least $1 - 1/2^2 = 3/4$ of the data points will fall in the interval. All six of the data points actually fall within the boundaries. The Chebychev Inequality provides us with a lower bound on the fraction of the data falling within k standard deviations of the mean. Often data is more closely concentrated than this rule suggests.

An Empirical Rule for Analytic Studies. In enumerative studies we often have the opportunity to know something more about the shape of our data. The data points may tend to follow one of a great variety of distributional forms such as the normal, chi square, exponential, uniform, binomial, Poisson, and triangular.

Unfortunately in analytic studies we usually don't have the luxury of such information for future output. Even if we did know the distributional form of historical data, there's no guarantee that the form will remain operational in the future. Nevertheless, it's unnecessary to assume that data are normally or otherwise distributed in order that control charts work effectively.

W. A. Shewhart demonstrated that the means and standard deviations computed from 25 samples of four observations each drawn from normal, uniform, and triangular distributions all were within three standard errors of their respective means. In Shewhart's words, "In each case all of the points are within the limits as we should expect them to be under the controlled conditions supposed to exist in drawing these samples."[2]

Wheeler and Chambers expand on this point, arguing "Control charts work well even if the data are not normally distributed."[3] They state that when data from a process is stable "no matter how the data 'behave,' virtually all of the data will fall within three sigma units of the average."[4] Their Empirical Rule states that for a stable process, "Approximately 99% to 100% of the data will be located within a distance of three sigma units on either side of the average."[5] They go on to show that for six widely differing distributions, virtually all of the means and ranges computed from samples of sizes 2, 4, and 10 fall within three standard errors of their respective means. Their results indicate that

> While the distribution of the measurements does affect the percentage falling outside the control limits, the actual effects are quite small. For Subgroup Averages the percentage outside the limits remains about 1% or less. For the Subgroups Ranges the percentage outside the control limits goes above 2% only for the Chi-Square and Exponential Distributions.[6]

This provides a very compelling argument for the use of control charts without the necessity for the assumptions regarding any underlying distributional forms. We'll discuss this further in the next chapter.

Summary

This chapter presented the classical definition and relative frequency definition of probability. The former allows us to calculate the probability of an event for enumerative studies, while the latter is used in analytic studies, where there's no frame from which classical probabilities can be computed because we're dealing with a process that has a past, present, and (unknown) future.

We've also looked at how to describe the characteristics of data taken from a population or process. We begin with a classification of data into two types: variables and attribute. Data may be characterized by both visual and numerical means. We've seen that frequency distributions provide a tabular display of numerical data grouped into non-overlapping intervals. Graphical displays offer a composite picture of the relationships and/or patterns in the data at a glance. Frequency polygons, histograms, bar charts, and run charts are useful visual tools for characterizing data.

Numerical measures summarize important characteristics of a population or process in a more quantitative way. We've seen that both grouped and ungrouped data can vary with respect to central tendency, variability, and shape, and we've discussed some important measures of these characteristics. The mean, median, and mode are used to describe the central tendency of variables data; the proportion is a useful measure for attribute data. The range and standard deviation provide useful characterizations of the variability and the skewness and kurtosis of a frequency distribution are measures of the shape.

For stable processes, we can further characterize the variability and conclude that virtually all process output will be contained within a three-sigma interval, providing us with a basis for constructing control charts in the next three chapters.

Exercises

4.1 You've just been appointed plant manager in a sock factory. On your first day you're presented with the following production history.

Grade	Number of Socks Produced January 1991–December 1993	
Firsts (good)	73,197	(72.21%)
Seconds (sell at lower price)	17,877	(17.64%)
Scrap (junk)	10,286	(10.15%)
Total	101,360	

a. What does the table tell you about the factory's capability to produce good socks (firsts) in the future?

b. Would more information about the table's figures improve their usefulness? What information?

4.2 The accompanying table provides seven years of departmental accident data for an industrial laundry. The data have been analyzed chronologically (by day) and found to exhibit stable patterns of variation with respect to accidents per day.

| | Number of Accidents | | | | | |
Department	Mon.	Tues.	Wed.	Thurs.	Fri.	Total
Receiving/Shipping	12	8	9	8	13	50
Washroom	30	19	21	27	33	130
Drying	10	10	11	9	9	49
Ironing/Folding	22	20	21	19	20	102
Total	74	57	62	63	75	331

    a. What's the probability that an accident will occur in the Washroom on Monday? In Ironing/Folding on Thursday?

    b. What's the probability that an accident will occur on Friday? In Receiving/Shipping?

4.3 An accounting clerk is responsible for computing the effective number of hours worked daily (regular time + (1.5 × overtime)) for each of 500 production workers and for recording the number of days absent each week on the time card for each worker. The preceding two functions (computing effective daily hours per worker and recording number of days absent each week per worker) are the job characteristics upon which she's judged. The accounting supervisor draws weekly random samples of 50 time sheets to monitor and help improve the clerk's performance. The following chart reflects 30 weeks of analysis.

| Recorded Number of Days Absent per Week | Computation of Effective Hours per Week | | Total |
	Correct (no mistakes)	Incorrect (one or more mistakes)	
Correct (recorded properly)	1,235	190	1,425
Incorrect (recorded improperly)	65	10	75
Total	1,300	200	1,500

    a. Can the accounting supervisor use this information to help the clerk improve her performance? Why?

    b. What assumptions are required for the supervisor to be able to use those figures in supervisory efforts?

    c. Assuming that the accounting clerk's work is in statistical control in relation to both quality characteristics:

        1. What's the probability that the effective hours will be computed correctly? That the number of days absent will be recorded correctly?

        2. What's the probability that a time card will be completed correctly? Incorrectly?

4.4 A consumer protection testing agency wants to study the life expectancy of a particular job lot of a new radial tire. Ten tires were randomly selected from the job lot. Mileages these tires achieved before the minimum tread depth was reached (and the tires were declared worn out) were: 32,800; 41,700; 35,200; 39,000; 36,200; 35,600; 35,700; 45,200; 42,800; and 35,700.

    a. Construct a frequency distribution and cumulative frequency distribution for these data.

    b. Using that frequency distribution, draw a histogram of the recorded mileages for the 10 new radial tires.

    c. Is a run chart an appropriate display for these data? Explain why or why not, and if so, construct a run chart.

    d. Calculate the mean life expectancy for this sample of 10 tires.

    e. Calculate the median life expectancy for this sample of 10 tires.

    f. Calculate the modal life expectancy for this sample of 10 tires.

    g. Estimate the standard deviation of the life expectancy of all new radial tires from which this sample of 10 tires has been drawn.

    h. Calculate the range for these data.

4.5 A machine shop manager wishes to study the time it takes an assembler to complete a given small subassembly. Measurements, in minutes, are made at 15 consecutive half-hour intervals. The times to complete the task are 12, 10, 18, 16, 4, 16, 11, 15, 15, 13, 19, 10, 15, 17, and 11.

    a. Construct a frequency distribution and cumulative frequency distribution for these data.

    b. Using the preceding frequency distribution, draw a frequency polygon for the times to complete the task. Comment on the skewness of the data.

    c. Is a run chart an appropriate display for these data? Explain why or why not, and if so, construct a run chart.

    d. Calculate the mean time to complete the task based upon this sample of 15 observations.

    e. Calculate the median time to complete the task based upon this sample of 15 observations.

    f. Calculate the modal time to complete the task based upon this sample of 15 observations.

    g. Estimate the standard deviation of the time to complete the task.

    h. Calculate the range for these data.

    i. Calculate Pearson's coefficient of skewness for these data.

4.6 A buyer for a large chain of restaurants is about to purchase a large number of chicken breasts. According to the supplier, the breasts weigh an average of one pound each. The buyer selects a random sample of 10 breasts and weighs them, revealing the following weights, in pounds: 1.04, 1.00, 0.94, 1.10, 1.02, 0.90, 0.97, 1.03, 1.05, and 0.95.

    a. Calculate the mean weight for this sample of 10 chicken breasts.

    b. Calculate the median weight for this sample of 10 chicken breasts.

    c. Estimate the standard deviation of the weight of chicken breasts from this supplier.

    d. Calculate the range of weights of chicken breasts, based upon the sample of 10 chicken breasts.

    e. Calculate the kurtosis for these data.

4.7 The ABC Company is planning to analyze the average weekly wage distribution of its 58 employees during fiscal year 1994. The 58 weekly wages are available as raw data corresponding to the alphabetic order of the employees' names:

241	253	312	258	264	265
316	242	257	251	282	305
298	276	284	304	285	307
263	301	262	272	271	265
249	229	253	285	267	250
288	248	276	280	252	258
262	314	241	257	250	275
275	301	283	249	288	275
281	276	289	228	275	
170	289	262	282	260	

    a. Calculate the range of the data.

    b. Calculate the mean, median, and mode of the data.

    c. Construct a frequency distribution for the data.

4.8 The following frequency distribution shows the distance covered (in miles) by a sample of 80 trucks belonging to a long-distance moving company during 1994.

Distance Covered (*in miles*)	*Number of Trucks*
30,000 but less than 40,000	2
40,000 but less than 50,000	3
50,000 but less than 60,000	7
60,000 but less than 70,000	12
70,000 but less than 80,000	18
80,000 but less than 90,000	24
90,000 but less than 100,000	14

    a. Represent this frequency distribution on a histogram.

    b. Represent this frequency distribution on a frequency polygon.

4.9 The following frequency distribution shows the number of minutes spent on an elementary task in an assembly line operation performed by 35 employees over a one-month period:

Number of Minutes	Number of Employees
2 but less than 4	3
4 but less than 6	8
6 but less than 8	20
8 but less than 10	4

    a.   Represent this information on a frequency polygon.

    b.   Represent this information on a histogram.

Endnotes

1. The kth moment about the mean is $\Sigma(x - \bar{x})^k/n$, where the first moment is always 0; the second moment is the variance; the third moment is an absolute measure of skewness, and the fourth moment is an absolute measure of kurtosis.
2. W. A. Shewhart, *Economic Control of Quality of Manufactured Product* (New York: Van Nostrand, 1931), p. 318.
3. Donald J. Wheeler and David S. Chambers, *Understanding Statistical Process Control,* 2d ed. (Knoxville, Tenn.: SPC Press, 1992), p. 65.
4. Ibid.
5. Ibid., p. 61.
6. Ibid., p. 76.

PART III Tools and Methods for Analytic Studies

This section discusses the tools and methods needed to conduct analytic studies. These techniques, when used in conjunction with the PDSA cycle, form a powerful arsenal that can be used to pursue continuous, never-ending improvement of any system (process) or system of systems.

Chapter 5 provides an overview of procedures for stabilizing and improving a documented and defined process. The chapter includes a section on the general theory and purpose of statistical control charts as well as a section on rational subgroups.

Chapter 6 discusses attribute control charts: p charts (for both constant and variable sample sizes), np charts, c charts, and u charts. Many examples of attribute control charts are included.

Chapter 7 examines variables control charts: x-bar and R charts, x-bar and s charts, median charts, and single measurement and moving range charts for variables data. Many examples are provided.

Chapter 8 presents specific control chart patterns that we can use to detect special sources of variation. It also provides statistical rules for detecting special sources of variation.

Chapter 9 discusses other tools and methods for determining the causes of special and common sources of variation: brainstorming, cause-and-effect diagrams, check sheets, Pareto diagrams, and stratification. These techniques provide information that may lead to plans for action resulting in resolving special sources of variation and reducing common sources of variation.

5 Stabilizing and Improving a Process with Control Charts

Introduction

A process that has been defined and documented can be stabilized and then improved. In great measure this can be accomplished through the use of statistical control charts, as well as other techniques that will be introduced in Chapter 9.

These tools and methods must be used in an environment that provides a positive atmosphere for process improvement; and top management must sincerely desire real process improvement. W. E. Deming points out that "any attempt to use statistical techniques under conditions that rob the hourly worker of his pride in workmanship will lead to disaster."[1] With this caveat clearly in sight, we may begin to consider the issues of stabilizing and improving a documented and defined process.

The Deming Cycle

The Deming cycle is a method that can aid management in stabilizing a process and pursuing continuous, never-ending process improvement.[2] Recall from Chapter 1 that the Deming cycle is composed of four basic stages: planning, doing, studying, and acting. Hence the Deming cycle is sometimes referred to as the PDSA cycle (*Plan-Do-Study-Act* cycle). A plan is developed (*plan*); the plan is tested on a trial basis using planned experiments (*do*); the effects of the experiments are monitored (*study*); and appropriate actions are taken on the process (*act*). These actions can lead to a new or modified plan, so the PDSA cycle continues forever in an uphill cycle of never-ending improvement. Figure 5.1 illustrates the Deming (PDSA) cycle.

The Deming cycle operates by recognizing that problems (opportunities for improvement) in a process are determined by the difference between customer (internal and/or external) needs and process performance. If the difference is large, customer dissatisfaction may be high, but there's great opportunity for

FIGURE 5.1   The Deming (PDSA) Cycle

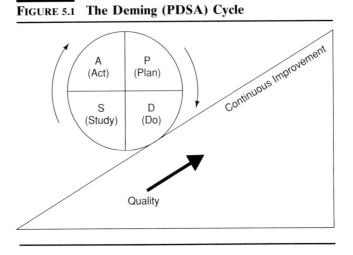

Quality

improvement. If the difference is small, the consequent opportunity for improvement is diminished.[3] Nevertheless, it's always economical to continually attempt to decrease the difference between customer needs and process performance, as discussed later in this chapter. Now let's review the four stages of the Deming cycle.

Stage 1: Plan

The collection of data about process variables is critical when determining a plan of action for what must be accomplished to decrease the difference between customer needs and process performance. In Chapter 1 we discussed the three types of quality that must be understood to optimize the interdependent system of stakeholders of an organization's progress toward its aim. Data concerning customer needs are collected via *quality-of-performance* studies. These data and more are evaluated and transformed into product/process characteristics (variables) that may be acted upon during *quality-of-design* studies. Data concerning the process's ability to surpass customer needs are collected via *quality-of-conformance* studies. These data should be collected on product/process characteristics (variables) for which process improvement action can be taken.

A plan must be developed to determine the effect(s) of manipulating process variables upon the difference between process performance and customer needs. The plan must be tested through experimentation in the Do phase.

All relevant variables must be operationally defined, as discussed in Chapter 3, and all procedures must be standardized so that everyone conducts a given procedure using the same format. This standardization increases communication and decreases the possibility of error.

Stage 2: Do

The plan established in the first stage is set into motion on a trial basis in the Do stage. This is accomplished through a three-part process. First, the organization must educate everyone involved with the planned experiment so that they understand the relationship between the manipulated variables and the proposed decrease in the difference between customer needs and process performance. Second, the organization must train everyone involved with the trial plan so that they understand how the plan will affect and modify their jobs (Deming's point 6). After the first two steps have been accomplished, the plan can be set into motion on a trial basis as the third part of the Do stage. The planned experiments should be conducted in a laboratory, production setting or office setting, or on a small scale with customers (both internal and external).

Stage 3: Study

The results of the planned experiment conducted in Stage 2 (Do) must be analyzed (Study) to answer two questions. First, are the manipulated process variables behaving according to the plan and causing a decrease in the difference between customer needs and process performance? And second, are the downstream effects of the plan creating problems or improvements? The results of statistical studies in this Study stage lead to the Act stage.

Stage 4: Act

The Act stage's purpose is to implement modifications to the plan discovered in the Study stage or to make improvements to the process. This leads to further narrowing of the difference between customer needs and process performance. Hence the PDSA cycle continues forever to promote never-ending improvement.

The Act stage operates by considering whether the manipulated process variables have effectively diminished the difference between customer needs and process performance. If at the Act stage we learn that they haven't been effective, the PDSA cycle returns to the Plan stage to search for other process variables that may decrease the difference. However, if the manipulation of process variables has produced the desired results, then the Act stage leads back to the Plan stage to determine the optimal levels at which to set the manipulated process variables. Finally, if at the Act stage we find that the plan is optimal in decreasing the difference between customer needs and process performance, we implement and standardize the process improvements specified by the plan through training relevant personnel and updating all training materials.

The Deming (PDSA) cycle is the basic method used to stabilize and continuously improve a process.[4] The next sections consider control charts, which are important components of the quality studies required by the Deming cycle.

Using Control Charts to Stabilize and Improve a Process

Control charts are statistical tools used to analyze and understand process variables, to determine a process's capability to perform with respect to those variables, and to monitor the effect of those variables on the difference between customer (internal and/or external) needs and process performance. Control charts accomplish this by allowing a manager to understand the sources of variation in a process and hence to manipulate and control those sources to decrease the difference between customer needs and process performance. This decrease can be managed only if the process under study is stable and capable of improvement.

Process Variation

All processes exhibit variation. It's unavoidable, yet, like a "wild beast," it must be controlled. For example, one-inch bolts will vary over several thousandths of an inch; the proportion of data entry errors will vary from day to day; a random sample of 100 two-liter bottles of cola will vary in contents from container to container; the time it takes to serve a customer will vary from customer to customer. Despite the omnipresence of process variation, we must always endeavor to control and reduce it.

We can classify process variation as the result of either *common causes* or *special causes*.

Common Causes of Variation

Common causes of variation are inherent in a process. Common variation is comprised of a myriad of small sources that are always present in a process and affect all elements of the process. Examples of possible causes of common variation[5] are

Procedures not suited to requirements.

Poor product design.

Machines out of order.

Machines not suited to requirements.

Barriers that rob the worker of the right to do a good job and take pride in his or her work.

Poor instruction and/or supervision of workers.

Poor lighting.

Incoming materials not suited to requirements.

Vibration.

Failure to provide hourly workers with statistical information to help them improve performance and reduce variation.

Humidity too high or too low.

Unpleasant working conditions, such as noise, dirt, extremes of heat or cold, or poor ventilation.

Management should not hold workers responsible for such problems of the system; the system is management's responsibility. If management is unhappy with the amount of common variation in the system, it must act to remove it. Professionals estimate that common variation causes at least 85 percent of the problems in a process, the remaining problems being caused by special variation.

Special Causes of Variation

Variations created by special causes lie outside the system. Frequently their detection, possible avoidance, and rectification are the responsibility of the people directly involved with the process. But sometimes management must try to find these special causes. When found, policy must be set so that if these special causes are undesirable, they won't recur. If, on the other hand, these special causes are desirable, policy must be set so that they do recur.

Control Charts and Variation

In this chapter we'll see that control charts are used to identify and differentiate between these two different causes of variation. When a process no longer exhibits special variation, but only common variation, it's said to be stable.

When only common causes of variation are present in a process, management must take action to reduce the difference between customer needs and process performance by endeavoring to move the centerline of the process closer to a desired level (*nominal*) and/or by reducing the level of common variation. These types of changes will aid in the quest for never-ending improvement.

The Need for the Continual Reduction of Variation

During the past two centuries, most mass production concerned itself with meeting engineering specifications most of the time; variation wasn't the central focus. As long as an item or a part served its intended purpose, it was classified as "good" and passed on to its next operation or final use. When excessive variation caused an item to be nonconforming, it was classified as "bad" and downgraded, reworked, discarded, or somehow removed from the mainstream of output. Little if any effort was made to investigate the causes of such variation; it was accepted as a way of life. High output was maintained by overproducing and then sorting the output into items that met specifications and items that didn't. It's still common to find firms increasing output by relaxing engineering specifications to include marginally defective items with good ones.[6]

Deming has written, "It is good management to reduce the variation in any quality characteristic, whether this characteristic be in a state of control or not, and even when few or no defectives are being produced."[7] When variation is reduced, parts will be more nearly alike, and services rendered will be more predictable. Finished products and services will work better and be more reliable. Customer satisfaction will increase because customers will know what to expect. Process output and capability will be known with greater certainty, and the results of any changes to the process will be more predictable.

Therefore, management must constantly attempt to reduce process variation around desired characteristic specification levels (nominal levels) to achieve the degree of uniformity required to get products and services to function during their life cycle as promised to the customer.

As discussed in Chapter 1, the belief that there's no loss associated with products and services that conform to specifications regardless of the size of the deviation from nominal is fallacious.[8] Figure 5.2 shows the traditional view of losses arising from deviations from nominal. Losses are zero until the specification limit is reached; and suddenly they become positive and constant, regardless of the size of the deviation from nominal.

Again, as seen in Chapter 1, Figure 5.3 shows a more realistic loss function: losses begin to accrue as soon as products deviate from nominal. As we can see, the view represented in this figure requires the never-ending reduction of process variation around nominal to be maximally cost-efficient and provide the degree of customer satisfaction demanded in today's marketplace.[9]

FIGURE 5.2   Traditional View of Losses Arising from Deviations from Nominal

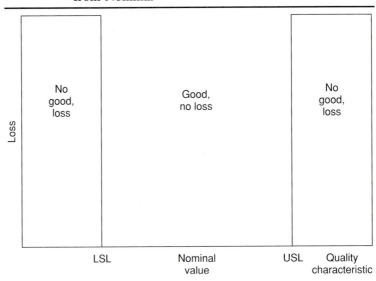

FIGURE 5.3   Realistic View of Losses Arising from Deviations from Nominal

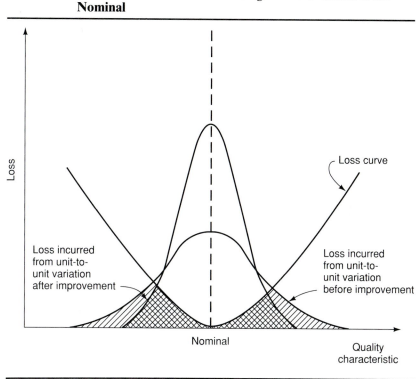

To illustrate, consider a car battery charged by an alternator. The alternator has a voltage regulator that controls the charge to the battery. The alternator-voltage regulator assembly must put out a charge of 13.2 volts to keep the battery's charge at 12 volts.

If the alternator produces a charge of less than 13.2 volts, the electrolyte (acid) in the battery will gradually turn into water, resulting in the battery failing. The lower the alternator output, the more quickly this will happen. If the alternator output is more than 13.2 volts, excessive heat will build up in the battery, causing the battery plates to warp, which will also result in failure of the battery. As the alternator output increases, this effect will occur more quickly.[10]

Any deviation from nominal clearly causes a loss. The greater the deviation, the greater the loss. Consequently we see the logic in the necessity for the continual reduction of process variation.

Donald J. Wheeler and David S. Chambers discuss another rationale for the continual reduction of process variation.[11] A process can be described as existing in one of four states: chaos, the brink of chaos, the threshold state, and the ideal state.

When a process is in a state of *chaos*, it's producing some nonconforming product and it's not in a state of statistical control. (That is, special causes of variation are present.) There's no way to know or predict the percentage of nonconforming product that the process will generate.

A process on the *brink of chaos* produces 100 percent conforming product; however, the process isn't stable. (There's variation resulting from special causes.) Hence there's no guarantee that the process will continue to produce 100 percent conforming product indefinitely. As it's unstable, the process may wander and the product's characteristics may change at any time, entering a state of chaos.

The *threshold state* describes a stable process that produces some nonconforming product; process variation results from common causes that are an inherent part of the system. The only way to reduce this variation is to change the process itself.

The *ideal state* describes a stable process producing 100 percent conforming product. The ideal state isn't a natural state; forces will always exist to push the process away from the ideal state. Wheeler and Chambers liken this phenomenon to entropy, in that there's similarly a trend toward disorder in the universe. It may help to visualize a process in the ideal state as a perfectly swept lawn. There will always be winds to mar its perfect appearance by depositing leaves, twigs, or other debris. Keeping the lawn perfectly swept is a never-ending challenge. In the same way, striving toward an ideal state for a process requires constant attention on management's part.

Control charts are statistical tools that make possible the distinction between common and special causes of variation. Consequently, control charts permit management to relentlessly pursue the continuous reduction of process variation and strive toward the ideal state for a process.

The Structure of Control Charts

All control charts have a common structure. As Figure 5.4 shows, they have a centerline (representing the process average) and upper and lower control limits that provide information on the process variation.

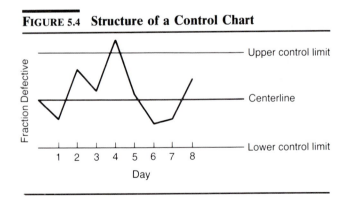

FIGURE 5.4   Structure of a Control Chart

Control charts are constructed by drawing samples and taking measurements of a process characteristic. Each set of measurements is called a subgroup. Control limits are based on the variation that occurs within the sampled subgroups. In this way, variation between the subgroups is intentionally excluded from the computation of the control limits; the common process variation becomes the variation on which we calculate the control limits. The control limit computations assume that there are no special causes of variation affecting the process. If a special cause of variation is present, the control chart, based solely on common variation, will highlight when and where the special cause occurred. Consequently, the control chart makes possible the distinction between common and special variation and provides management and workers with a basis on which to take corrective action on a process.

Control limits are often called *three-sigma limits,* where the lowercase Greek letter *sigma* (σ) is used in enumerative studies to denote the population standard deviation, as described in Chapter 4. In analytic studies, this notation may be used to denote a process standard deviation.

When Walter Shewhart described creating a range for allowable variation (common variation), he proposed using a statistic (the mean of the process characteristic of interest) plus and minus three times the standard deviation of that statistic (where the standard deviation of a statistic is called the *standard error*) as an acceptable economic value.[12] In practice, as pointed out in Chapter 4, virtually all of the process output will be located within a three-sigma interval of the process mean, provided that the process is stable. Further, virtually all of the statistics for a given sample size will be located within a three–standard-error interval around the process mean, provided that the process is stable. This provides us with a basis for distinguishing between common and special variation for the process statistics to be discussed here and in Chapters 6 and 7.

In general, the centerline of a control chart is taken to be the estimated mean of the process; the upper control limit is the mean plus three times the estimated standard error, and the lower control limit is the mean minus three times the estimated standard error. These are computed from the process output, assuming that no special sources of variation are present. Subgroup means that behave nonrandomly with respect to these control limits will be said to be indications of the presence of special causes of variation.

Stabilizing a Process with Control Charts

As an example of the use of control charts to detect special variation, consider a data entry operation that makes numerous entries daily.[13] On each of 24 consecutive days, subgroups of 200 entries are inspected. Figure 5.5 illustrates the resulting raw data; Figure 5.6 plots the fraction of defective entries as a function of time. Figure 5.6 seems to indicate that on days 5, 6, 10, and 20 something unusually good happened (0 percent defectives), and on days 8 and 22 something unusually bad happened. A simple control chart will help to determine whether these points were caused by common or special variation.

FIGURE 5.5   **Formulation of a Control Chart for Data Entry Operation**

Raw Data for Construction of Control Chart

Day	Number of Entries Inspected	Number of Defective Entries	Fraction of Defective Entries
1	200	6	.030
2	200	6	.030
3	200	6	.030
4	200	5	.025
5	200	0	.000
6	200	0	.000
7	200	6	.030
8	200	14	.070
9	200	4	.020
10	200	0	.000
11	200	1	.005
12	200	8	.040
13	200	2	.010
14	200	4	.020
15	200	7	.035
16	200	1	.005
17	200	3	.015
18	200	1	.005
19	200	4	.020
20	200	0	.000
21	200	4	.020
22	200	15	.075
23	200	4	.020
24	200	1	.005
Total	4,800	102	

When the data consist of a series of fractions defective, the appropriate control chart is a p chart. This is a depiction of the fraction of process data that has some attribute of interest—in our example, the fraction defective.

The centerline for a p chart is the mean of the fraction defective, $\bar{p}$, which we calculate as

$$\bar{p} = \left[\frac{\text{Total number of defectives in all subgroups under investigation}}{\text{Total number of units examined in all subgroups under investigation}} \right] \quad (5.1)$$

Control limits are calculated as $\bar{p}$ plus and minus three times the standard error. The standard error for the average proportion, σ_p, is given by the expression

$$\sigma_p = \sqrt{\frac{\bar{p}(1 - \bar{p})}{n}}$$

FIGURE 5.6   Plot of Fraction of Defective Entries against Time

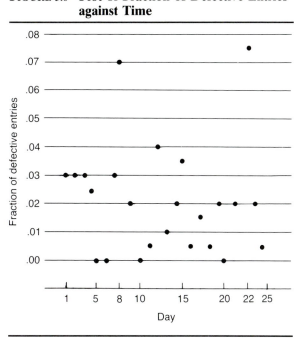

Using this value, the upper and lower control limits for a p chart are given by:

$$\text{UCL(p)} = \bar{p} + 3 \sqrt{\frac{\bar{p}(1 - \bar{p})}{n}} \tag{5.2}$$

$$\text{LCL(p)} = \bar{p} - 3 \sqrt{\frac{\bar{p}(1 - \bar{p})}{n}} \tag{5.3}$$

where n is the subgroup size.

We can now use Equations 5.1, 5.2, and 5.3 to find the numerical values for constructing our p chart:

$$\bar{p} = \frac{102}{4,800} = 0.02125$$

$$\text{Centerline(p)} = 0.021$$

$$\text{UCL(p)} = 0.02125 + 3 \sqrt{\frac{(0.02125)(1 - 0.02125)}{200}}$$

$$= 0.02125 + 3(0.010198)$$

$$= 0.05184$$

$$\text{Upper control limit} = 0.052$$

FIGURE 5.7   Control Chart for Fraction Defective

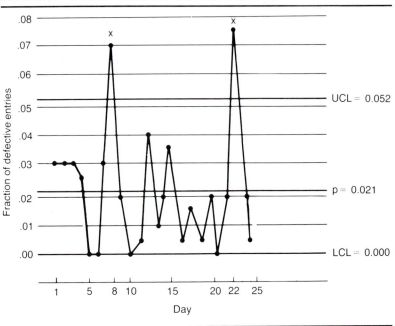

$$LCL(p) = 0.02125 - 3 \sqrt{\frac{(0.02125)(1 - 0.02125)}{200}}$$

$$= 0.02125 - 3(0.010198)$$

$$= -0.00934$$

Lower control limit $= 0.00$

Notice that since a negative fraction defective isn't possible, the lower control limit is set at 0.00.

Figure 5.7 shows the completed p chart. Clearly on days 8 and 22 there's some special variation. Notice, however, that the fractions defective on days 5, 6, 10 and 20 aren't beyond the control limits. Days with no defectives aren't out of control; we've merely observed that the process is capable of producing zero defectives some of the time. The observations on days 8 and 22 indicate that the process exhibits a lack of statistical control.

When a manager or worker who determines that the cause of variation is special, he should search for and resolve the causes that may be attributable to such factors as a specific machine, worker or group of workers, or a new batch of raw materials. After the causes of special variation have been identified and rectified, a stable process that's in statistical control will result.

In our example, to bring the process under control, management investigated the observations that were out of control (days 8 and 22) in an effort to discover and remove the special causes of variation in the process. In this case, management found that on day 8 a new operator had been added to the work force without any training. The logical conclusion was that the new environment probably caused the unusually high number of errors. To ensure that this special cause wouldn't recur, the company added a one-day training program in which data entry operators would be acclimated to the work environment.

A team of managers and workers conducted an investigation of the circumstances occurring on day 22. Their work revealed that on the previous night one of the data entry terminals malfunctioned and was replaced with a standby unit. The standby unit was older and slightly different from the ones currently used in the department. Repairs on the regular terminal weren't expected to be completed until the morning of day 23. To correct this special source of variation, the team recommended developing a proactive program of preventative maintenance on the terminals to decrease the likelihood of future breakdowns. Employees then implemented the solution with the policy commitment of management.

Negative special causes of variation can be eliminated from a process, or positive special causes of variation can be incorporated into a process, by setting and enforcing policy changes. Once this has been done, the process has, in essence, been changed.

The action taken on the process stemming from investigations of days 8 and 22 should change the process so that the special causes of variation will be eliminated. Consequently data from days 8 and 22 may now be deleted. After eliminating the data for the days in which the special causes of variation are found, the control chart statistics are recomputed.

$$\bar{p} = \frac{73}{4,400} = 0.01659 \cong 0.017$$

$$UCL(p) = 0.01659 + 3\sqrt{\frac{(0.01659)(1 - 0.01659)}{200}} = 0.04369 \cong 0.044$$

$$LCL(p) = 0.01659 - 3\sqrt{\frac{(0.01659)(1 - 0.01659)}{200}} = -0.01051 \cong -0.011$$

Lower control limit = 0.00

Figure 5.8 shows the revised control chart. The process appears to be stable and in statistical control. Notice that the revised control chart has somewhat narrower control limits than the original. When special causes have been eliminated, the narrower limits that occur may reveal other points that are now out of control. It will then be necessary to again search for special causes. Several such iterations may be required until the process is stable and in statistical control. At least 20 subgroups should remain to determine that the process is indeed stable. If the elimination of subgroups has left fewer than 20 subgroups remaining, additional data should be collected to ensure that the process is in a state of statistical control.

FIGURE 5.8   Control Chart for Fraction Defective

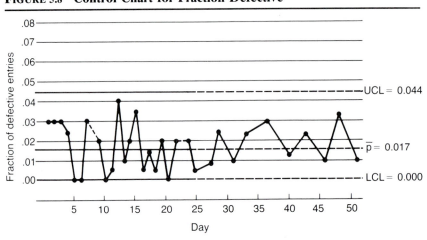

Chart of the process after special causes on days 8 and 22 are found by management and removed.

In this example, the control limits were extended into the future (see the dashed lines on the right side of Figure 5.8), and more data were collected and compared to the revised control limits. The process was found to be stable and consequently capable of being improved.

Advantages of a Stable Process

A *stable process* is a process that exhibits only common variation or variation resulting from inherent system limitations. The advantages of achieving a stable process are

1. Management knows the process capability and can predict performance, costs, and quality levels.
2. Productivity will be at a maximum, and costs will be minimized.
3. Management will be able to measure the effects of changes in the system with greater speed and reliability.
4. If management wants to alter specification limits, it will have the data to back up its decision.

A stable process is a basic requirement for process improvement efforts.[14]

Improving a Process with Control Charts

Once a process is stable, it has a known capability. A stable process may, nevertheless, produce an unacceptable number of defects (threshold state) and continue to do so as long as the system, as currently defined, remains the same. Management owns the system and must assume the ultimate responsibility for changing the system to reduce common variation as well as to reduce the difference between customer needs and process performance.

There are two areas for action to reduce the difference between customer needs and process performance. First, action may be taken to change the process average. This might include action to reduce the level of defects or process changes to increase production or service. Second, management can act to reduce the level of common variation with an eye toward never-ending improvement of the process. Procedures and inputs (such as composition of the work force, training, supervision, materials, tools and machinery, and operational definitions) are the responsibility of management. The workers can only suggest changes; they can not effect changes to the system.

In our example of the data entry firm, an employee-suggested training program was instituted. The program was aimed at reducing the average fraction of errors and the common variation, which would result in narrower control limits. Figure 5.9 shows the data entry control chart after management instituted the new training program. The average proportion of entries with errors decreased from 0.017 to 0.008, and the process variation decreased as well.

FIGURE 5.9   Control Chart for Fraction Defective after Institution of a New Training Program

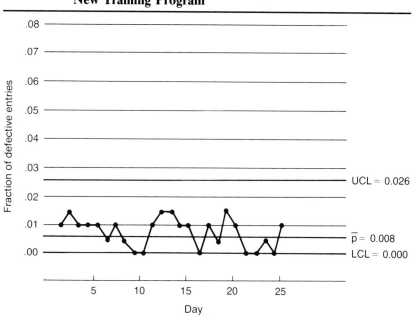

Chart of the process after training and procedure changes have been implemented by management.

Quality Consciousness and Types of Control Charts

Being conscious of quality typically follows a logical pattern. It begins with defect detection, moves to defect prevention, moves through the never-ending improvement of current products and services, and leads ultimately to the realization that the only path to continued prosperity must include innovation in future products and services.

Defect Detection

The purpose of defect detection is to sort conforming and nonconforming products or services through mass inspection. Defect detection assumes that defects will be produced; they are expected. In this first stage of quality consciousness no feedback loops or tools are available for correcting the factors that created the defectives in the first place. Once a defect is produced, it's too late to do anything but remove it from the process output.

Defect Prevention: Attribute Control Charts

The purpose of defect prevention is to achieve "zero defects." This stage of quality consciousness assumes that if all products and services are within specification limits, then all output will meet customers' needs and wants. The initial entry into this phase generally involves the use of control charts based on attribute data, such as conforming versus nonconforming with respect to some specification.

The most common types of attribute control charts are

- p chart: used to control the fraction of units with some characteristic (such as the fraction defective).
- np chart: used to control the number of units with some characteristic (such as the number of defectives per batch).
- c chart: used to control the number of events (such as defects) in some fixed area of opportunity (such as a single unit).
- u chart: used to control the number of events (such as defects) in a changeable area of opportunity (such as square yards of paper drawn from an operational paper machine).

Attribute control charts can help move the firm's processes toward a zero percent defective rate. However attribute control charts don't provide specific information on the cause of the defectives; and as the percent defective approaches zero, larger and larger sample sizes will be needed to detect defective output. For example, if a process is generating an average of one defective in every million units produced, then the average sample size needed to find one defective is one million units. Hence attribute control charts become ineffective as the proportion of defective output approaches zero. Control charting must continue, but in the face of the limitations of attribute control charts, a better means of process improvement and control is required. This will lead management to the next level of quality consciousness—never-ending improvement.

Never-Ending Improvement: Variables Control Charts

The purpose of never-ending improvement is to modify current products and processes to continuously reduce the difference between customer needs and process performance. Never-ending improvement necessitates using control charts based on variables data. These types of control charts allow for the never-ending reduction of unit-to-unit variation, even within specification limits. For example, steel rods from a process may all conform to specifications; an attribute control chart would show zero percent defective in every sample. However, by taking actual measurements on rod lengths (variables data), management can collect information that will enable them to consistently strive for the reduction of unit-to-unit variation.

The most common types of variables control charts are

- x-bar chart: used to control the process average.
- R chart: used to control the process range.
- s chart: used to control the process standard deviation.
- median chart: used as a simple alternative to the combination of an x-bar and R chart.
- Individuals chart: used to control subgroups of size one drawn from a process; frequently used when sampling is expensive or only one observation is available per subgroup (e.g., production per month).

Using variables control charts, management may continuously seek to reduce variation, center a process on nominal, and decrease the difference between customer needs and process performance. Chapters 6 and 7 extensively discuss the uses and applications of each of these different types of control charts.

Innovation (Quality Creation)

The purpose of innovation is twofold: first, to create a dramatic breakthrough in decreasing the difference between customer needs and process performance; and second, to discover the customer's future needs. Ideas for innovation with respect to the customer's future needs can't come from direct queries to customers; rather, they must come from the producer. In this regard, consumer research is backward looking; that is, asking customers what they want can only help producers improve existing products or services; it can't help producers anticipate the customer's future needs. As the Konica camera example discussed in Chapter 1 illustrated, consumers don't know what innovations they'll want in the future; for example, a consumer couldn't tell you he or she wants a facsimile machine or an automatic loading camera before such things existed. Breakthroughs must be discovered by the producer studying the problems customers have when using products and services. Recall that, in 1974 the camera market was saturated with cameras that satisfied customers' current needs; cameras were reliable, were relatively inexpensive to use, and produced good pictures. This created a nightmare for the camera industry. Consequently Konica decided to ask consumers what more they'd like in a camera. Consumers replied that they were satisfied with their cameras. Unfortunately asking consumers what more they would like in a camera didn't yield the information Konica needed to create a breakthrough. In response to this situation, Konica studied negatives at film processing laboratories and discovered that often the first few pictures on rolls of film were overexposed, indicating that users had difficulty in loading cameras. This presented an opportunity to innovate camera technology. The customer couldn't have been expected to think of this innovation. In response to this analysis, Konica developed the automatic loading camera. This is an excellent example of innovation of a current product or service. The same type of procedure could have been used to discover consumer needs for a completely new product or service.

An Example of an Attribute Control Chart

As an example of a situation in which an attribute control chart is appropriate, consider a new can-forming process used in the orange juice industry. After encountering severe quality problems, management decides to select periodic samples of size 50 from 30 consecutive trial production runs. The data are attribute in nature because each can is either defective or not. Each set of 50 comprises a subgroup that may contain defective units. The number of defectives in each subgroup can take on a value from 0 to 50. The fraction defective in each subgroup of 50 cans will be recorded and charted. The proper control chart in this case is the p chart (or, alternatively, an np chart, which Chapter 6 discusses in detail). Data derived from the subgroups are shown in Figure 5.10.

Equation 5.1 yields a centerline value of

$$\bar{p} = 347/1{,}500 = 0.2313$$

Then from Equations 5.2 and 5.3 the UCL and the LCL values are

$$UCL(p) = 0.2313 + 3\sqrt{\frac{(0.2313)(1 - 0.2313)}{50}} = 0.4102 \cong 0.41$$

and

$$LCL(p) = 0.2313 - 3\sqrt{\frac{(0.2313)(1 - 0.2313)}{50}} = 0.0524 \cong 0.05$$

Figure 5.11 illustrates the control chart for this process.

Two points indicate a lack of control because they fall above the UCL. There may be other grounds for believing that this process is exhibiting signs of special causes of variation; we discuss some of these later in this chapter. Chapter 8 discusses other methods for determining whether special sources of variation are present in a process.

FIGURE 5.10   Number of Defectives in 30 Subgroups of Size 50

Production Run	Subgroup Size	Number of Defectives	Subgroup Fraction Defective
1	50	12	0.24
2	50	15	0.30
3	50	8	0.16
4	50	10	0.20
5	50	4	0.08
6	50	7	0.14
7	50	16	0.32
8	50	9	0.18
9	50	14	0.28
10	50	10	0.20
11	50	5	0.10
12	50	6	0.12
13	50	17	0.34
14	50	12	0.24
15	50	22	0.44
16	50	8	0.16
17	50	10	0.20
18	50	5	0.10
19	50	13	0.26
20	50	11	0.22
21	50	20	0.40
22	50	18	0.36
23	50	24	0.48
24	50	15	0.30
25	50	9	0.18
26	50	12	0.24
27	50	7	0.14
28	50	13	0.26
29	50	9	0.18
30	50	6	0.12
	1,500	347	

Another Example of an Attribute Control Chart

A large metropolitan hospital must fill out a particular state-mandated form for each new patient admitted. The form requires information to be entered in 78 different fields, many of which have several subfields. The finished form thus presents many opportunities for clerical errors. To study the reasons that errors are made, 10 forms are selected each week and carefully examined by the clerical supervisor. The count of total number of errors found can take on nonnegative integer values. Figure 5.12 shows the result for the past 25 weeks.

As we'll see in the next chapter, when the data are a sequence of counts that have occurred in a series of identical areas of opportunity, the appropriate control chart may be a c chart.

FIGURE 5.11 Control Chart for Fraction Defective

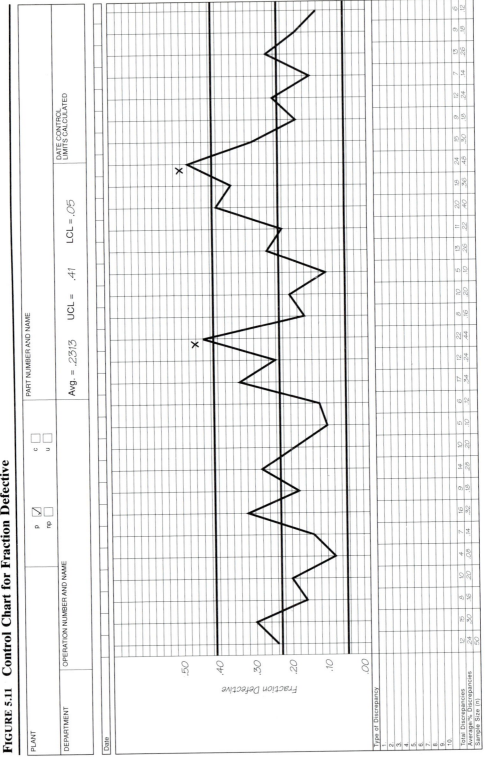

FIGURE 5.12   Number of Errors on New Patient Admission Forms

Week	Number of Errors (c)	Week	Number of Errors (c)
1	17	14	11
2	15	15	8
3	21	16	13
4	16	17	4
5	5	18	22
6	9	19	14
7	23	20	5
8	25	21	10
9	6	22	24
10	17	23	22
11	7	24	19
12	12	25	25
13	18		Total 368

The centerline for the c chart is the average count observed in the data:

$$\bar{c} = \frac{\Sigma c}{k} \qquad (5.4)$$

where

$$c = \text{number of errors per admission form, and}$$

$$n = \text{number of admission forms in the sample.}$$

Here that would take on a value of

$$\bar{c} = \frac{368}{25} = 14.72$$

The UCL and LCL values are found, as we'll see in the next chapter, by adding and subtracting three times the square root of the average value from the centerline:

$$UCL(c) = \bar{c} + 3\sqrt{\bar{c}} \qquad (5.5)$$

and

$$LCL(c) = \bar{c} - 3\sqrt{\bar{c}} \qquad (5.6)$$

These yield control limit values of

$$UCL(c) = 14.72 + 3\sqrt{14.72} = 14.72 + 3(3.84) = 26.24$$

and

$$LCL(c) = 14.72 - 3\sqrt{14.72} = 14.72 - 3(3.84) = 3.21$$

Figure 5.13 illustrates the control chart for this process. There are no indications of the presence of special variation in this process. It's stable and predictable. If

FIGURE 5.13 Control Chart for Errors on New Patient Admission Forms

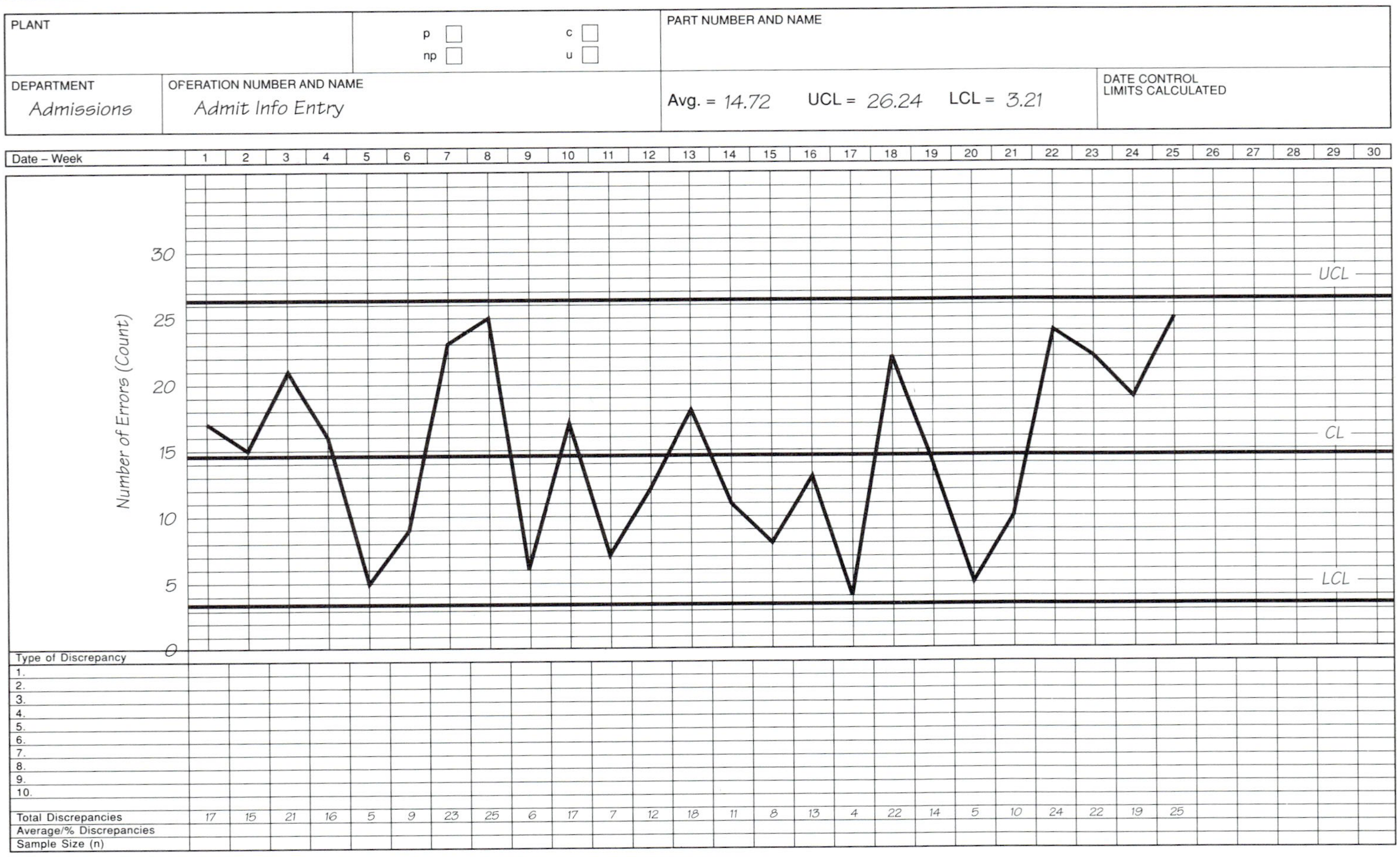

nothing happens to change this process, it will continue to produce an average of 14.72 clerical errors for every 10 forms processed.

An Example of a Variables Control Chart

A large pharmaceutical firm provides vials filled to a nominal value (specification) of 52.0 grams. The firm's management has embarked on a program of statistical process control and has decided to use variables control charts for this filling process to detect special causes of variation. Samples of six vials are selected every five minutes during a 105-minute period. Each set of six measurements makes up a subgroup.

As will be discussed more fully in Chapter 7, the appropriate control chart in this instance is an x-bar and R chart. The purpose of this chart is to see whether the process output is stable with regard to its variability and its average value. The output of each subgroup is summarized by its sample average and range. x-bar is the average for each of the subgroups, as given by Equation 4.3, while R is the range, or the largest data value in each subgroup minus the smallest data value in that subgroup, for each subgroup, calculated from Equation 4.8.

For our example, the subgroup ranges are shown in Figure 5.14. They begin with a range at 9:30 of

$$R = 53.10 - 52.22 = 0.88$$

FIGURE 5.14   Vial Weights

Date: 1-10-93				Measurement (gm)				Nominal Fill: 52.00 gms.	
Obs. #	Time	1	2	3	4	5	6	R	$\bar{x}$
1	9:30	52.22	52.85	52.41	52.55	53.10	52.47	0.88	52.60
2	9:35	52.25	52.14	51.79	52.18	52.26	51.94	0.47	52.09
3	9:40	52.37	52.69	52.26	52.53	52.34	52.81	0.55	52.50
4	9:45	52.46	52.32	52.34	52.08	52.07	52.07	0.39	52.22
5	9:50	52.06	52.35	51.85	52.02	52.30	52.20	0.50	52.13
6	9:55	52.59	51.79	52.20	51.90	51.88	52.83	1.04	52.20
7	10:00	51.82	52.12	52.47	51.82	52.49	52.60	0.78	52.22
8	10:05	52.51	52.80	52.00	52.47	51.91	51.74	1.06	52.24
9	10:10	52.13	52.26	52.00	51.89	52.11	52.27	0.38	52.11
10	10:15	51.18	52.31	51.24	51.59	51.46	51.47	1.13	51.54
11	10:20	51.74	52.23	52.23	51.70	52.12	52.12	0.53	52.02
12	10:25	52.38	52.20	52.06	52.08	52.10	52.01	0.37	52.14
13	10:30	51.68	52.06	51.90	51.78	51.85	51.40	0.66	51.78
14	10:35	51.84	52.15	52.18	52.07	52.22	51.78	0.44	52.04
15	10:40	51.98	52.31	51.71	51.97	52.11	52.10	0.60	52.03
16	10:45	52.32	52.43	53.00	52.26	52.15	52.36	0.85	52.42
17	10:50	51.92	52.67	52.80	52.89	52.56	52.23	0.97	52.51
18	10:55	51.94	51.96	52.73	52.72	51.94	52.99	1.05	52.38
19	11:00	51.39	51.59	52.44	51.94	51.39	51.67	1.05	51.74
20	11:05	51.55	51.77	52.41	52.32	51.22	52.04	1.19	51.89
21	11:10	51.97	51.52	51.48	52.35	51.45	52.19	0.90	51.83
22	11:15	52.15	51.67	51.67	52.16	52.07	51.81	0.49	51.92

and continue for all 22 subgroups to the last range at 11:15 of

$$R = 52.16 - 51.67 = 0.49$$

The average of the R values is called $\overline{R}$, and it is computed by taking the simple arithmetic average of the R values:

$$\overline{R} = \Sigma R/k \qquad (5.7)$$

where k is the number of subgroups.

Thus the centerline can be computed as

$$\overline{R} = \frac{16.28}{22} = 0.740$$

Using a multiple of three standard errors to construct the upper control limit and the lower control limit, we have

$$UCL(R) = \overline{R} + 3\sigma_R$$

Assuming that the distribution of process output was, and will be, stable and approximately normally distributed, we can derive control limits. Note, however, that due to the empirical rule discussed in Chapter 4, the assumption of normality isn't necessary to interpret the control limits.

Consequently, when the characteristic of interest is stable, and the subgroup size is small ($2 \leq n \leq 9$), then

$$\overline{R} = d_2\sigma$$

and

$$\sigma_R = d_3\sigma$$

where the values of d_2 and d_3 are tabulated as a function of subgroup size in Table 1.

We can thus replace $\overline{R}$ and σ_R with these functions of the process standard deviation, σ:

$$UCL(R) = d_2\sigma + 3d_3\sigma = d_2\sigma(1 + 3d_3/d_2)$$
$$= \overline{R}(1 + 3d_3/d_2)$$

Similarly

$$LCL(R) = \overline{R} - 3\sigma_R$$
$$= d_2\sigma - 3d_3\sigma = d_2\sigma(1 - 3d_3/d_2)$$
$$= \overline{R}(1 - 3d_3/d_2)$$

A more convenient way to represent these control limits is by defining two new constants, D_3 and D_4:

$$D_3 = 1 - 3d_3/d_2$$
$$D_4 = 1 + 3d_3/d_2$$

so that

$$UCL(R) = D_4\overline{R} \tag{5.8}$$

and

$$LCL(R) = D_3\overline{R} \tag{5.9}$$

where D_3 and D_4 are tabulated as a function of subgroup size in Table 1. For our example,

$$UCL(R) = 2.004(0.740) = 1.48$$

and

$$LCL(R) = 0(0.740) = 0$$

Figure 5.15(a) illustrates the resulting average range and upper and lower control limits for the subgroup ranges. The R chart is then examined for signs of special variation. None of the points on the R chart is outside of the control limits, and there are no other signals indicating a lack of control. Thus there are no indications of special sources of variation on the R chart. More will be said later in this chapter and in Chapter 8 about indications of special causes of variation.

After analyzing the R chart, we construct the x-bar chart. This control chart depicts variations in the averages of the subgroups. To find the average for each subgroup, we add the data points for each subgroup and divide by the number of entries in the subgroup, as given by Equation 4.3. For the pharmaceutical company, the average of the 9:30 subgroup is

$$\frac{52.22 + 52.85 + 52.41 + 52.55 + 53.10 + 52.47}{6} = 52.60$$

This calculation is repeated for each of the subgroups. The x-bar results for this example can be found in the last column of Figure 5.14.

The centerline of an x-bar control chart is found by taking the average of the subgroup averages, $\overline{\overline{x}}$, calculated from Equation 4.4. In our example, the average of the 22 subgroup averages is

$$\overline{\overline{x}} = \frac{1146.55}{22} = 52.12$$

Using a multiple of three standard errors to construct the control limits, we have

$$UCL(\overline{x}) = \overline{\overline{x}} + 3\sigma_{\overline{x}}$$

Assuming that the distribution of process output was, and will be, stable and approximately normally distributed, we can derive control limits. Note, however, that due to the empirical rule discussed in Chapter 4, the assumption of normality isn't necessary to interpret the control limits.

Consequently, when the characteristic of interest is stable,

FIGURE 5.15 x-bar and R Charts for Filling Operation

PLANT	DEPT	OPERATION	DATE CONTROL LIMITS CALCULATED	ENGINEERING SPECIFICATION	PART NO.
MACH. NO.	DATES	CHARACTERISTIC		SAMPLE SIZE/FREQUENCY	PART NAME

AVERAGES (X BAR CHART)

$\bar{\bar{X}} = \text{Average } \bar{X} = 52.116 \quad UCL = \bar{\bar{X}} + A_2\bar{R} = 52.47 \quad LCL = \bar{\bar{X}} - A_2\bar{R} = 51.76$

(b)

UCL — Centerline — LCL

52.80 52.60 52.40 52.20 52.00 51.80 51.60

RANGES (R CHART)

$\bar{R} = \text{Average } R = .74 \quad UCL = D_4\bar{R} = 1.48 \quad LCL = D_3\bar{R} = 0$

(a)

UCL — Centerline — LCL

1.5 1.0 .5

DATE		1	2	3	4	5	6	7	8	9	10	11	12	13	14	15	16	17	18	19	20	21	22
TIME																							
R 1																							
E 2																							
A 3																							
D 4																							
I N G S 5																							
SUM																							
$\bar{X}$ = SUM / NO OF READINGS		52.6	52.1	52.5	52.2	52.1	52.2	52.2	52.2	52.1	51.5	52.0	52.1	51.8	52.0	52.0	52.4	52.5	51.7	52.4	52.0	51.8	51.9
R = HIGHEST - LOWEST		.9	.5	.6	.4	.5	1.0	.8	1.1	.4	1.1	.5	.4	.7	.4	.6	.9	1.0	1.1	1.1	1.2	.9	.5

$$\bar{\bar{x}} = \frac{\Sigma \bar{x}}{\text{Number of subgroups}} \qquad (5.10)$$

$$\sigma_{\bar{x}} = \sigma/\sqrt{n} = (\bar{R}/d_2)/\sqrt{n}$$

where, since $\bar{R} = d_2\sigma$, $\sigma = \bar{R}/d_2$ and the values of d_2 are tabulated as a function of subgroup size in Table 1.

We can thus rewrite the control limits as

$$\text{UCL}(\bar{x}) = \bar{\bar{x}} + 3[(\bar{R}/d_2)/\sqrt{n}]$$

or, more conveniently,

$$\text{UCL}(\bar{x}) = \bar{\bar{x}} + A_2\bar{R} \qquad (5.11)$$

and

$$\text{LCL}(\bar{x}) = \bar{\bar{x}} - 3\sigma_{\bar{x}}$$

$$= \bar{\bar{x}} - 3(\sigma/\sqrt{n})$$

$$= \bar{\bar{x}} - 3[(\bar{R}/d_2)/\sqrt{n}]$$

or

$$\text{LCL}(\bar{x}) = \bar{\bar{x}} - A_2\bar{R} \qquad (5.12)$$

where $A_2 = 3/(d_2\sqrt{n})$ is tabulated as a function of subgroup size in Table 1.

For the pharmaceutical company, the upper and lower control limits can now be computed as

$$\text{UCL}(\bar{x}) = 52.12 + (0.483)(0.740) = 52.47$$

and

$$\text{LCL}(\bar{x}) = 52.12 - (0.483)(0.740) = 51.76$$

Figure 5.15(b) illustrates the x-bar chart. Notice that a total of five points on the x-bar chart are outside of the control limits and therefore indicate a lack of control. Further investigation is warranted to determine the source(s) of these special variations.

We must realize that an x-bar chart can't be meaningfully analyzed if its corresponding R chart isn't in statistical control. This is because x-bar chart control limits are calculated from $\bar{R}$ (i.e., $\bar{\bar{x}} \pm A_2\bar{R}$), and if the range isn't stable, no calculations based on it will be accurate.

Two Possible Mistakes in Using Control Charts

There are two different types of mistakes that the user of a control chart may make: overadjustment and underadjustment. Proper use of control charts will minimize the total economic consequences of making either of these types of errors.

Overadjustment

The *overadjustment* error occurs when the user reacts to swings in the process data that are merely the result of common variation, such as adjusting a process downward if its past output is above average or adjusting a process upward if its past output is below average. When a process is overadjusted, it resembles a car being oversteered, veering back and forth across the highway. In general, processes should be adjusted not on the basis of time-to-time observations, but on the basis of information provided by a statistical control chart.

The effects of overadjustment can be seen by examining a frequently used demonstration device known as a Quincunx board (Figure 5.16). The Quincunx board is a rectangular box with an upper chamber containing a large number of beads. A horizontal sliding bar feeds one or more beads at a time into a triangular hopper that then allows the beads to fall at a specified lateral point directly above 10 rows of pegs. Each time a bead hits a peg, it will bounce right or left so that its position after falling through the 10 rows of pegs is a result of 10 random events. It doesn't seem unreasonable to expect that many beads would tend to fall almost directly beneath the point at which they were released. But some beads may tend to wander a bit and end up to the right or left of their release point.

Notice that in Figure 5.16 the beads will fall into a series of slots after passing through the rows of pegs. The slots have been numbered from 1 to 10 for purposes of illustration. Note also that the opening of the triangular hopper is set to release the beads directly above the number 5 slot. Provided that the hopper isn't moved, the process output (i.e., slot position of the beads) will be stable.

The slots themselves are divided into two portions: a short upper and a longer lower one. The short portion is used to observe subgroups, while the longer portion is used for the accumulation of subgroups. Figures 5.16 through 5.19 viewed as a sequence show the use of the Quincunx board.

The collection of data may be simulated by allowing small groups of beads to pass through the triangular hopper and fall into the upper portion of the slots. Each small group of beads represents a subgroup of data points. In Figure 5.17 we can see the result of allowing a subgroup of five beads to pass through the hopper. The bottom of the hopper is centered directly above slot number 9, and beads have fallen into slots 7 through 12.

When the data has been examined and recorded, the small group of beads are allowed to fall into the lower chamber where all beads will be accumulated. Figure 5.18 shows not only this group but the results of several subgroups accumulated in the bottom chamber. If the position of the hopper isn't changed, the accumulated beads in the lower chamber will usually follow an approximately normal curve. In Figure 5.18 we can see that the process average seems to be about 8.5, and the range of the process is 8.

Now suppose that, instead of having the good sense to leave the hopper alone, we were to adjust it after each bead dropped. This is *adjusting for common variation*.

After each bead passes through the pegs, we count the number of slots above or below the target of 5, and adjust the hopper that many slots in the opposite

FIGURE 5.16   A Quincunx Board

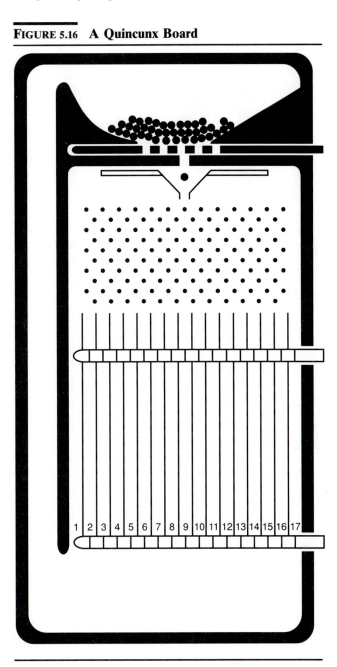

FIGURE 5.17   Quincunx Bead Falling

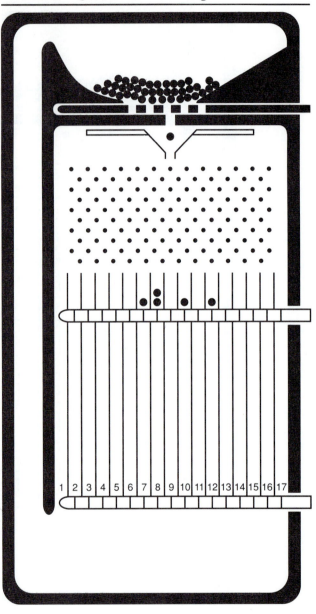

Beads being released from the hopper of a Quincunx board. Notice that most beads end up below the release point.

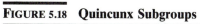

FIGURE 5.18   Quincunx Subgroups

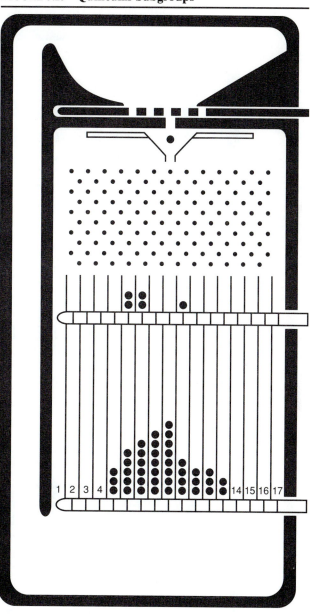

FIGURE 5.19   Results of Compensating for Bead Motion from Trial to Trial

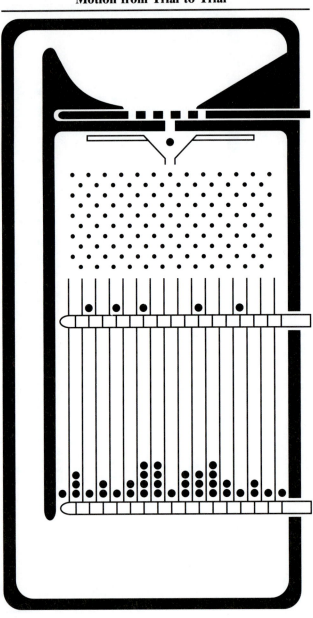

direction in an attempt to compensate. This will result in a dramatic increase in the process variation indicated by a greatly increased range. Figure 5.19 shows the results for the collection of several subgroups with the hopper being moved in this manner. While the process average still appears to be approximately the same, the range is now 16. This is the penalty for adjusting on the basis of common variation. This dangerous situation is illustrated by Rule 2 of the Funnel Experiment discussed in Chapter 14.

Underadjustment

Underadjustment, or lack of attention, results when a process is out of control and no effort is made to provide the necessary regulation. The process swings up and down in response to one or more special causes of variation, which may have compounding effects.

Avoiding both of these mistakes all of the time is an impossible task. That is, never adjusting the process—so that we never make the mistake of overadjusting—could result in severe underadjustment. On the other hand, if we made very frequent adjustments to avoid the problem of underadjustment, we would probably be overadjusting. Control charts provide an economical means to minimize the total loss that results from these two errors. Consequently control charts provide management with information on when to take action on a process and when to leave it alone.

Three Uses of Control Charts

As we've seen, control charts fall into two broad categories: attribute and variables. In both cases, a particular quality characteristic is measured and then examined. That examination can be used (1) to evaluate the history of the process, (2) to evaluate the present state of the process, or (3) to predict the near future state of a process in conjunction with the opinion of a process expert.

Evaluating the Past

The retrospective examination of the process's completed output using a control chart answers the question of whether the process has been in statistical control. A lack of control, or the presence of special causes of variation, is indicated when one or more of the control chart points is beyond the control limits or is otherwise in violation of one of the several rules introduced later in this chapter. Chapter 8 more fully discusses patterns indicating a lack of control. When no special causes of variation are present, the characteristic measured is said to be *in statistical control* or stable.

Evaluating the Present

Another use of control charts is to maintain a state of statistical control during a process's operation. Control charts can be used to generate a signal in a real time sense. The signal might, for example, call attention to tool wear or changes in humidity that might require intervention in the process. In this sense, control charts are useful in maintaining an existing state of process stability.

Predicting the Near Future

Finally, control charts can be used to predict the near future condition of a process, based on statistical evidence of a process's stability and process knowledge concerning future conditions that could affect the process. For example, if a process is stable and a process expert foresees no future sources of special variation, then the expert can predict that the process will remain stable in the near future.

Some Out-of-Control Evidence

As mentioned earlier, a process exhibits a lack of statistical control if a subgroup statistic falls beyond either of the control limits. But it's possible for all subgroup statistics to be within the control limits while there are other factors that indicate a lack of control in the process. Stable processes always exhibit random patterns of variation. Accordingly, most data points will tend to cluster about the mean value, or centerline, with an approximately equal number of points falling above and below the mean. A few of the values will lie close to the control limits. Points rarely fall beyond a control limit. Also there will seldom be prolonged runs upward or downward for a number of subgroups. If one or more of these conditions is missing in a control chart, the chart does not exhibit statistical control. Hence, for a process that is out of control, there will be an absence of points near the centerline, an absence of points near the control limits, one or more points located beyond the control limits, or runs or nonrandom patterns among the points.

So that we can examine patterns indicating a lack of control, the area between the control limits is divided into six bands, each band one standard error wide. As Figure 5.20 shows, bands within one standard error of the centerline are called the *C zones;* bands between one and two standard errors from the centerline are called *B zones;* and the outermost bands, which lie between two and three standard errors from the mean, are *A zones.*

Seven simple rules based on these bands are commonly applied to determine if a process is exhibiting a lack of statistical control. Any out-of-control points found are marked with an X directly on the control chart.

Rule 1. *A process exhibits a lack of control if any subgroup statistic falls outside of the control limits.*

As we've already seen, this is the first criterion—and the most obvious one. Figure 5.7 exhibits points (indicated by an X) that are out of control by virtue of this rule.

FIGURE 5.20   A, B, and C Zones for a Control Chart

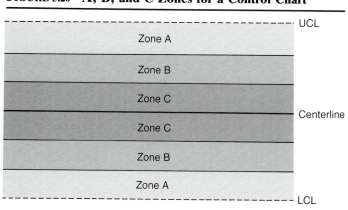

FIGURE 5.21   Lack of Control by Virtue of Rule 2

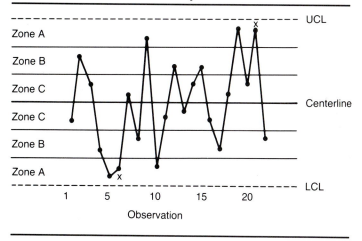

Rule 2. *A process exhibits a lack of control if any two out of three consecutive subgroup statistics fall in one of the A zones or beyond on the same side of the centerline.*

This means that if any two of three consecutive subgroup statistics are in the A zone or beyond, the process exhibits a lack of control when the second A zone point occurs. The two points must be in zone A or beyond on the same side of the centerline; the third point can be anywhere. Figure 5.21 illustrates a process exhibiting a lack of control by virtue of Rule 2 at two points in time: at observations 6 and 21.

When applying Rule 2 or any other indicator of a lack of control, it's always best to look for patterns demonstrating evidence of a lack of control by looking backward along the control chart. This makes any patterns or trends more obvious and makes it easier to find the beginning of a pattern or trend.

Consider subgroup statistics 21, 20, and 19. Point 21 is in the upper zone A; point 20 in zone C; and point 19 in the upper zone A again. Point 21 is thus the second of two out of three consecutive points in zone A, so it's marked as demonstrating a lack of control. Next, notice that observation number 6 is in zone A, preceded by observation number 5 (also in zone A) and then observation number 4 (in zone B). This makes observation number 6 the second of two out of three consecutive points in the A zone. Therefore observation number 6 is marked with an X as evidence of a lack of control. The analysis began by looking from right to left on the control chart, so more recent indications of a lack of control are the first to be found.

Rule 3. *A process exhibits a lack of control if four out of five consecutive subgroup statistics fall in one of the B zones or beyond on the same side of the centerline.*

This means that if any four out of five consecutive subgroup statistics are in either one of the B zones or beyond on the same side of the centerline while the fifth is not, the fourth point in the B zone or beyond is deemed to be providing evidence of a lack of control. It should be marked with an X. Figure 5.22 illustrates several possible patterns by which a process may be out of control via Rule 3. Observation number 15 is in zone B; number 14 is in zone B; number 13 is also in zone B; number 12 is in zone A (beyond zone B), while number 11 is in zone C. This means that observation number 15 is an indication of a lack of control. Notice that observation 16 lies in zone B also so it too should be considered out of control because it's the fourth of a different set of five consecutive points (points 12, 13, 14, 15, and 16) that lie in zone B or beyond.

FIGURE 5.22   Lack of Control by Virtue of Rule 3

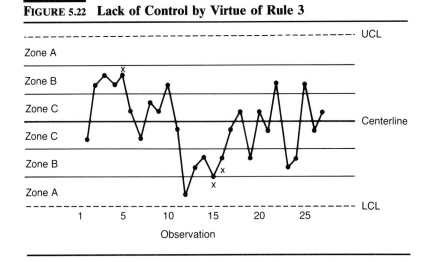

Subgroup statistics 5, 4, 3, and 2 all lie in zone B. They constitute four points in a row in this zone (point 1 can be considered the fifth point because it's in zone C); so the fourth point in zone B, point number 5, is marked with an X.

Rule 4. *A process exhibits a lack of control if eight or more consecutive subgroup statistics lie on one side of the centerline.*

The eighth and subsequent subgroup statistics are said to provide evidence of a lack of control by virtue of this rule. Figure 5.23 shows a process exhibiting a lack of control by virtue of this rule.

Notice that observation number 10 is the eighth of a string of points on one side of the centerline and can therefore be considered evidence of lack of control.

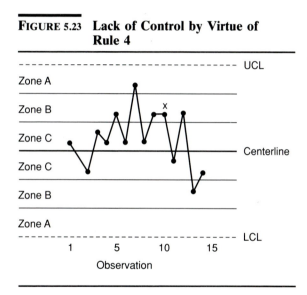

FIGURE 5.23   Lack of Control by Virtue of Rule 4

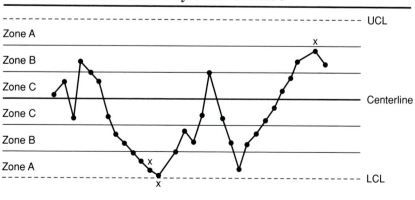

FIGURE 5.24   Lack of Control by Virtue of Rule 5

A note on rules 1 through 4: Rules 1 through 4 assume that the distribution of the control chart statistic is continuous, stable, and normally distributed. However, we do not require the assumption of normality to apply rules 1 through 4, due to the empirical rule discussed earlier in Chapter 4. For example, if the control chart statistic is continuous, stable, and nonsymetrically distributed, as is frequently the case with the range chart, the empirical rule can still be used to detect out-of-control points.

Rule 5. *A process exhibits a lack of control if eight or more consecutive subgroup statistics move upward in value or if eight or more consecutive subgroup statistics move downward in value.*

The eighth and subsequent subgroup statistics that continue moving up (or down) are said to provide evidence of a lack of control. Figure 5.24 shows a process exhibiting a lack of control by virtue of this rule.

Rule 6. *A process exhibits a lack of control if an unusually small number of runs above and below the centerline are present (a saw-tooth pattern).*

Figure 5.25 shows a process exhibiting a lack of control by virtue of this rule.

Rule 7. *A process exhibits a lack of control if 13 consecutive points fall within zone C on either side of the centerline.*

The thirteenth and subsequent subgroup statistics are said to provide evidence of a lack of control by virtue of this rule. Figure 5.26 shows such a process.

Note that rules 6 and 7 are used to determine whether a process is unusually noisy (high variability) or unusually quiet (low variability).

A note on discrete statistics: One possible problem using rules of thumb with respect to discrete quantities on c charts (or u charts, as we will see in Chapter 6) occurs when the process average for the discrete variable is small. As discussed in Chapter 6, this can create a situation where a zone can be entirely between two consecutive whole numbers, giving that zone a zero probability of occurring. For

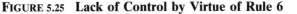

FIGURE 5.25   Lack of Control by Virtue of Rule 6

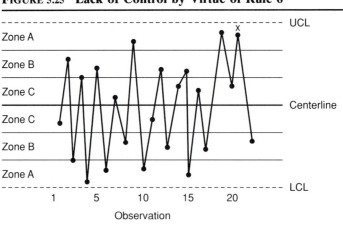

FIGURE 5.26   Lack of Control by Virtue of Rule 7

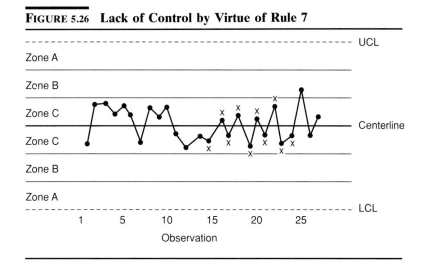

FIGURE 5.27   Zones for Number of Defects per Unit

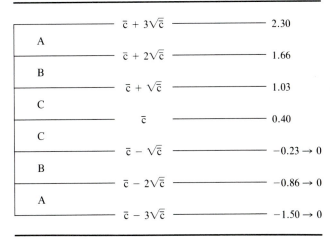

example, a firm makes radios with an average of 0.4 defects per unit and the number of defects per unit is stable. Consequently, zones A, B, and C would range between the values shown in Figure 5.27. Note that zone B is entirely between one and two defects per unit, and hence has no probability of occurring. Consequently, if the average value of a discrete variable is small, only the first and fourth rules of thumb make sense:

Rule 1. *A process exhibits a lack of control if any subgroup statistic falls outside of a control limit.*

Rule 4. *A process exhibits a lack of control if eight or more consecutive subgroup statistics lie on one side of the centerline.*

Collecting Data: Rational Subgrouping

Proper organization of the data to be control-charted is critical if the control chart is to be helpful in process improvement. We must be certain that we're asking the right questions. In other words, the data must be organized in such a way as to permit examination of variation productively and in a manner that will reveal special sources of variation. The organization of the data defines the question the control chart is addressing to achieve process improvement. Each subgroup should be selected from a small area so that relatively homogeneous conditions exist within each subgroup; this is called *rational subgrouping*. An excellent example can be found in *Understanding Statistical Process Control* by Wheeler and Chambers.[15]

Consider the case of a manufacturer of industrial paints. One-gallon cans are filled four at a time, each one by a separate filling head. The department manager is interested in learning if the weight of the product is stable and within specification and has decided to use statistical process control as an aid.

The supervisor is asked to take five successive cans from each of the four filling heads every hour. The gross weight (in kilograms) of each can is recorded for 20 measurements each hour. The supervisor continues observations for eight consecutive hours, yielding 160 individual observations (Figure 5.28).

How these observations are arranged may reveal variation from one of three sources: variations over *time,* variations between *measurements,* or variations between *filling heads.* Variation over time (hour to hour in this case) is represented by the differences in the groups of 20 cans; variation between measurements is represented by the difference between the five cans selected at each hour regardless of filling head; and variation between filling heads is represented by the differences between the results of the filling heads for each of the five cans selected per head, per hour.

The manager must decide on the proper arrangement of these data; their arrangement will dictate the variation that might be revealed. Let's consider an arrangement of the data for each possible source of variation.

Arrangement 1

If the basic subgroup consists of the four head readings for a given measurement and hour (Figure 5.29), then the variation within the subgroups will be the variation from head to head. That is, our estimate of the process standard error will be based on the measurements taken across all four filling heads. The process is being studied on a measurement-to-measurement and hour-to-hour basis, which means that the process variation is allocated as follows:

Source of Variation	Allocation
Hour-to-hour	Between subgroups
Measurement-to-measurement	Between subgroups
Head-to-head	Within subgroups

FIGURE 5.28   Fill Data for 160 Paint Cans

		Time: 8 AM Measurement							Time: 9 AM Measurement				
		1	*2*	*3*	*4*	*5*			*1*	*2*	*3*	*4*	*5*
H	1	6.09	6.10	6.09	6.09	6.09	H	1	6.13	6.13	6.14	6.13	6.11
E	2	6.09	6.09	6.10	6.09	6.09	E	2	6.12	6.12	6.11	6.13	6.10
A	3	6.10	6.11	6.12	6.11	6.11	A	3	6.11	6.13	6.13	6.14	6.14
D	4	6.16	6.16	6.17	6.17	6.17	D	4	6.20	6.20	6.20	6.17	6.17

		Time: 10 AM Measurement							Time: 11 AM Measurement				
		1	*2*	*3*	*4*	*5*			*1*	*2*	*3*	*4*	*5*
H	1	6.14	6.13	6.12	6.13	6.13	H	1	6.10	6.10	6.08	6.15	6.11
E	2	6.11	6.10	6.11	6.10	6.14	E	2	6.08	6.12	6.10	6.11	6.10
A	3	6.13	6.13	6.11	6.13	6.14	A	3	6.13	6.13	6.15	6.15	6.09
D	4	6.16	6.16	6.19	6.19	6.21	D	4	6.20	6.18	6.21	6.21	6.20

		Time: 12 Noon Measurement							Time: 1 PM Measurement				
		1	*2*	*3*	*4*	*5*			*1*	*2*	*3*	*4*	*5*
H	1	6.14	6.15	6.14	6.13	6.16	H	1	6.12	6.09	6.11	6.10	6.10
E	2	6.10	6.12	6.15	6.13	6.13	E	2	6.16	6.13	6.13	6.09	6.10
A	3	6.13	6.13	6.14	6.14	6.13	A	3	6.16	6.11	6.11	6.13	6.10
D	4	6.17	6.18	6.18	6.18	6.17	D	4	6.19	6.21	6.21	6.19	6.16

		Time: 2 PM Measurement							Time: 3 PM Measurement				
		1	*2*	*3*	*4*	*5*			*1*	*2*	*3*	*4*	*5*
H	1	6.07	6.07	6.08	6.07	6.08	H	1	6.11	6.12	6.13	6.13	6.13
E	2	6.07	6.08	6.07	6.07	6.09	E	2	6.10	6.11	6.13	6.10	6.13
A	3	6.09	6.09	6.09	6.09	6.10	A	3	6.13	6.16	6.14	6.13	6.13
D	4	6.15	6.15	6.16	6.16	6.14	D	4	6.18	6.19	6.20	6.19	6.21

FIGURE 5.29   Arrangement 1

		Time: 8:00 AM Measurement				
		1	*2*	*3*	*4*	*5*
H	1	6.09	6.10	6.09	6.09	6.09
E	2	6.09	6.09	6.10	6.09	6.09
A	3	6.10	6.11	6.12	6.11	6.11
D	4	6.16	6.16	6.17	6.17	6.17

Arrangement 1 asks the following questions:

1. Is there a systematic difference in hour-to-hour observations?
2. Is there a systematic difference in measurement-to-measurement observations?
3. Are the head-to-head differences consistent?

Consequently the first arrangement of the data looks for hour-to-hour and measurement-to-measurement special causes of variation.

For the eight hours, there will be a total of 40 subgroups. The average and range for each subgroup have been computed and appear in Figure 5.30. The average of the averages, $\bar{\bar{x}}$, has been calculated to be 6.13. The average of the range values, $\bar{R}$, is 0.08. When the data are analyzed via this arrangement, they show some evidence of a lack of control, as Figure 5.31 shows. The evidence is the long string of points above the centerline on the x-bar portion of the control chart (Rule 4). This, in all likelihood, results from some special source of variation that is unclear at this time.

Arrangement 2

If the basic subgroup consists of five measurements for a given head and hour (Figure 5.32), then the variation from measurement to measurement is the basis for our estimate of the standard error. Our estimate of the process standard error is based on the observations taken across all five measurements for each one of the heads for each hour. The process is being studied on a filling-head–to–filling-head and hour-to-hour basis. The process variation is allocated as follows:

Source of Variation	Allocation
Hour-to-hour	Between subgroups
Measurement-to-measurement	Within subgroups
Head-to-head	Between subgroups

The second arrangement has 32 subgroups each with five measurements. (We still have the same 160 measurements.) These five are the measurements taken each hour on each of the four filling heads. The questions now being asked of the data are

1. Is there a systematic difference in hour-to-hour observations?
2. Is there a systematic difference in head-to-head observations?
3. Are measurement-to-measurement differences consistent?

Consequently the second arrangement of the data looks for hour-to-hour and head-to-head special sources of variation.

The centerline, $\bar{\bar{x}}$, remains 6.13, while the average range, $\bar{R}$, is now computed as 0.03. Figure 5.33 shows a control chart illustrating this.

FIGURE 5.30   Arrangement 1: 40 Subgroups

		Time: 8 AM Measurement								Time: 9 AM Measurement				
		1	*2*	*3*	*4*	*5*				*1*	*2*	*3*	*4*	*5*
H	1	6.09	6.10	6.09	6.09	6.09		H	1	6.13	6.13	6.14	6.13	6.11
E	2	6.09	6.09	6.10	6.09	6.09		E	2	6.12	6.12	6.11	6.13	6.10
A	3	6.10	6.11	6.12	6.11	6.11		A	3	6.11	6.13	6.13	6.14	6.14
D	4	6.16	6.16	6.17	6.17	6.17		D	4	6.20	6.20	6.20	6.17	6.17
$\bar{x}$		6.11	6.12	6.12	6.12	6.12		$\bar{x}$		6.14	6.15	6.15	6.14	6.13
R		0.07	0.07	0.08	0.08	0.08		R		0.09	0.08	0.09	0.04	0.07

		Time: 10 AM Measurement								Time: 11 AM Measurement				
		1	*2*	*3*	*4*	*5*				*1*	*2*	*3*	*4*	*5*
H	1	6.14	6.13	6.12	6.13	6.13		H	1	6.10	6.10	6.08	6.15	6.11
E	2	6.11	6.10	6.11	6.10	6.14		E	2	6.08	6.12	6.10	6.11	6.10
A	3	6.13	6.13	6.11	6.13	6.14		A	3	6.13	6.13	6.15	6.15	6.09
D	4	6.16	6.16	6.19	6.19	6.21		D	4	6.20	6.18	6.21	6.21	6.20
$\bar{x}$		6.14	6.13	6.13	6.14	6.16		$\bar{x}$		6.13	6.13	6.14	6.16	6.13
R		0.05	0.06	0.08	0.09	0.08		R		0.12	0.08	0.13	0.10	0.11

		Time: 12 Noon Measurement								Time: 1 PM Measurement				
		1	*2*	*3*	*4*	*5*				*1*	*2*	*3*	*4*	*5*
H	1	6.14	6.15	6.14	6.13	6.16		H	1	6.12	6.09	6.11	6.10	6.10
E	2	6.10	6.12	6.15	6.13	6.13		E	2	6.16	6.13	6.13	6.09	6.10
A	3	6.13	6.13	6.14	6.14	6.13		A	3	6.16	6.11	6.11	6.13	6.10
D	4	6.17	6.18	6.18	6.18	6.17		D	4	6.19	6.21	6.21	6.19	6.16
$\bar{x}$		6.14	6.15	6.15	6.15	6.15		$\bar{x}$		6.16	6.14	6.14	6.13	6.12
R		0.07	0.06	0.04	0.05	0.04		R		0.07	0.12	0.10	0.10	0.06

		Time: 2 PM Measurement								Time: 3 PM Measurement				
		1	*2*	*3*	*4*	*5*				*1*	*2*	*3*	*4*	*5*
H	1	6.07	6.07	6.08	6.07	6.08		H	1	6.11	6.12	6.13	6.13	6.13
E	2	6.07	6.08	6.07	6.07	6.09		E	2	6.10	6.11	6.13	6.10	6.13
A	3	6.09	6.09	6.09	6.09	6.10		A	3	6.13	6.16	6.14	6.13	6.13
D	4	6.15	6.15	6.16	6.16	6.14		D	4	6.18	6.19	6.20	6.19	6.21
$\bar{x}$		6.10	6.10	6.10	6.10	6.10		$\bar{x}$		6.13	6.15	6.15	6.14	6.15
R		0.08	0.08	0.09	0.09	0.06		R		0.08	0.08	0.07	0.09	0.08

Notice that the process can now be seen as being wildly out of control, with many points beyond the control limits. Grouping the measurements by fill head reduced the within group variation so that the average range was lowered. This subsequently tightened the control limits and revealed the out-of-control points. Undoubtedly special sources of variation are present, and the control chart indi-

FIGURE 5.31 **Control Chart for Arrangement 1**

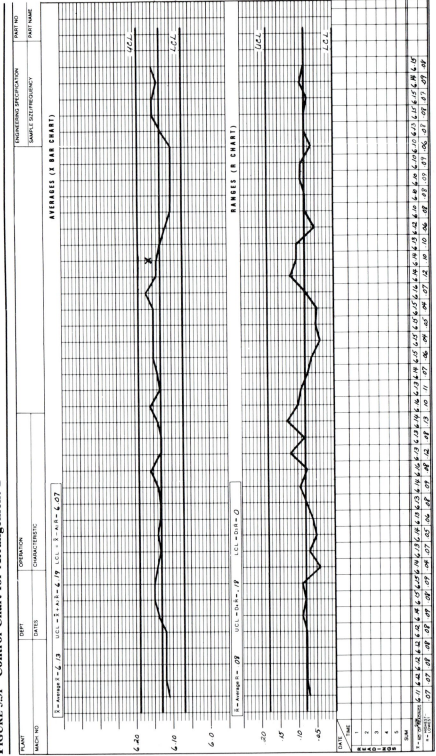

150

FIGURE 5.32   Arrangement 2

Time: 8 AM

		Measurement						
		1	*2*	*3*	*4*	*5*	$\bar{x}$	*R*
H	1	6.09	6.10	6.09	6.09	6.09	6.09	0.01
E	2	6.09	6.09	6.10	6.09	6.09	6.09	0.01
A	3	6.10	6.11	6.12	6.11	6.11	6.11	0.02
D	4	6.16	6.16	6.17	6.17	6.17	6.17	0.01

Time: 9 AM

		Measurement						
		1	*2*	*3*	*4*	*5*	$\bar{x}$	*R*
H	1	6.13	6.13	6.14	6.13	6.11	6.13	0.03
E	2	6.12	6.12	6.11	6.13	6.10	6.12	0.03
A	3	6.11	6.13	6.13	6.14	6.14	6.13	0.03
D	4	6.20	6.20	6.20	6.17	6.17	6.19	0.03

Time: 10 AM

		Measurement						
		1	*2*	*3*	*4*	*5*	$\bar{x}$	*R*
H	1	6.14	6.13	6.12	6.13	6.13	6.13	0.02
E	2	6.11	6.10	6.11	6.10	6.14	6.11	0.04
A	3	6.13	6.13	6.11	6.13	6.14	6.13	0.03
D	4	6.16	6.16	6.19	6.19	6.21	6.18	0.05

Time: 11 AM

		Measurement						
		1	*2*	*3*	*4*	*5*	$\bar{x}$	*R*
H	1	6.10	6.10	6.08	6.15	6.11	6.11	0.07
E	2	6.08	6.12	6.10	6.11	6.10	6.10	0.04
A	3	6.13	6.13	6.15	6.15	6.09	6.13	0.06
D	4	6.20	6.18	6.21	6.21	6.20	6.20	0.03

Time: 12 Noon

		Measurement						
		1	*2*	*3*	*4*	*5*	$\bar{x}$	*R*
H	1	6.14	6.15	6.14	6.13	6.16	6.14	0.03
E	2	6.10	6.12	6.15	6.13	6.13	6.13	0.05
A	3	6.13	6.13	6.14	6.14	6.13	6.13	0.01
D	4	6.17	6.18	6.18	6.18	6.17	6.18	0.01

Time: 1 PM

		Measurement						
		1	*2*	*3*	*4*	*5*	$\bar{x}$	*R*
H	1	6.12	6.09	6.11	6.10	6.10	6.10	0.03
E	2	6.16	6.13	6.13	6.09	6.10	6.12	0.07
A	3	6.16	6.11	6.11	6.13	6.10	6.12	0.06
D	4	6.19	6.21	6.21	6.19	6.16	6.19	0.05

FIGURE 5.32   *(concluded)*

		Time: 2 PM Measurement						
		1	*2*	*3*	*4*	*5*	$\bar{x}$	*R*
H	1	6.07	6.07	6.08	6.07	6.08	6.07	0.01
E	2	6.07	6.08	6.07	6.07	6.09	6.08	0.02
A	3	6.09	6.09	6.09	6.09	6.10	6.09	0.01
D	4	6.15	6.15	6.16	6.16	6.14	6.15	0.02
		Time: 3 PM Measurement						
		1	*2*	*3*	*4*	*5*	$\bar{x}$	*R*
H	1	6.11	6.12	6.13	6.13	6.13	6.12	0.02
E	2	6.10	6.11	6.13	6.10	6.13	6.11	0.03
A	3	6.13	6.16	6.14	6.13	6.13	6.14	0.03
D	4	6.18	6.19	6.20	6.19	6.21	6.19	0.03

cates where we should begin our investigation; that is, careful examination reveals that many of the out-of-control points on the x-bar portion correspond to the number 4 fill head. Its fill values are consistently above the upper control limit. Obviously fill head number 4 is putting more product on average into the containers than the other three fill heads.

The reason this wasn't revealed by the first arrangement of the data and the first control chart is that the first chart wasn't aimed at finding differences between the fill heads; it was aimed at examining measurement-to-measurement differences and hour-to-hour differences. The second arrangement was grouped to reveal differences between the fill heads and differences from hour to hour.

Arrangement 3

Most revealing at this point would be a third arrangement of the data that keeps separate control charts for each fill head. As each filling head has been separated with its own control chart, there's no longer any variation between the filling heads on our control chart. There are, in fact, now four distinct control charts, none of which can detect filling-head–to–filling-head variation. Computationally this third arrangement of the data is merely a slight reorganization of the data in Figure 5.32; it's shown in Figure 5.34. Using this arrangement, the process variation is allocated as follows:

Source of Variation for a Head	*Allocation*
Hour-to-hour	Between subgroups
Measurement-to-measurement	Within subgroups

FIGURE 5.33   Control Chart for Arrangement 2

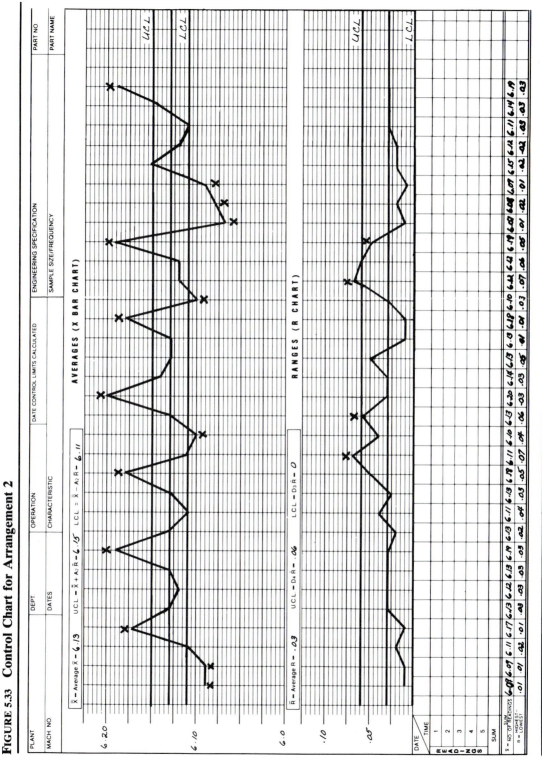

FIGURE 5.34   **Arrangement 3**

Filling Head 1
Measurement

Time	1	2	3	4	5	$\bar{x}$	R
8 AM	6.09	6.10	6.09	6.09	6.09	6.09	0.01
9 AM	6.13	6.13	6.14	6.13	6.11	6.13	0.03
10 AM	6.14	6.13	6.12	6.13	6.13	6.13	0.02
11 AM	6.10	6.10	6.08	6.15	6.11	6.11	0.07
12 NOON	6.14	6.15	6.14	6.13	6.16	6.14	0.03
1 PM	6.12	6.09	6.11	6.10	6.10	6.10	0.03
2 PM	6.07	6.07	6.08	6.07	6.08	6.07	0.01
3 PM	6.11	6.12	6.13	6.13	6.13	6.12	0.02

Filling Head 2
Measurement

Time	1	2	3	4	5	$\bar{x}$	R
8 AM	6.09	6.09	6.10	6.09	6.09	6.09	0.01
9 AM	6.12	6.12	6.11	6.13	6.10	6.12	0.03
10 AM	6.11	6.10	6.11	6.10	6.14	6.11	0.04
11 AM	6.08	6.12	6.10	6.11	6.10	6.10	0.04
12 NOON	6.10	6.12	6.15	6.13	6.13	6.13	0.05
1 PM	6.16	6.13	6.13	6.09	6.10	6.12	0.07
2 PM	6.07	6.08	6.07	6.07	6.09	6.08	0.02
3 PM	6.10	6.11	6.13	6.10	6.13	6.11	0.03

Filling Head 3
Measurement

Time	1	2	3	4	5	$\bar{x}$	R
8 AM	6.10	6.11	6.12	6.11	6.11	6.11	0.02
9 AM	6.11	6.13	6.13	6.14	6.14	6.13	0.03
10 AM	6.13	6.13	6.11	6.13	6.14	6.13	0.03
11 AM	6.13	6.13	6.15	6.15	6.09	6.13	0.06
12 NOON	6.13	6.13	6.14	6.14	6.13	6.13	0.01
1 PM	6.16	6.11	6.11	6.13	6.10	6.12	0.06
2 PM	6.09	6.09	6.09	6.09	6.10	6.09	0.01
3 PM	6.13	6.16	6.14	6.13	6.13	6.14	0.03

Filling Head 4
Measurement

Time	1	2	3	4	5	$\bar{x}$	R
8 AM	6.16	6.16	6.17	6.17	6.17	6.17	0.01
9 AM	6.20	6.20	6.20	6.17	6.17	6.19	0.03
10 AM	6.16	6.16	6.19	6.19	6.21	6.18	0.05
11 AM	6.20	6.18	6.21	6.21	6.20	6.20	0.03
12 NOON	6.17	6.18	6.18	6.18	6.17	6.18	0.01
1 PM	6.19	6.21	6.21	6.19	6.16	6.19	0.05
2 PM	6.15	6.15	6.16	6.16	6.14	6.15	0.02
3 PM	6.18	6.19	6.20	6.19	6.21	6.19	0.03

FIGURE 5.35 Control Charts for Arrangement 3

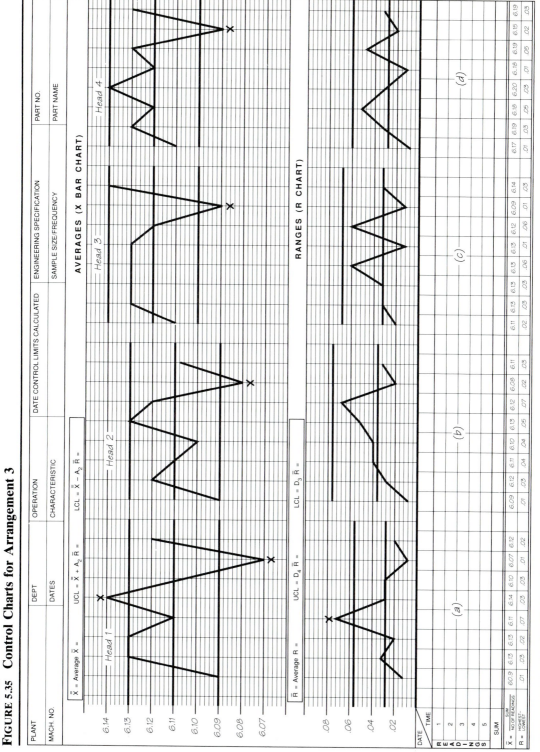

155

Note that head-to-head variation is no longer within the control chart and is only visible by comparing the different control charts.

The third arrangement asks the following questions for each filling head:

1. Is there a systematic difference in hour-to-hour observations?

2. Are measurement-to-measurement differences consistent?

Consequently the third arrangement of the data is set up to look for hour-to-hour special sources of variation for each head.

Arranging the data in this way permits the construction of individual sets of x-bar and R charts for each filling head. When the four head control charts are drawn on the same scale (Figure 5.35), the charts reveal things that may have been obscured earlier: Head 4 is significantly different from heads 1, 2, and 3. This information allows management to take appropriate action on the process (fix head 4).

Thus, proper subgrouping of the data to be control-charted is critical to process improvement efforts. Knowledge of the process under study is often the best guide to rational subgrouping of data.

Summary

This chapter has reviewed the Deming cycle and has demonstrated the use of control charts to stabilize and improve a process. The basic thrust of the Deming cycle is to decrease the difference between customer needs and process performance. Control charts are important vehicles for accomplishing this.

We took a first look at p charts, c charts, and x-bar and R charts (which the next two chapters will detail). We presented seven rules for determining whether a process is exhibiting a lack of control.

We next saw that rational subgrouping is important to ensure that our control charts focus on the proper sources of variation. We saw that when subgroups aren't rationally created, control charts may fail to reveal a lack of control and may not produce all of the useful information they are capable of producing.

In the next chapter we'll discuss control charts for attributes used to stabilize and improve a process. These charts include p charts, np charts, c charts and u charts. Chapter 7 focuses on the specifics of variables control charts: x-bar and R charts, x-bar and s charts, median charts, and single-measurement and moving range charts.

Exercises

The control charts in Exercises 5.1 through 5.6 have been divided into zones. In each exercise, identify any points that indicate a lack of control, and explain why you think the point does so.

5.1

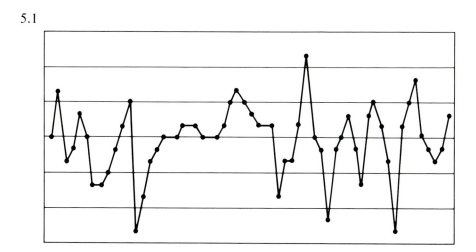

5.2

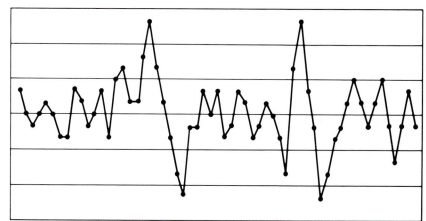

5.3

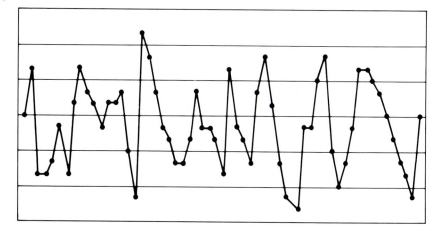

5.4

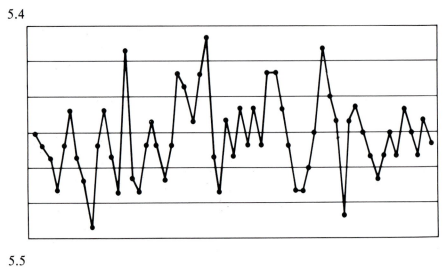

5.5

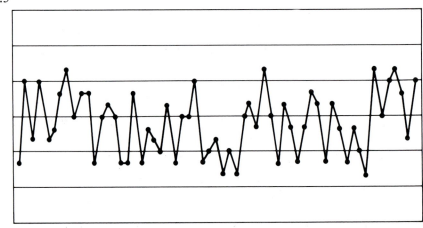

5.6

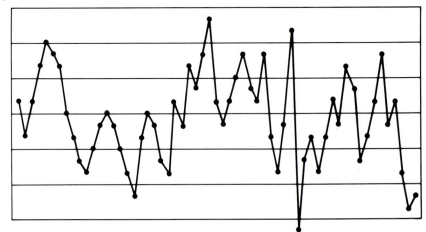

5.7 In the production of steel sheeting, five measurements are made once an hour at each of five indicator positions. The results of those 25 measurements for six consecutive hours are:

| | | Time: 9:00 AM Measurement | | | | |
		1	*2*	*3*	*4*	*5*
I	1	20.4	20.1	20.1	19.5	19.8
n	2	19.7	20.0	20.0	19.6	19.5
d	3	20.1	19.7	20.4	19.8	19.6
i	4	21.3	20.8	21.2	20.9	20.8
c.	5	19.6	20.1	19.9	20.0	19.9

| | | Time: 10:00 AM Measurement | | | | |
		1	*2*	*3*	*4*	*5*
I	1	20.2	19.9	19.7	20.3	20.3
n	2	20.1	19.7	20.0	19.7	19.9
d	3	19.5	20.0	19.9	20.1	19.5
i	4	20.9	20.9	21.4	21.1	21.1
c.	5	20.4	20.1	20.1	19.8	20.4

| | | Time: 11:00 AM Measurement | | | | |
		1	*2*	*3*	*4*	*5*
I	1	19.8	19.9	19.5	19.6	20.0
n	2	19.9	19.6	19.8	20.4	20.2
d	3	19.6	20.3	19.6	20.4	19.7
i	4	20.6	21.2	20.6	21.3	20.7
c.	5	20.4	19.5	20.1	20.3	20.3

| | | Time: 12:00 Noon Measurement | | | | |
		1	*2*	*3*	*4*	*5*
I	1	20.1	20.4	20.1	20.3	19.8
n	2	19.9	20.2	19.9	19.6	19.5
d	3	20.1	19.9	19.9	20.2	20.0
i	4	20.8	21.3	21.4	20.6	21.3
c.	5	20.1	20.3	19.8	20.1	19.5

| | | Time: 1:00 PM Measurement | | | | |
		1	*2*	*3*	*4*	*5*
I	1	20.2	19.6	20.1	20.4	20.0
n	2	20.0	20.2	19.7	19.8	19.7
d	3	20.1	19.5	19.6	19.5	20.3
i	4	21.0	21.2	20.8	20.8	20.5
c.	5	20.2	19.8	19.5	20.0	20.2

		Time: 2:00 PM Measurement				
		1	2	3	4	5
I	1	20.4	19.9	20.1	19.7	19.8
n	2	20.3	20.2	20.1	19.7	19.9
d	3	19.8	19.6	20.0	20.3	19.9
i	4	20.8	20.8	20.6	21.2	20.6
c.	5	19.6	20.2	19.5	19.7	19.6

a. Arrange the data to answer the following questions:
  1. Are there systematic differences in the hour-to-hour averages?
  2. Are there systematic differences in the measurement-to-measurement averages?
  3. Are the indicator-to-indicator differences consistent?
b. Use the results of part *a* to construct an $\bar{x}$ and R chart. Identify any indications of a lack of statistical control.
c. Arrange the data to answer the following questions:
  1. Are there systematic differences in the hour-to-hour averages?
  2. Are there systematic differences in the indicator-to-indicator averages?
  3. Are the measurement-to-measurement differences consistent?
d. Use the results of part *c* to construct an $\bar{x}$ and R chart. Identify any indications of a lack of statistical control.
e. Use the results of part *c* to allocate the three sources of variation: hour to hour, measurement to measurement, and indicator to indicator.
f. Arrange the data to answer the following questions:
  1. Is there a systematic difference in hour-to-hour observations?
  2. Are measurement-to-measurement differences consistent?
g. Construct the appropriate $\bar{x}$ and R charts with reference to your answer to part *f*.

5.8 Customers being served at a fast-food restaurant experience loss only when service takes too much time. Depict the loss functions for this one-sided situation from both traditional and realistic viewpoints.

5.9 Steel pails are manufactured at a high rate. Periodic samples of 50 pails are selected from the process. Results of that sampling are:

Sample Number	n	Number Defective	Sample Number	n	Number Defective
1	50	5	11	50	4
2	50	6	12	50	4
3	50	3	13	50	3
4	50	6	14	50	5
5	50	8	15	50	4
6	50	5	16	50	2
7	50	4	17	50	4
8	50	5	18	50	5
9	50	6	19	50	1
10	50	7	20	50	6

    a. Calculate the string of successive proportions of defective pails.

    b. Calculate the centerline and control limits for the p chart.

    c. Draw the p chart.

    d. Is the process stable? How do you know?

5.10 In studying the output of a process designed to produce pizza crusts, data have been collected on the finished diameter. Subgroups of size four are selected every half hour. After 23 subgroups have been amassed, the average of the $\bar{x}$ values is found to be 11.38 inches and the average of the ranges is found to be 0.16 inches. Determine the control limits for both the x-bar and range charts.

Endnotes

1. W. Edwards Deming, *Quality, Productivity and Competitive Position* (Cambridge, Mass.: Massachusetts Institute of Technology, Center for Advanced Engineering Study, 1982), p. 116.

2. W. Edwards Deming, *Out of the Crisis* (Cambridge, Mass.: Massachusetts Institute of Technology, Center for Advanced Engineering Study, 1986), pp. 88–89.

3. W. Scherkenbach, *The Deming Route to Quality and Productivity: Road Maps and Roadblocks* (Washington, D.C.: Ceepress Books, 1986), pp. 35–40.

4. Kaoru Ishikawa, *What Is Total Quality Control? The Japanese Way* (Englewood Cliffs, N.J.: Prentice-Hall, 1985), pp. 60–71.

5. Deming (1982), p. 123.

6. Donald J. Wheeler and David S. Chambers, *Understanding Statistical Process Control,* 2d ed. (Knoxville, Tenn.: SPC Press, 1992), pp. 1–7.

7. W. Edwards Deming, "On Some Statistical Aids toward Economic Production," *Interfaces* 5 (August 1975), pp. 1–15.

8. Genichi Taguchi and Yu-In Wu, *Off-Line Quality Control* (Nagoya, Japan: Central Japan Quality Control Association, 1980), pp. 7–9.

9. Ibid.

10. Howard S. Gitlow, "Definition of Quality," *Proceedings—Case Study Seminar—Dr. Deming's Management Methods: How They Are Being Implemented in the U.S. and Abroad* (Andover, Mass.: G.O.A.L., Nov. 6, 1984), pp. 4–18.

11. Wheeler and Chambers, pp. 1–7.

12. Walter A. Shewhart, *Economic Control of Quality of Manufactured Product* (Milwaukee, Wisc.: American Society for Quality Control, 1980), pp. 276–77.

13. Howard S. Gitlow and Paul Hertz, "Product Defects and Productivity," *Harvard Business Review,* Sept./Oct. 1983, pp. 131–41.

14. Deming (1982), p. 130.

15. Wheeler and Chambers, pp. 102–12.

CHAPTER 6 Attribute Control Charts

Introduction

Attribute data are data based on counts, or the number of times we observe a particular event. The events may be the number of nonconforming items, the fraction of nonconforming items, the number of defects, or any other distinct occurrence that's operationally defined. Attribute data may include such classifications as defective or conforming, go or no-go, acceptable or not acceptable, or number of defects per unit.

In Chapter 5 we saw that the first step on the ladder of quality consciousness consists of sorting conforming from nonconforming product; shipping conforming items; and discarding, reworking, or downgrading nonconforming ones. In the first step, efforts focus on defect detection and on trying to inspect quality in by removing defective items. This stage is characterized by dependence on mass inspection to achieve quality rather than statistical process control. Even today, many firms consider this quality control.

Attribute control charts represent the second step on the ladder of quality consciousness. The second step occurs when organizations move toward defect prevention in the drive toward zero percent defective. At the second step statistical process control is initiated using various attribute control charts. However, information that output failed to meet a given specification doesn't answer the question of *why* the specification wasn't met. Total (100 percent) conformance to specifications doesn't provide a mechanism for never-ending process improvement. As we'll see in Chapter 7, the third step on the ladder leads to continuous, never-ending improvement through the use of variables control charts.

Types of Attribute Control Charts

There are two basic classifications of attribute control charts: binomial count charts and area of opportunity charts.

162

Binomial Count Charts

Binomial count charts deal with either the fraction of items or the number of items in a series of subgroups that have a particular characteristic. The *p chart* is used to control the fraction of items with the characteristic. Subgroup sizes in a p chart may remain constant or may vary. A p chart might be used to control defective versus conforming, go versus no-go, or acceptable versus not acceptable. The *np chart* serves the same function as the p chart except that it's used to control the number rather than the fraction of items with the characteristic and is used only with constant subgroup sizes.

Area of Opportunity Charts

Area of opportunity charts deal with the number of times a particular characteristic appears in some given area of opportunity. An area of opportunity can be a radio, a crate of radios, a hospital room, an airline reservation, a roll of paper, a section of a roll of paper, a time period, a geographical region, a stretch of highway, or any delineated observable region in which one or more events may be observed. There are two types of area of opportunity control charts: c charts and u charts.

c Charts.   *c charts* are used to control the number of times a particular characteristic appears in a constant area of opportunity. A *constant area of opportunity* is one in which each subgroup used in constructing the control chart provides the same area or number of places in which the characteristic of interest may occur. For example, defects per air conditioner, accidents per workweek in a factory, and deaths per week in a city all provide an approximately constant area of opportunity for the characteristic of interest to occur. The area of opportunity is the subgroup, whether it be the air conditioner, the factory workweek, or the week in the city.

u Charts.   *u charts* serve the same basic function as c charts, but they're used when the area of opportunity changes from subgroup to subgroup. For example, we may examine varying square footage of paper selected from rolls for blemishes, or carloads of lumber for damage when the contents of the rail cars varies.

Attribute Control Chart Paper

Standard forms exist for the construction of attribute control charts. Although there may be some slight individualizing from firm to firm, certain standard areas are almost always provided on the forms. Figure 6.1 shows an example of an attribute control chart form.

In the upper left corner the plant/factory/office location is entered; then just to the right the type of control chart is noted. The next box requires information on the part name and number. Other identifying entries include the department

FIGURE 6.1 A Typical Attribute Control Chart Form

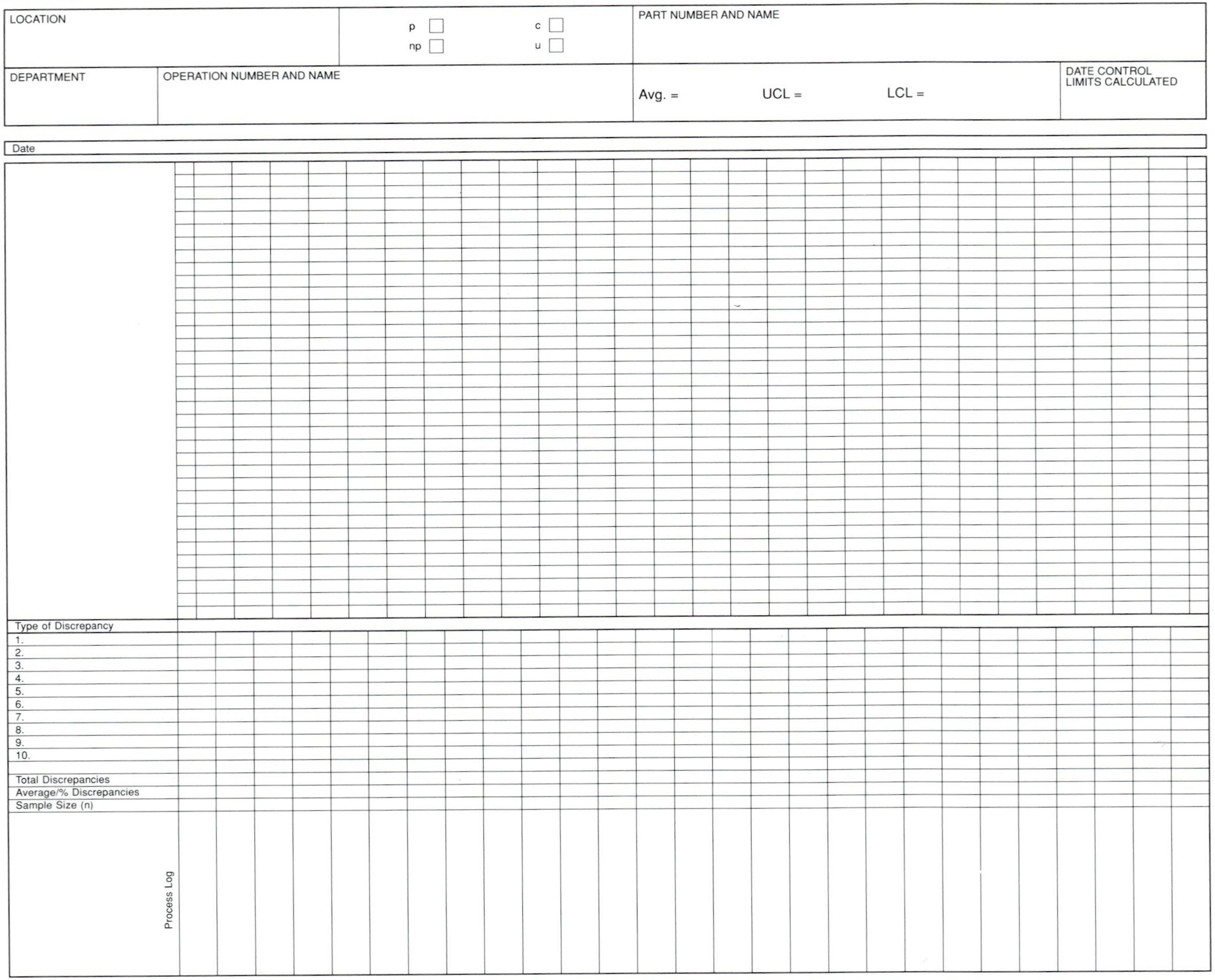
LOCATION
p ☐ np ☐ c ☐ u ☐
PART NUMBER AND NAME
DEPARTMENT
OPERATION NUMBER AND NAME
Avg. = UCL = LCL =
DATE CONTROL LIMITS CALCULATED
Date
Type of Discrepancy
1.
2.
3.
4.
5.
6.
7.
8.
9.
10.
Total Discrepancies
Average/% Discrepancies
Sample Size (n)
Process Log

and the operation number and name. The next two boxes provide space to enter the process average, UCL, and LCL plus the date on which they were calculated.

At the very bottom of the page are spaces for entering the total number of discrepancies (or defects); the percentage (or fraction) of discrepancies; and the sample (subgroup) size, n. Also included is a process log to help identify possible sources of variation. There are 10 cells directly above these for listing the type of discrepancy, usually by code number because of space constraints on the form.

The large open area on the left is for calibration and identification of the control chart's scale. The scale should be created to accommodate all observed and anticipated data entries. The control limits should fall well within the created scale so that there's room left for any points beyond the control limits to be entered on the graph.

Notice that the larger, upper portion of the cells (the ones on which the control chart will actually be drawn) is offset by exactly one half cell width from those below. This is to avoid any confusion as to which vertical bar corresponds to which data entry.

Just above this larger area is a single row of boxes for noting the date, time, or other identifying information for each observation.

Binomial Count Charts

Defect prevention, the second step in the journey toward quality consciousness, relies on the use of attribute control charts to help to begin to reduce the difference between customer specifications and process performance. When the data are in the form of binomial counts, either a p chart or np chart is used.

Conditions for Use

When each unit can be classified as either conforming or nonconforming (or having some characteristic of interest or not), a binomial count chart is appropriate. Samples of n items are periodically selected from process output. For these n distinct units in a subgroup:

1. Each unit must be classifiable as either possessing or not possessing the characteristic of interest. For example, each unit in a subgroup might be classified as either defective or nondefective, or conforming or nonconforming. The number of units possessing the characteristic of interest is called the *count,* X.

2. The probability that a unit possesses the characteristic of interest is assumed to be stable from unit to unit.

3. Within a given area of opportunity, the probability that a given unit possesses the characteristic of interest is assumed to be independent of whether any other unit possesses the characteristic.

For data satisfying these conditions, we may use a p chart or np chart. But we must exercise caution to avoid using either of these control charts inappropriately.

When Not to Use p Charts or np Charts

Occasionally data based on measurements (variables data) are downgraded into data in terms of conformance or nonconformance (attribute data). This isn't a good practice because the data based on measurements can provide more information than the data based on conformance or nonconformance.

It also is important that the denominator in the fraction being charted is the proper area of opportunity. If it's not, then the data aren't truly a proportion but a ratio. For example, the fraction of defectives found on the second shift will be a useful proportion only if it's computed by dividing the number of defectives found on the second shift by the proper area of opportunity, the number of units produced on the second shift. If a ratio is created using the number of defectives found on the second shift divided by the number of items shipped by the second shift, there's no way of knowing that the items shipped during the second shift were all produced on the second shift. Some items shipped on the second shift may have been produced during the first shift and therefore may be an inappropriate area of opportunity.

Last, we must exercise caution to ensure that the control chart is being created for a single process. Control charting output from combined different processes will result in irrational subgroups or subgroups that won't identify process problems. Little if anything can be learned from such charts, and the net effect may be a masking of special causes of variation.

Constructing Binomial Count Charts

A version of the Deming cycle, discussed in Chapter 5, may be used to construct and use a p chart or np chart.

   I. Plan
      A. The purpose of the chart must be determined.
      B. The characteristic to be charted must be selected and operationally defined.
      C. The manner, size, and frequency of subgroup selection must be established.
      D. The type of chart (i.e., p chart or np chart) must be established.
      E. If subgroup sizes are to vary, it must be decided whether new control limits will be computed for each subgroup or whether one of the approximate methods (to be discussed later in this chapter) will be used.
      F. Forms for recording and constructing the control chart must be established.
  II. Do
      A. Data must be recorded.
      B. The fraction of items with the characteristic of interest must be calculated for each of the subgroups.
      C. The average value must be calculated.
      D. The control limits and zone boundaries must be calculated and plotted onto the control chart.

      E.   The data points must be entered on the control chart.

III.   Study

      A.   The control chart must be examined for indications of a lack of control.

      B.   All aspects of the control chart must be reviewed periodically and appropriate changes made when required.

IV.   Act

      A.   Actions must be undertaken to bring the process under control by eliminating any special causes of variation.

      B.   Actions must be undertaken to reduce the causes of common variation for the purpose of never-ending improvement of the process.

      C.   Specifications must be reviewed in relation to the capability of the process.

      D.   The purpose of the control chart must be reconsidered by returning to the Plan stage.

The Plan Stage.   The first step in the Plan stage is to determine the purpose of the chart. A p chart or np chart may be created to:

1. Search for special causes of variation indicating a lack of statistical control.
2. Help determine and monitor the fraction or number of items with some characteristic of interest, such as the fraction or number of output that is nonconforming or defective. This will help management search for the common causes of variability in a stable system.

The second step in the Plan stage is to select and operationally define the characteristic for control charting. Very often a single item possesses several characteristics, any of which may cause the item to be considered defective or nonconforming. Most often a single chart will be kept for the entire item, but frequently separate charts will be kept for individual characteristics. It's usually efficient to concentrate initial efforts on control charts for the characteristics that cause problems for the customer and are within control of the process owner studying the problem. Some of the techniques to be discussed in Chapter 9, such as brainstorming, may be useful in selecting the characteristics to be charted.

The third step in the Plan stage is to determine the manner, size, and frequency for the selection of subgroups. As Chapter 5 said, rational subgroups should be selected to minimize the within subgroup variation. Frequently subgroups are selected in the order of production. The decisions concerning the method of selection and the factors to be isolated will require careful planning by those individuals with knowledge of, and experience with, the process. Early efforts may need revision as a result of unexpected factors that may be revealed while developing the control chart. This may lead to the creation of several charts where only one was initially contemplated, but this may be of use in resolving special causes of variation and reducing common variation in the areas charted.

The necessary size of subgroups will be discussed in a later section of this chapter.

The frequency with which the subgroups are selected is generally specific to each situation and depends upon factors such as the rate of production, elapsed time, and shift duration. The frequency should be logical in terms of shifts, time periods, or any other rational grouping. The shorter the intervals between subgroups, the more quickly information may be fed back for possible action. Cost will naturally be a factor; but after process stability has been established, frequency of subgroup selection can often be decreased and efforts focused elsewhere.

The fourth step in the Plan stage is to decide whether to use a p chart or np chart. There's no substantive difference between these two charts. The information portrayed is essentially the same; only the form is different. The p chart displays the *fraction* with the characteristic of interest, while the np chart displays the *number* of items with that characteristic of interest. From a technical standpoint, they may be used interchangeably. Nevertheless, as the np chart permits the data to be entered as whole numbers (rather than as the ratio of the number of nonconforming items to the subgroup size), the np chart may be preferable. Still, as we'll discuss later, if subgroup size varies from subgroup to subgroup, a p chart is typically used.

If subgroup sizes will vary, it must be decided whether the control limits and zone boundaries will be recalculated for each subgroup. Control limits depend on subgroup size, so varying subgroup size implies varying control limits and zone boundaries; however, we'll explore several techniques that we may employ to overcome this complication.

The final step in the Plan stage is to select the control chart form. Standard forms are available from the American Society for Quality Control for attribute control charts.[1] Many firms have developed their own forms, such as the one in Figure 6.1.

Occasionally, supplemental forms (checksheets) are used merely to collect data (Figure 6.2). Data are then transferred to a control chart. This technique may be especially convenient if the control chart is to be drawn at another time or with the aid of a computer—or if the work environment isn't suitable for drawing the chart.

The Do Stage. The Do stage begins with the recording of the required data for each subgroup on either the data collection sheet or directly on the control chart paper. Any abnormalities or unusual occurrences should be recorded in the space provided for comments. This may help provide clues to special causes of variation should a lack of control be found.

If the chart is a p chart, the fraction of items with the characteristic of interest must be calculated for each subgroup. After data for each subgroup have been collected (using no fewer than 20 subgroups), the average value for p is calculated using Equation 5.1. This value provides a centerline for the control chart and is the basis for the calculation of the standard error used to determine the control limits and zone boundaries.

Next, the control limits and zone boundaries are computed—using the equations introduced in Chapter 5 and discussed later in this chapter—and are then drawn onto the control chart.

FIGURE 6.2   Sample Data Collection Form for p Charts and np Charts

DATA SHEET FOR p CHART OR np CHART

Department: _____

Part Name: _____   Part Number: _____

Date	Time	Inspected by	Number Inspected	No. of Defectives	Fraction Defective	Comments

Last in the Do stage, the p values (or np values for the np chart) are plotted onto the control chart.

It's usually desirable to complete the control chart promptly and display it for those individuals working with the process. It's not unusual for such a display to have some beneficial results, especially if those involved have been educated about the purpose and meaning of control charts.

The Study Stage.   Using the rules introduced in Chapter 5 and detailed in Chapter 8, we then examine the control chart for indications of a lack of control. The data should be examined from right to left (looking backward in time), and any indications appropriately noted.

Periodically the centerline, control limits, and zone boundaries should be reviewed. Timing of the review, of course, depends on the process and its history. Typically p charts and np charts are kept for long periods of time. Any change in any aspect of the process is cause to consider a review of chart parameters.

The Act Stage.   Indications of special sources of variation may be revealed during the Study stage. If any special variation is found, steps must be initiated to guarantee that these sources of special variation are removed from the process if they're bad, or are incorporated into the process if they're good. It's not uncommon for supervisors or foremen to already be aware of problem areas; the control chart helps to discover the cause and reinforce arguments for improvement. Furthermore, control charts help focus attention on areas needing immediate help.

If it appears there's a lack of control on the desirable side of the chart, it's a good practice to examine the inspection procedures; faulty inspection procedures may be to blame. Or perhaps a special cause is responsible for points on the desirable side of the chart that should be formally incorporated into the process— that is, improvements may have spontaneously occurred in the process that, once discovered, should be incorporated.

Although a control chart may reveal no indications of special causes of variation, the overall level of the fraction or number of items with the characteristic of interest may not be at a satisfactory level (threshold state). Other tools and techniques, such as cause-and-effect diagrams and Pareto analysis (to be discussed in Chapter 9), may be used in an attempt to reduce the high fraction of nonconforming items as the Deming cycle rolls as a wheel up the hill of never-ending process improvement.

In the drive toward never-ending improvement, *no* level of defectives is low enough; nevertheless, as the proportion of defectives shrinks as a result of efforts at process improvement, subgroups will often contain no defectives at all. This will make the use of p charts or np charts difficult if not impossible because of the large subgroup sizes needed to detect even a single defective item. This leads to the use of variables control charts, which we discuss in Chapter 7.

Last in the Act stage is the reconsideration of the purpose of the control chart. We return to the beginning of the Plan stage, where the cycle begins again.

The p Chart for Constant Subgroup Sizes

A p chart with constant subgroup size is a relatively straightforward control chart. Constant subgroup size implies that the same number of items is sampled and then classified for each subgroup on the chart. We use a discrete, countable characteristic of output to construct a p chart; for example, the fraction of customers who pay their bills in fewer than 30 days, the fraction of correspondence sent electronically, or the fraction of an airline's flights that arrive within 15 minutes of their scheduled arrival time.

The Centerline and Control Limits.   Categorization of data into two classes suggests a binomial process. That is, for a stable process, every item has approximately the same probability of being in one of the two categories. We say *approximately* because even stable processes exhibit variation.

In Chapter 5 we saw one example of this type of control chart; the centerline for the p chart is established at $\bar{p}$, the overall fraction of output that was nonconforming, as given by Equation 5.1. The upper control limit and the lower control limit are found by adding and subtracting three times the standard error from the centerline value using Equations 5.2 and 5.3.

To reiterate,

$$\text{Centerline(p)} = \bar{p} = \left[\frac{\text{Total number of defectives in all subgroups under investigation}}{\text{Total number of units examined in all subgroups under investigation}} \right] \qquad (6.1)$$

$$\text{UCL(p)} = \bar{p} + 3 \sqrt{\frac{\bar{p}(1 - \bar{p})}{n}} \qquad (6.2)$$

$$\text{LCL(p)} = \bar{p} - 3 \sqrt{\frac{\bar{p}(1 - \bar{p})}{n}} \qquad (6.3)$$

Construction of a p Chart: An Example.　　As an illustration, consider the case of an importer of decorative ceramic tiles. Some tiles are cracked or broken before or during transit, rendering them useless scrap. The fraction of cracked or broken tiles is naturally of concern to the firm. Each day a sample of 100 tiles is drawn from the total of all tiles received from each tile vendor. Figure 6.3 presents the sample results for 30 days of incoming shipments for a particular vendor.

FIGURE 6.3　Daily Cracked Tiles

Day	Sample Size	Number Cracked or Broken	Fraction
1	100	14	0.14
2	100	2	0.02
3	100	11	0.11
4	100	4	0.04
5	100	9	0.09
6	100	7	0.07
7	100	4	0.04
8	100	6	0.06
9	100	3	0.03
10	100	2	0.02
11	100	3	0.03
12	100	8	0.08
13	100	4	0.04
14	100	15	0.15
15	100	5	0.05
16	100	3	0.03
17	100	8	0.08
18	100	4	0.04
19	100	2	0.02
20	100	5	0.05
21	100	5	0.05
22	100	7	0.07
23	100	9	0.09
24	100	1	0.01
25	100	3	0.03
26	100	12	0.12
27	100	9	0.09
28	100	3	0.03
29	100	6	0.06
30	100	9	0.09
Totals	3,000	183	

The average fraction of cracked or broken tiles can be calculated from this data using Equation 6.1. This is the centerline for the control chart.

$$\text{Centerline(p)} = p = 183/3000 = 0.061$$

The upper and lower control limits can then be computed using Equations 6.2 and 6.3.

$$\text{UCL(p)} = 0.061 + 3 \sqrt{\frac{(0.061)(1 - 0.061)}{100}} = 0.133$$

$$\text{LCL(p)} = 0.061 - 3 \sqrt{\frac{(0.061)(1 - 0.061)}{100}} = -0.011$$

Recall from our discussion in Chapter 5 that a negative lower control limit in a p chart is meaningless. Instead we use a value of 0 for the lower control limit. Figure 6.4 illustrates this control chart.

For a stable process the probability that any subgroup fraction will be outside the three-sigma limits is small. Also, if the process is stable, the probability is small that the data will demonstrate any other indications of the presence of special causes of variation by virtue of the other rules discussed in Chapters 5 and 8. But if the process isn't in a state of statistical control, the control chart provides an economical basis upon which to search for and identify indications of this lack of control.

For this p chart—or, in fact, for any of the attribute control charts—the exact probabilities that a stable process will generate points indicating a lack of control are impossible to calculate because even a stable process exhibits variation in its mean, dispersion, and shape. Nevertheless, the exact value of these probabilities isn't too important for ordinary applications; the fact that they're small is important. Therefore, if a point does lie beyond the upper or lower control limits, we'll infer that it indicates a lack of control. Additionally, for p charts, the six other rules for out-of-control points described in Chapter 5 can all be applied. In order to apply all of these rules, we need to compute the boundaries for the A, B, and C zones.

Recall from Chapter 5 that the width of each zone is one standard error, or one third of the distance between the upper control limit and the centerline. Thus the boundaries between zones B and C are one standard error on either side of the centerline. Here they are found by adding and subtracting the quantity $\sqrt{\bar{p}(1 - \bar{p})/n}$ from the centerline, $\bar{p}$.

$$\sqrt{\frac{\bar{p}(1 - \bar{p})}{n}} = \sqrt{\frac{0.061(1 - 0.061)}{100}} = 0.024$$

so that

$$\begin{matrix}\text{Boundary between} \\ \text{upper zones B and C}\end{matrix} = \bar{p} + \sqrt{\frac{\bar{p}(1 - \bar{p})}{n}} \qquad (6.4)$$

FIGURE 6.4 p Chart for Fraction of Cracked or Broken Tiles

PLANT				PART NUMBER AND NAME	

p ☒ c ☐
np ☐ u ☐

DEPARTMENT	OPERATION NUMBER AND NAME					DATE CONTROL LIMITS CALCULATED

Avg. = .061   UCL = .133   LCL = 0

Fraction cracked or broken

Zone A — UCA
Zone B
Zone C
— P̄
Zone C
Zone B
Zone A — LCL

Date	2	3	4	5	6	7	8	9	10	11	12	13	14	15	16	17	18	19	20	21	22	23	24	25	26	27	28	29	30

Type of Discrepancy
1.
2.
3.
4.
5.
6.
7.
8.
9.
10.

		2	3	4	5	6	7	8	9	2	3	8	4	15	3	8	4	2	5	7	9	1	3	12	9	3	6	9
Total Discrepancies	14	2	11	9	7	7	8	3	2	2	3	8	4	15	3	8	4	2	5	7	9	1	3	12	9	3	6	9
Average/% Discrepancies	.14	.02	.11	.04	.09	.07	.06	.08	.03	.02	.03	.08	.04	.15	.03	.08	.04	.02	.05	.07	.09	.01	.03	.12	.09	.03	.03	.09
Sample Size (n)	100	100	100	100	100	100	100	100	100	100	100	100	100	100	100	100	100	100	100	100	100	100	100	100	100	100	100	100

In our example this value is $0.061 + 0.024 = 0.085$ and

$$\text{Boundary between lower zones B and C} = \bar{p} - \sqrt{\frac{\bar{p}(1 - \bar{p})}{n}} \qquad (6.5)$$

In our example this value is $0.061 - 0.024 = 0.037$.

These boundaries appear in Figure 6.4.

We find the upper and lower boundaries between zones A and B by adding and subtracting, respectively, two standard errors from the centerline, $\bar{p}$.

$$\text{Boundary between upper zones A and B} = \bar{p} + 2\sqrt{\frac{\bar{p}(1 - \bar{p})}{n}} \qquad (6.6)$$

and

$$\text{Boundary between lower zones A and B} = \bar{p} - 2\sqrt{\frac{\bar{p}(1 - \bar{p})}{n}} \qquad (6.7)$$

Using these in our example,

$$0.061 + 2\sqrt{\frac{(0.061)(0.939)}{100}} = 0.109$$

and

$$0.061 - 2\sqrt{\frac{(0.061)(0.939)}{100}} = 0.013$$

Figure 6.4 shows the completed control chart. Examining it, we find a process that lacks control. On day 1, the mean fraction value is above the upper control limit. The sample mean for day 14 is also above the upper control limit, another indication of lack of control. None of the other rules presented in Chapter 5 seems to be violated. That is, there are no instances when two out of three consecutive points lie in zone A on one side of the centerline; there are no instances when four out of five consecutive points lie in zone B or beyond on one side of the centerline; there are no instances when eight consecutive points move upward or downward; nor are there eight consecutive points on one side of the centerline. There does not appear to be a lack of runs; there are no instances of 13 consecutive points in zone C.

Nevertheless, the incoming flow of ceramic tiles needs further examination. The special causes of these erratic shifts in the fraction of cracked or broken tiles should be eliminated so that expectations for usable portions can be stabilized. Only after this is done can improvements be made in the process.

Further study reveals that on both day 1 and day 14 the regular delivery truck operator was absent because of illness. Another employee loaded and drove the delivery truck on those days. That individual had never been instructed in the proper care and handling of the product, which needs special handling and treatment. To solve this problem and eliminate this special cause of variation, management created and implemented a training program using the regular driver's experience for three other employees. Any one of these three employees can now properly fill in and perform satisfactorily. Thus the system has been changed to eliminate this special cause of variation.

After the process has been changed so that special causes of variation have been removed, the out-of-control points are removed from the data. The points are removed from the control chart, and the graph merely skips over them.

Removing these points also changes the process average and standard error. Therefore the centerline, control limits, and zone boundaries must be recalculated.

The new centerline and control limits are

$$\bar{p} = 154/2800 = 0.055$$

$$\text{UCL(p)} = 0.055 + 3 \sqrt{\frac{(0.055)(0.945)}{100}} = 0.123$$

$$\text{LCL(p)} = 0.055 - 3 \sqrt{\frac{(0.055)(0.945)}{100}} = -0.013 \text{ (use 0.000)}$$

The new upper and lower boundaries between zones B and C are calculated using Equations 6.4 and 6.5:

$$\text{Boundary between upper zones B and C} = 0.055 + \sqrt{\frac{(0.055)(0.945)}{100}} = 0.078$$

$$\text{Boundary between lower zones B and C} = 0.055 - \sqrt{\frac{(0.055)(0.945)}{100}} = 0.032$$

The new upper and lower boundaries between zones A and B are calculated using Equations 6.6 and 6.7:

$$\text{Boundary between upper zones A and B} = 0.055 + 2 \sqrt{\frac{(0.055)(0.945)}{100}} = 0.101$$

$$\text{Boundary between lower zones A and B} = 0.055 - 2 \sqrt{\frac{(0.055)(0.945)}{100}} = 0.009$$

The entire control chart is redrawn (Figure 6.5). Notice that the data points for day 1 and day 14 don't appear; the chart has been drawn without these points. Figure 6.5 is the revised control chart. None of the seven rules discussed in Chapter 5 is violated, so there doesn't appear to be a lack of control. The process now appears to be stable and in a state of statistical control. Management may now look for ways to reduce the overall process average of the number of cracked or broken tiles to raise the usable number of tiles per shipment and effectively increase the process output.

Iterative Reevaluations.   It's possible—and not altogether uncommon—that by changing the process, removing points that were out of control, and recomputing the control limits and zone boundaries, points that initially exhibited only common variation will now indicate a lack of control. If and when this happens, the

FIGURE 6.5 Revised p Chart for Fraction of Cracked or Broken Tiles

PLANT

DEPARTMENT

OPERATION NUMBER AND NAME

PART NUMBER AND NAME

Avg. = .055   UCL = .123   LCL = 0

DATE CONTROL LIMITS CALCULATED

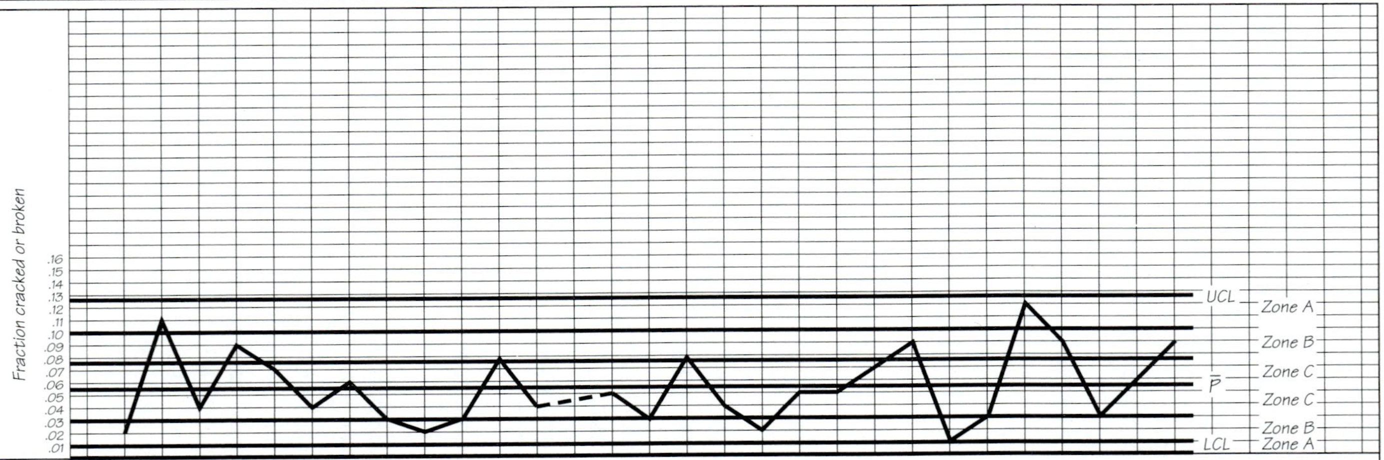

p ☒   c ☐
np ☐ u ☐

Fraction cracked or broken

UCL — Zone A
Zone B
Zone C
P̄ — Zone C
Zone B
LCL — Zone A

DROP POINT

Date	1	2	3	4	5	6	7	8	9	10	11	12	13	14	15	16	17	18	19	20	21	22	23	24	25	26	27	28	29	30

.16
.15
.14
.13
.12
.11
.10
.09
.08
.07
.06
.05
.04
.03
.02
.01

Type of Discrepancy
1
2
3
4
5
6
7
8
9
10

	1	2	3	4	5	6	7	8	9	10	11	12	13	14	15	16	17	18	19	20	21	22	23	24	25	26	27	28	29	30
Total Discrepancies	14	2	11	4	9	7	6	8	3	2	3	8	4	14	15	5	3	8	4	2	5	7	9	1	3	9	3	6	9	
Average% Discrepancies	14	.02	.11	.04	.09	.07	.06	.08	.03	.02	.03	.08	.04	.14	.15	.05	.03	.08	.04	.02	.05	.07	.09	.01	.03	.09	.03	.06	.09	
Sample Size (n)	100	100	100	100	100	100	100	100	100	100	100	100	100	100	100	100	100	100	100	100	100	100	100	100	100	100	100	100	100	

DROP POINT

system must again be reevaluated to eliminate the newly revealed special causes of variation.

This may once again uncover even more indications of a lack of control, which also must be removed from the system. Analysis of the process will continue to iterate in this manner until there no longer appears to be a lack of control. Keep in mind that in the course of these iterations, some of the data will be discarded. Hence the data base will shrink, and the control chart will be based on fewer and fewer subgroups. Furthermore, as changes are made, the process may no longer resemble the original process.

We must also keep in mind that if control limits are recalculated too frequently (as might be the temptation with automatic data processing available with many computer control routines), it becomes possible to mistake common variation for special variation. This effect parallels the oversteering many new drivers experience when first learning to drive a car. Knowledge and experience with the process are the best guides here.

At some point a decision must be made to stop analyzing the original data and collect new data. There's no clear rule for the point at which this should be done. Only knowledge and experience can dictate when to stop analyzing previous data and begin collecting and analyzing new data.

Subgroup Size.   When constructing a p chart, the subgroup size (or the number of sample items to be observed at each inspection to determine the fraction conforming or nonconforming) is much larger than that required for variables control charts. This is because the sample size must be large enough so that some nonconforming items are likely to be included in the subgroup. If, for example, a process produces 1.0 percent defectives, sample subgroups of size 10 will only occasionally contain a nonconforming item. As a general rule of thumb, control charts based on binomial count data should have sample sizes large enough so that the average count per subgroup is at least 2.00. This allows the A, B, and C zones to be wide enough to provide a reasonable working region into which data points may fall for analysis. This is true for both the p chart and the np chart (which we discuss later in this chapter).

Consider, for example, a process producing 5 percent of its output with a characteristic of interest. Subgroups of size 20 would be able to show these items only in counts of 0, 1, 2, and so on, yielding fractions in increments of 5 percent. The centerline would be at 0.050, the lower control limit would be at 0.000, and the upper control limit would be at

$$0.05 + 3 \sqrt{\frac{(0.05)(1 - 0.05)}{20}} = 0.196$$

Only counts of 0, 1, 2, or 3 would fall within the control limits. Examining patterns such as runs up or down wouldn't be practical; finding eight points moving upward or downward would almost always be redundant because the beginning or end of the run would be beyond the upper control limit and would indicate a lack of control for that reason. Clearly we wouldn't be able to learn too much from a p chart, based on a subgroup size of 20 items, with its centerline at 0.05 or an np chart with its centerline at 1.0. Samples taken from a process producing nonconforming items at a rate of 1 percent would require samples of 200 to have an average count of 2.00. Even with samples of size 200, samples would provide counts of 0, 1, 2, and so on for subgroup fractions in increments of 0.005. With a centerline at 0.01, the p chart wouldn't be very detailed and might not provide satisfactory indications of a lack of control.

Average subgroup counts of fewer than 2.00 present problems that can become extreme, especially if the average count per subgroup falls below 1.00. Hence, subgroups must be made large enough so that the average number of events is at least 2.00.

Ideally, subgroup sizes should remain the same for all subgroups, but occasionally circumstances require variations in subgroup size. Whether the subgroup size for a p chart varies or remains constant, the larger the subgroup size, the narrower the control limits will be. This is because the subgroup size, n, appears in the denominator of the expression for the standard error; the larger the value for n, the narrower the width of the control limits and zones A, B, and C around the process average will be.

Number of Subgroups. Every process goes through physical cycles, such as shifts and ordering sequences. Control chart calculations must be based on a sufficient number of subgroups to encompass all of the cycles of a process to include all possible sources of variation. As a rule of thumb, the number of subgroups should be at least 25.

It's possible to construct control charts for rational subgroups that don't represent chronological events. For example, a p chart for the fraction defective produced by a battery of 100 machines performing the same task (such as spot welding) might be kept on a single control chart. In these situations the number of subgroups must encompass all machines to encompass all possible sources of variation. Additionally, the rules concerning indications of a lack of control by virtue of trends in the data—such as two out of three consecutive points in zone A or four out of five consecutive points in zone B or beyond—should be ignored.

Another Example. An injection molding process provides a bracket to be used on aircraft passenger seats. Daily samples of 500 brackets are selected from the production output and examined carefully for cracks, splits, or other imperfections that will render them defective. Figure 6.6 lists the results. The centerline, control limits, and zone boundaries are calculated from Equations 6.1 through 6.7.

$$\bar{p} = 504/12500 = 0.040$$

$$UCL(p) = 0.040 + 3\sqrt{\frac{(0.040)(0.960)}{500}} = 0.066$$

$$LCL(p) = 0.040 - 3\sqrt{\frac{(0.040)(0.960)}{500}} = 0.014$$

$$\text{Boundary between lower zones A and B} = 0.040 - 2\sqrt{\frac{(0.040)(0.960)}{500}} = 0.022$$

FIGURE 6.6   Defective Aircraft Seat Brackets

Date	Sample Size	No. of Defectives	Fraction Defective
July 5	500	12	0.024
6	500	9	0.018
7	500	8	0.016
8	500	10	0.020
9	500	17	0.034
12	500	33	0.066
13	500	15	0.030
14	500	46	0.092
15	500	22	0.044
16	500	13	0.026
19	500	9	0.018
20	500	15	0.030
21	500	4	0.008
22	500	37	0.074
23	500	20	0.040
26	500	15	0.030
27	500	14	0.028
28	500	18	0.036
29	500	45	0.090
30	500	25	0.050
Aug. 2	500	27	0.054
3	500	33	0.066
4	500	17	0.034
5	500	28	0.056
6	500	12	0.024
Totals	12,500	504	

$$\text{Boundary between lower zones B and C} = 0.040 - \sqrt{\frac{(0.040)(0.960)}{500}} = 0.031$$

$$\text{Boundary between upper zones B and C} = 0.040 + \sqrt{\frac{(0.040)(0.960)}{500}} = 0.049$$

$$\text{Boundary between upper zones A and B} = 0.040 + 2\sqrt{\frac{(0.040)(0.960)}{500}} = 0.058$$

Figure 6.7 illustrates the p chart for these data. Many points indicate that this process wasn't in a state of statistical control. The operator running the molding process initiates a study of the molding process that reveals that the mold is poorly designed, so consistent parts can't be formed. The operator suggests a redesign of the mold that may eliminate most of the special causes of variation and reduce the average fraction defective.

The p Chart for Variable Subgroup Sizes

Sometimes subgroups of observations vary in size. This makes the construction of a p chart somewhat more difficult, but occasionally circumstances make this

FIGURE 6.7 p Chart for Fraction of Defective Aircraft Seat Brackets

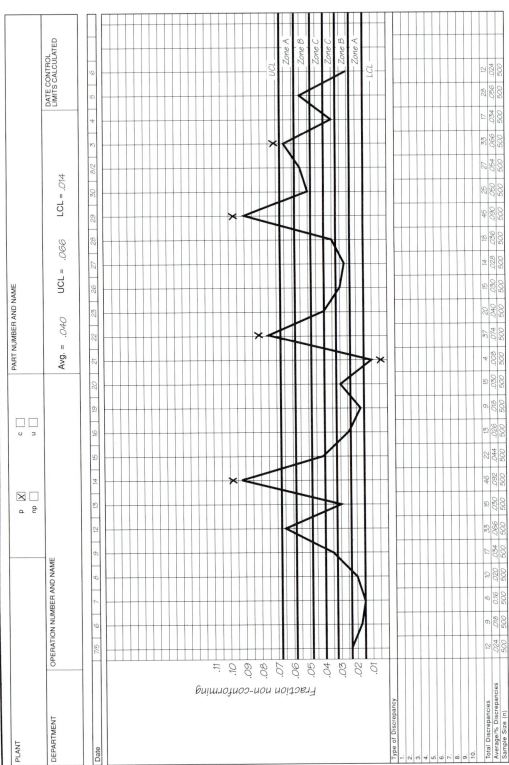

situation unavoidable. Common among these is when data initially collected for some purpose other than the creation of a control chart are later used to construct a control chart.

The standard error, $\sqrt{\overline{p}(1 - \overline{p})/n}$, varies inversely with the sample size. That is, as the sample size increases, the standard error decreases, and vice versa. Because control limits and zone boundaries are calculated based on this value, as the sample size changes so will the control limits and the zone boundaries.

Three Alternatives for Coping with Variable Subgroup Sizes. There are three different ways around the problem of variable subgroup sizes: The first is to compute the average subgroup size and to use this value for the control chart. The second is to compute new control limits and class boundaries for every subgroup based on that subgroup's size. The third is to compute both a wide and a narrow set of control limits based upon the smallest and largest possible values for n, respectively.

When the first method, an average subgroup size, is used, and a point falls in the outer portion of zone A or indicates a lack of control by falling outside the control limits, the true value for the subgroup size should be used to calculate the control limits and the zone boundaries. The point may then be properly evaluated regarding lack of control. If the point actually indicates a lack of control, appropriate steps must then be taken to find the sources of the special variation and to change the system so that they're rectified.

Additionally, when this method is used, the average value for n should be periodically recalculated to ensure it hasn't drifted too far from the value calculated initially. The first method works best when the subgroup sizes haven't been too different in the recent past and won't be too different in the immediate future. The terms *recent* and *immediate* must be carefully considered and defined to make them relevant to the particular application. Many statisticians feel that ±25 percent in subgroup size is permissible. There's no good substitute for knowledge and experience.

The second method, where control limits are recalculated for each subgroup, is the most accurate but is also the most tedious and time-consuming. Further, p charts created with variable control limits and zone boundaries may be difficult to read and interpret (especially for a relatively inexperienced user) so they're often undesirable.

The third method utilizes an inner and an outer set of control limits. Of the two sets of control limits, the outer set of control limits is based on the smallest anticipated value for n, and is wider since a small value of n in the denominator results in a large standard error. Points that fall outside these limits—assuming they're based on subgroup sizes greater than those used for the calculation of the outer limits—clearly indicate a lack of statistical control. The inner set of control limits is based on the largest subgroup's value of n. Points that fall within these limits (and don't otherwise indicate a lack of control) represent points under control. Exact values for control limits will need to be calculated for those points falling between the inner and outer control limits. The use of zone boundaries with this method is extremely cumbersome, and they generally aren't used unless a pattern (or some other indication of a lack of control) exists.

Using Average Subgroup Size: An Example.   Consider the case of a small manufacturer of low-tension electric insulators. The insulators are sold to wholesalers who subsequently sell them to electrical contractors. Each day during a one-month period the manufacturer inspects the production of a given shift; the number inspected varies somewhat. Based on carefully laid out operational definitions, some of the production is deemed nonconforming and is downgraded. Figure 6.8 illustrates the results for 25 weekdays beginning on September 2.

As the subgroup sizes don't vary by more than 25 percent, the centerline and control limits can be calculated using an average value for n:

$$\text{Centerline(p)} = \bar{p} = 594/9769 = 0.061$$

$$\text{Average value of n} = 9769/25 = 390.76$$

$$\text{UCL(p)} = 0.061 + 3 \sqrt{\frac{(0.061)(0.939)}{390.76}} = 0.097$$

$$\text{LCL(p)} = 0.061 - 3 \sqrt{\frac{(0.061)(0.939)}{390.76}} = 0.025$$

FIGURE 6.8   Nonconforming Electric Insulators

Date	Number Inspected	Number Nonconforming	Fraction Nonconforming
9/2	350	22	0.063
3	420	27	0.064
4	405	20	0.049
5	390	12	0.031
6	410	23	0.056
9/9	384	23	0.060
10	392	25	0.064
11	415	26	0.063
12	364	24	0.066
13	377	29	0.077
9/16	409	12	0.029
17	376	36	0.096
18	399	23	0.058
19	355	21	0.059
20	410	26	0.063
9/23	414	21	0.051
24	366	24	0.066
25	377	22	0.058
26	404	24	0.059
27	387	26	0.067
9/30	402	27	0.067
10/1	358	30	0.084
2	411	28	0.068
3	404	17	0.042
4	390	26	0.067
Totals	9,769	594	

Equations 6.4, 6.5, 6.6, and 6.7 may be used to compute the zone boundaries:

$$\text{Boundary between lower zones A and B} = 0.061 - 2 \sqrt{\frac{(0.061)(0.939)}{390.76}} = 0.037$$

$$\text{Boundary between lower zones B and C} = 0.061 - \sqrt{\frac{(0.061)(0.939)}{390.76}} = 0.049$$

$$\text{Boundary between upper zones A and B} = 0.061 + 2 \sqrt{\frac{(0.061)(0.939)}{390.76}} = 0.085$$

$$\text{Boundary between upper zones B and C} = 0.061 + \sqrt{\frac{(0.061)(0.939)}{390.76}} = 0.073$$

Figure 6.9 illustrates the control chart generated from these data. The process output doesn't indicate a lack of control. At this point, efforts should focus on reducing the fraction of nonconforming production and the variation in that fraction. As the fraction nonconforming is decreased, larger subgroups are needed so that the average count remains at least 2.00. Efforts must be made to look for measurable variables to use for a variables control chart to continue the pursuit of never-ending process improvement.

Using Varying Control Limits: An Example. When sample sizes do vary by more than ±25 percent we may either compute two sets of control limits or calculate new zone boundaries and control limits for each subgroup. Although the calculations required for new zone boundaries for each subgroup are more tedious, the technique is more sensitive.

Consider, for example, the case of an automatic toll barrier with two types of toll collection mechanisms: automatic and manned. The automatic lanes require exact change while the manned lanes don't. The fraction of vehicles arriving with exact change is examined using a control chart for a series of rush hour intervals on consecutive weekdays. As the number of vehicles passing through the toll varies by more than 25 percent, the control limits change day-to-day. One-hour periods (7:30 to 8:30 AM) for 20 consecutive weekdays yield the data in Figure 6.10.

Using these data, $\bar{p}$, the centerline, can be calculated from Equation 6.1 as

$$\text{Centerline(p)} = \bar{p} = 2569/6421 = 0.400$$

We can also calculate each UCL, LCL, and zone boundary using Equations 6.2 through 6.7.

For example, for the first data point,

$$\text{UCL(p)} = 0.40 + 3 \sqrt{\frac{(0.40)(1 - 0.40)}{465}} = 0.468$$

$$\text{LCL(p)} = 0.40 - 3 \sqrt{\frac{(0.40)(1 - 0.40)}{465}} = 0.332$$

$$\text{Boundary between lower zones A and B} = 0.40 - 2 \sqrt{\frac{(0.40)(1 - 0.40)}{465}} = 0.355$$

FIGURE 6.9 p Chart for Fraction of Nonconforming Electric Insulators: Average Subgroup Size Used

PLANT

DEPARTMENT

OPERATION NUMBER AND NAME

PART NUMBER AND NAME

p ☒ c ☐
np ☐ u ☐

Avg. = .061   UCL = .097   LCL = .025

DATE CONTROL LIMITS CALCULATED

Fraction non-conforming

Zone A — Zone B — Zone C — Zone C — Zone B — Zone A

.10 .09 .08 .07 .06 .05 .04 .03 .02 .01

Date	9/2	3	4	5	6	9	10	11	12	13	16	17	18	19	20	23	24	25	26	27	30	10/1	2	3	4

	22	27	20	12	23	23	25	26	24	29	12	36	23	21	26	23	24	22	26	27	30	28	17	26

Type of Discrepancy
1.
2.
3.
4.
5.
6.
7.
8.
9.
10.

	Total Discrepancies	22	27	20	12	23	23	25	26	24	29	12	36	23	21	26	23	24	22	26	27	30	28	17	26
Average/% Discrepancies		.063	.064	.049	.031	.056	.060	.064	.063	.066	.077	.029	.096	.058	.059	.063	.058	.059	.066	.058	.063	.067	.084	.068	.042	.067
Sample Size (n)		350	420	405	390	410	384	392	415	364	377	409	376	399	355	410	377	366	404	387	402	358	411	404	390

FIGURE 6.10 Number of Vehicles Using Exact Change

Day	n	Number with Exact Change	Day	n	Number with Exact Change
1	465	180	11	406	186
2	123	38	12	415	149
3	309	142	13	379	90
4	83	20	14	341	148
5	116	35	15	258	107
6	306	108	16	270	84
7	333	190	17	480	185
8	265	106	18	350	184
9	354	94	19	433	210
10	256	116	20	479	197
			Totals	6,421	2,569

$$\text{Boundary between lower zones B and C} = 0.40 - \sqrt{\frac{(0.40)(1-0.40)}{465}} = 0.377$$

$$\text{Boundary between upper zones A and B} = 0.40 + 2\sqrt{\frac{(0.40)(1-0.40)}{465}} = 0.445$$

$$\text{Boundary between upper zones B and C} = 0.40 + \sqrt{\frac{(0.40)(1-0.40)}{465}} = 0.423$$

Since subgroup sizes vary, these control limits and zone boundaries are only valid for the first observation, where n = 465. Each subgroup will have its own control limits and zone boundaries. Figure 6.11 shows the results of calculating these in the same manner as for the first point.

We use these values to draw the control limits and zone boundaries in Figure 6.12. The process indicates many instances of a lack of control. Fully 25 percent of the subgroup proportions are out of control, and the data seems to be behaving in an extremely erratic pattern. Days 19, 18, 13, 9, and 7 are all beyond the control limits. Day 5 also indicates a lack of control because it's the second of three consecutive points falling in zone C or beyond on the same side of the centerline.

Careful study is warranted to determine the cause or causes of this special variation. Removing the special cause(s) of variation may require some fundamental changes in the way this system operates. Nevertheless, we must eliminate the special sources of variation before attempting to reduce the common causes of variation in the process.

Changing the Process.   Management decides it would be advantageous to remove the erratic patterns in the preceding process. This would enable it to better serve the public by having adequate toll lanes of either the automatic or manned type available during rush hours. As a result of brainstorming, management institutes the sale of tokens that can be used in the exact change lanes. These are sold at a

---

FIGURE 6.11   Control Limits and Zone Boundaries for Vehicles with Exact Change

Subgroup Number	n	Fraction Defective	UCL	LCL	Upper Zone C	Lower Zone C	Upper Zone B	Lower Zone B
1	465	0.387	0.468	0.332	0.423	0.377	0.446	0.354
2	123	0.309	0.533	0.267	0.444	0.356	0.488	0.312
3	309	0.460	0.484	0.316	0.428	0.372	0.456	0.344
4	83	0.241	0.561	0.239	0.454	0.346	0.508	0.292
5	116	0.302	0.536	0.264	0.445	0.355	0.491	0.309
6	306	0.353	0.484	0.316	0.428	0.372	0.456	0.344
7	333	0.571	0.481	0.319	0.427	0.373	0.454	0.346
8	265	0.400	0.490	0.310	0.430	0.370	0.460	0.340
9	354	0.266	0.478	0.322	0.426	0.374	0.452	0.348
10	256	0.453	0.492	0.308	0.431	0.369	0.461	0.339
11	406	0.458	0.473	0.327	0.424	0.376	0.449	0.351
12	415	0.359	0.472	0.328	0.424	0.376	0.448	0.352
13	379	0.237	0.475	0.325	0.425	0.375	0.450	0.350
14	341	0.434	0.480	0.320	0.427	0.373	0.453	0.347
15	258	0.415	0.491	0.309	0.430	0.370	0.461	0.339
16	270	0.311	0.489	0.311	0.430	0.370	0.460	0.340
17	480	0.385	0.467	0.333	0.422	0.378	0.445	0.355
18	350	0.526	0.479	0.321	0.426	0.374	0.452	0.348
19	433	0.485	0.471	0.329	0.424	0.376	0.447	0.353
20	479	0.411	0.467	0.333	0.422	0.378	0.445	0.355

discount (to encourage their purchase) to motorists in the manned lanes. Because the process has now been changed, a new set of observations is made. After a period of two months, to allow for transient effects to die down, the same sample selection method is again employed. Results for those subgroups appear in Figure 6.13. Their corresponding control limits and zone boundaries are shown in Figure 6.14.

Using Equation 6.1 the centerline is calculated as

$$\bar{p} = 3,293/7,190 = 0.458$$

Figure 6.15 illustrates the control chart for the revised process. The process now appears stable, with no indications of a lack of control. Furthermore, the process is now better. Not only is the proportion of motorists using the exact change lanes stable and predictable, but the proportion of those motorists has risen from .400 to .458, which results in a smoother flow of traffic at the toll barrier.

Using Two Sets of Control Limits: An Example. Because the process now appears to be operating in a stable manner, management decides to continue the control chart with less frequent sampling. Based upon the early analysis, the process is deemed sufficiently stable to use the easier-to-read third method. Instead of computing exact control limits for each of the subgroups, we compute

FIGURE 6.12 p Chart for Fraction of Vehicles with Exact Change: Subgroup Sizes Vary

PLANT	p ☒ np ☐ c ☐ u ☐	PART NUMBER AND NAME	
DEPARTMENT	OPERATION NUMBER AND NAME	Avg. = 0.40 UCL = LCL =	DATE CONTROL LIMITS CALCULATED

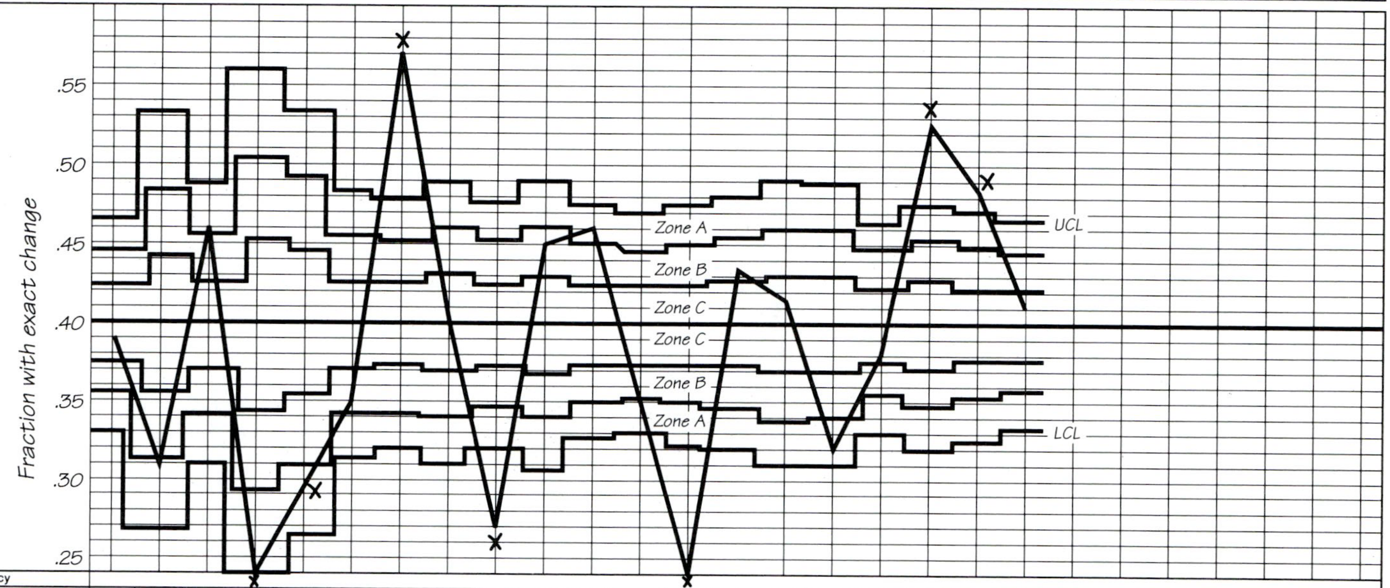

Date	1	2	3	4	5	6	7	8	9	10	11	12	13	14	15	16	17	18	19	20
Type of Discrepancy																				
1.																				
2.																				
3.																				
4.																				
5.																				
6.																				
7.																				
8.																				
9.																				
10.																				
Total Discrepancies	180	38	142	20	35	108	190	106	94	116	186	149	90	148	107	84	185	184	210	197
Average/% Discrepancies	.387	.309	.460	.241	.302	.353	.571	.400	.266	.453	.458	.359	.237	.434	.415	.311	.385	.526	.485	.411
Sample Size (n)	465	123	309	83	116	306	333	265	354	256	406	415	379	341	258	270	480	350	433	479

FIGURE 6.13   Number of Vehicles Using Exact Change or Tokens

Day	n	Number with Exact Change	Day	n	Number with Exact Change
1	421	171	11	401	199
2	466	197	12	384	165
3	389	192	13	428	213
4	254	107	14	352	149
5	186	89	15	444	193
6	456	189	16	357	158
7	411	211	17	283	147
8	322	139	18	424	207
9	287	136	19	337	143
10	262	131	20	326	157
			Totals	7,190	3,293

FIGURE 6.14   Control Limits and Zone Boundaries for Vehicles with Exact Change

Subgroup Number	n	Fraction Defective	UCL	LCL	Upper Zone C	Lower Zone C	Upper Zone B	Lower Zone B
1	421	0.406	0.531	0.385	0.482	0.434	0.507	0.409
2	466	0.423	0.527	0.389	0.481	0.435	0.504	0.412
3	389	0.494	0.534	0.382	0.483	0.433	0.509	0.407
4	254	0.421	0.552	0.364	0.489	0.427	0.521	0.395
5	186	0.479	0.568	0.348	0.495	0.421	0.531	0.385
6	456	0.415	0.528	0.388	0.481	0.435	0.505	0.411
7	411	0.513	0.532	0.384	0.483	0.433	0.507	0.409
8	322	0.432	0.541	0.375	0.486	0.430	0.514	0.402
9	287	0.474	0.546	0.370	0.487	0.429	0.517	0.399
10	262	0.500	0.550	0.366	0.489	0.427	0.520	0.396
11	401	0.496	0.533	0.383	0.483	0.433	0.508	0.408
12	384	0.430	0.534	0.382	0.483	0.433	0.509	0.407
13	428	0.498	0.530	0.386	0.482	0.434	0.506	0.410
14	352	0.423	0.538	0.378	0.485	0.431	0.511	0.405
15	444	0.435	0.529	0.387	0.482	0.434	0.505	0.411
16	357	0.443	0.537	0.379	0.484	0.432	0.511	0.405
17	283	0.519	0.547	0.369	0.488	0.428	0.517	0.399
18	424	0.488	0.531	0.385	0.482	0.434	0.506	0.410
19	337	0.424	0.539	0.377	0.485	0.431	0.512	0.404
20	326	0.482	0.541	0.375	0.486	0.430	0.513	0.403

two sets of control limits: a narrow set and a wide set. Zone A, B, and C boundaries aren't generally included because they would make the control chart confusing. Management calculates the exact control limits and zone boundaries for any suspicious-looking points or patterns, and any patterns that seem to suggest a lack of control are investigated further. Care must be exercised because the zone

FIGURE 6.15 p Chart for Revised Process for Fraction of Vehicles with Exact Change

PLANT

PART NUMBER AND NAME

DEPARTMENT | OPERATION NUMBER AND NAME | DATE CONTROL LIMITS CALCULATED

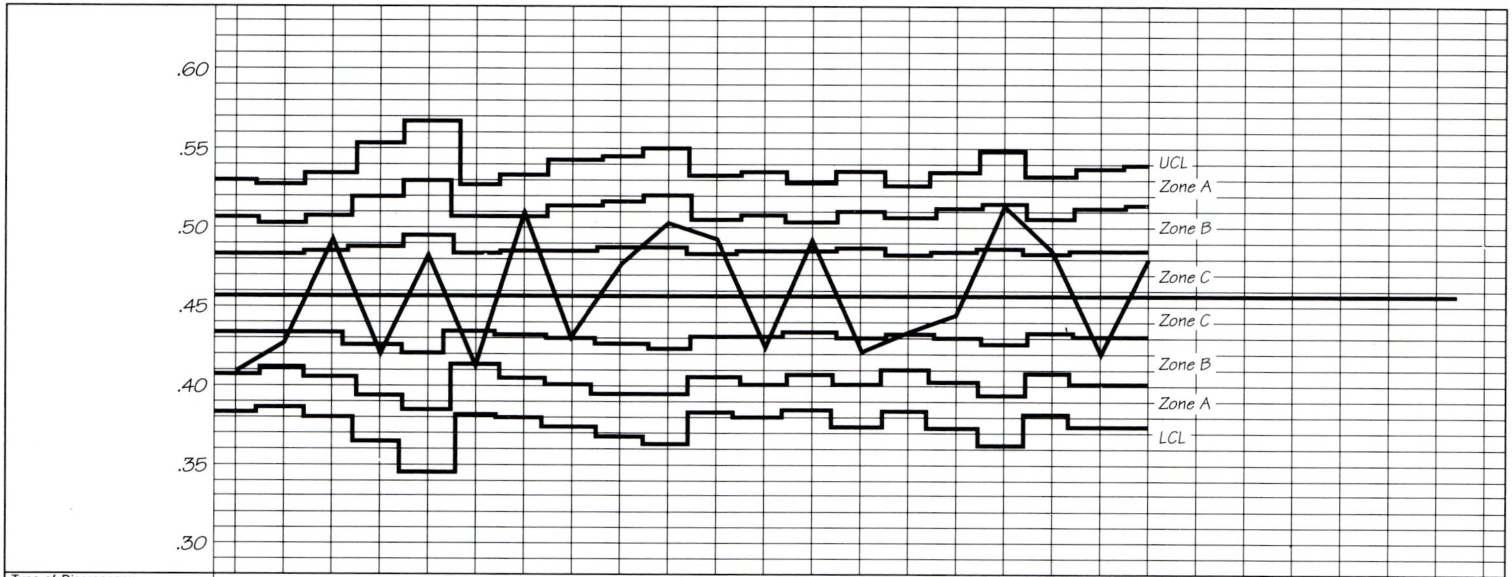

p [X] np [ ] c [ ] u [ ]

Avg. = .458   UCL =   LCL =

Zone labels: UCL, Zone A, Zone B, Zone C, Zone C, Zone B, Zone A, LCL

Date	1	2	3	4	5	6	7	8	9	10	11	12	13	14	15	16	17	18	19	20
Total Discrepancies	171	197	192	107	89	189	211	139	136	131	199	165	213	149	193	158	147	207	143	157
Average/% Discrepancies	.406	.423	.494	.421	.479	.415	.513	.432	.474	.500	.496	.430	.498	.423	.435	.443	.519	.488	.424	.482
Sample Size (n)	421	466	389	254	186	456	411	322	287	262	401	384	428	352	444	357	283	424	337	326

Type of Discrepancy
1.
2.
3.
4.
5.
6.
7.
8.
9.
10.

Y-axis values: .60, .55, .50, .45, .40, .35, .30

FIGURE 6.16   Vehicles Using Exact Change or Tokens

Day	n	Number with Exact Change	Fraction	Day	n	Number with Exact Change	Fraction
1	387	182	0.470	11	422	182	0.431
2	404	190	0.470	12	356	160	0.449
3	342	160	0.468	13	431	182	0.422
4	394	177	0.449	14	280	118	0.421
5	411	181	0.440	15	291	139	0.478
6	344	164	0.477	16	248	121	0.488
7	387	177	0.457	17	345	148	0.429
8	390	175	0.449	18	408	199	0.488
9	312	136	0.436	19	388	161	0.415
10	433	205	0.473	20	432	190	0.440
				Totals	7,405	3,347	

boundaries don't appear; patterns indicating a lack of control may be overlooked. For this reason, it's best to change to a sampling plan utilizing fixed subgroup sizes so that one set of control limits and zone boundaries can be used.

Subgroups at the toll are selected less frequently; Figure 6.16 illustrates results for 20 nonconsecutive days.

$$\bar{p} = 3,347/7,405 = 0.452$$

Minimum and maximum values for n are needed to construct the two sets of control limits. The minimum and maximum values for n should be based upon the smallest and largest subgroup sizes that can reasonably be anticipated to occur. In this case, a decision is made to use 75 as the minimum and 500 as the maximum. As these are slightly beyond the range of past experience, they're likely to provide conservative control limits. The smaller subgroup size will yield the wider set of control limits, and the larger will yield the narrow set of limits.

Recall that points falling outside of the outer control limits are considered to be sure signs of a lack of control; points inside the inner control limits aren't considered further (except if they form part of some other nonrandom pattern); and for those points falling between the two sets of control limits, exact control limits are calculated using the actual subgroup size to determine whether they're true indications of a lack of control.

The outer set of control limits are computed as

$$LCL(p) = 0.452 - 3 \sqrt{\frac{(0.452)(1 - 0.452)}{75}} = 0.280$$

$$UCL(p) = 0.452 + 3 \sqrt{\frac{(0.452)(1 - 0.452)}{75}} = 0.624$$

and the inner set of control limits are

$$LCL(p) = 0.452 - 3 \sqrt{\frac{(0.452)(1 - 0.452)}{500}} = 0.385$$

$$UCL(p) = 0.452 + 3 \sqrt{\frac{(0.452)(1 - 0.452)}{500}} = 0.519$$

The control chart is shown in Figure 6.17. No points are beyond the inner control limits and within the outer control limits. If there were, exact values for the control limits at those points would have to be calculated using the subgroup size actually observed and Equations 6.2 and 6.3. Then it could be determined whether those points truly indicated the presence of special causes of variation. Neither are there any other indications of a lack of control, such as runs up or down or too many consecutive points on one side of the centerline. If there were any signs whatsoever of special causes of variation, the procedure would be the same as it would be in the other situations we've discussed in this chapter. That is, we would attempt to identify and correct the special sources of variation using tools such as those we'll present in Chapter 9. After the process has been stabilized, we would recalculate the centerline and control limits and continue our efforts at eliminating common causes of variation.

The np Chart

Binomial count data can sometimes be more easily understood if the data appear as counts rather than fractions. This is especially true when using attribute control charts to introduce control charting and encountering reluctance by some members of the affected community to deal with fractions rather than whole numbers such as the number of defects.

The quantity np is the number of units in the subgroup with some particular characteristic, such as the number of nonconforming units. Traditionally, np charts are used only when subgroup sizes are constant. As the information used is the same as for p charts with constant subgroup sizes, these two charts are interchangeable.

Just as in the p chart, the categorization of data into two classes suggests a binomial process. Here too, for a stable process, every item must have approximately the same probability of being in one of the two categories. In a series of subgroups of constant size n, the mean or expected number of nonconforming items is approximated by np, and the associated standard error is given by $\sqrt{n\bar{p}(1 - \bar{p})}$. This enables us to construct the np chart.

Constructing the np Chart.   Data collected for an np chart will be a series of integers, each representing the number of nonconforming items in its subgroup. Computations for the centerline, the control limits, and the zone boundaries are quite similar to those of the p chart with constant sample sizes.

The centerline is the overall average number of nonconforming items found in each subgroup of the data. For the ceramic tile importer discussed earlier in this chapter (the data appear in Figure 6.3), there are a total of 183 cracked or broken

FIGURE 6.17 p Chart for Fraction of Vehicles with Exact Change: Two Sets of Control Limits Used

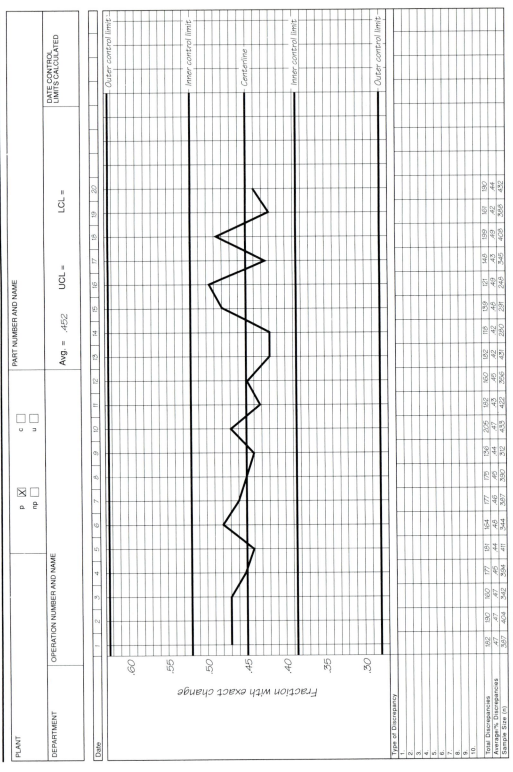

tiles in the 30 subgroups examined; this represents an average count of 183/30 = 6.1 tiles per day. In general, finding the average count is easier than computing $n\bar{p}$ directly as

$$n\bar{p} = (100) \left[\frac{183}{3000} \right] = 6.100$$

The standard error is

$$\sqrt{n\bar{p}(1 - \bar{p})} = \sqrt{(100)(0.061)(1 - 0.061)} = 2.393$$

The upper and lower control limits are found by adding or subtracting three times the standard error from the centerline, respectively:

$$\text{UCL}(np) = n\bar{p} + 3 \sqrt{n\bar{p}(1 - \bar{p})} \tag{6.8}$$

$$\text{LCL}(np) = n\bar{p} - 3 \sqrt{n\bar{p}(1 - \bar{p})} \tag{6.9}$$

For the tile importer this yields values of

$$\text{UCL}(np) = (100) \, (0.061) + 3\sqrt{(100) \, (0.061) \, (1-0.061)} = 13.280$$

$$\text{LCL}(np) = (100) \, (0.061) - 3\sqrt{(100) \, (0.061) \, (1-0.061)} = -1.080$$

As the LCL value is negative (-1.08), and, just as with the p chart, a negative value is meaningless, a value of 0 is used instead.

As with the p chart, the upper and lower boundaries between zones B and C are found by adding and subtracting one standard error from the centerline, $n\bar{p}$:

$$\begin{array}{l}\text{Boundary between}\\\text{upper zones B and C}\end{array} = n\bar{p} + \sqrt{n\bar{p}(1-\bar{p})} \tag{6.10}$$

and

$$\begin{array}{l}\text{Boundary between}\\\text{lower zones B and C}\end{array} = n\bar{p} - \sqrt{n\bar{p}(1-\bar{p})} \tag{6.11}$$

The upper boundary between zones B and C for this example is given by

$$6.1 + \sqrt{(100) \, (0.061) \, (1-0.061)} = 8.493$$

and the lower boundary between zones B and C is given by

$$6.1 - \sqrt{(100) \, (0.061) \, (1-0.061)} = 3.707$$

Upper and lower boundaries between zones A and B can be found by adding and subtracting two standard errors from the centerline, $n\bar{p}$:

$$\begin{array}{l}\text{Boundary between}\\\text{upper zones A and B}\end{array} = n\bar{p} + 2 \sqrt{n\bar{p}(1-\bar{p})} \tag{6.12}$$

and

$$\begin{array}{l}\text{Boundary between}\\\text{lower zones A and B}\end{array} = n\bar{p} - 2 \sqrt{n\bar{p}(1-\bar{p})} \tag{6.13}$$

The results for this example are

$$6.1 + 2 \sqrt{(100)\,(0.061)\,(1-0.061)} = 10.887$$

$$6.1 - 2 \sqrt{(100)\,(0.061)\,(1-0.061)} = 1.313$$

Figure 6.18 illustrates the np chart for this process. A comparison of Figures 6.4 and 6.18 reveals that these two control charts are mathematically equivalent and present the same information in a slightly different light. The only reason that one is preferred to the other is the form in which the data are presented, or the way in which the user prefers to visualize the control chart. Subsequent actions to stabilize the process are identical to those for the p chart.

Area of Opportunity Charts

A *defective* item is a nonconforming unit. It must be discarded, reworked, returned, sold, scrapped, or downgraded. It is unusable for its intended purpose in its present form. A *defect,* on the other hand, is an imperfection of some type that doesn't necessarily render the entire product unusable, yet is undesirable. One or more defects may not make an entire good or service defective. For example, we wouldn't scrap or discard a computer, a washing machine, or an air conditioner because of a small scratch in the paint.

An assembled piece of machinery such as a car, dishwasher, or air conditioner may have one or more defects that may not render the entire unit defective but may cause it to be downgraded or may necessitate its being reworked. Additionally, any product produced in sheets or rolls, such as paper, fabric, or plastic, may have several defects in a sheet or roll and still need not be scrapped as totally defective. A hotel room or airline reservation may have one or more defects, but may serve most of their intended purposes. In fact, there are many instances where more than one defect is the norm rather than the exception. This has created situations where products or services may not even be downgraded as a result of having several flaws. Naturally, in the quest for improvement, we prefer no defects in our output. Control charting is one of the tools to help achieve this end.

When there are multiple opportunities for defects or imperfections in a given unit (such as a large sheet of fabric), we call each such unit an *area of opportunity;* each area of opportunity is a subgroup. When areas of opportunity are discrete units and a single defect will make the entire unit defective, a p chart or np chart is appropriate. But when areas of opportunity are continuous or very nearly so, and more than one defect may occur in a given area of opportunity, a c chart or u chart should be used. The c chart is used when the areas of opportunity are of constant size, while the u chart is used when the areas of opportunity are not of constant size.

Conditions for Use

Area of opportunity charts have wide applicability. If we're counting defects, the enamel on an appliance represents a continuous area of opportunity; a roll of cloth

FIGURE 6.18 np Chart for Fraction of Cracked or Broken Tiles

PLANT	p ☐ np ☒ c ☐ u ☐	PART NUMBER AND NAME
DEPARTMENT	OPERATION NUMBER AND NAME	Avg. = 6.10 UCL = 13.28 LCL = 0
		DATE CONTROL LIMITS CALCULATED

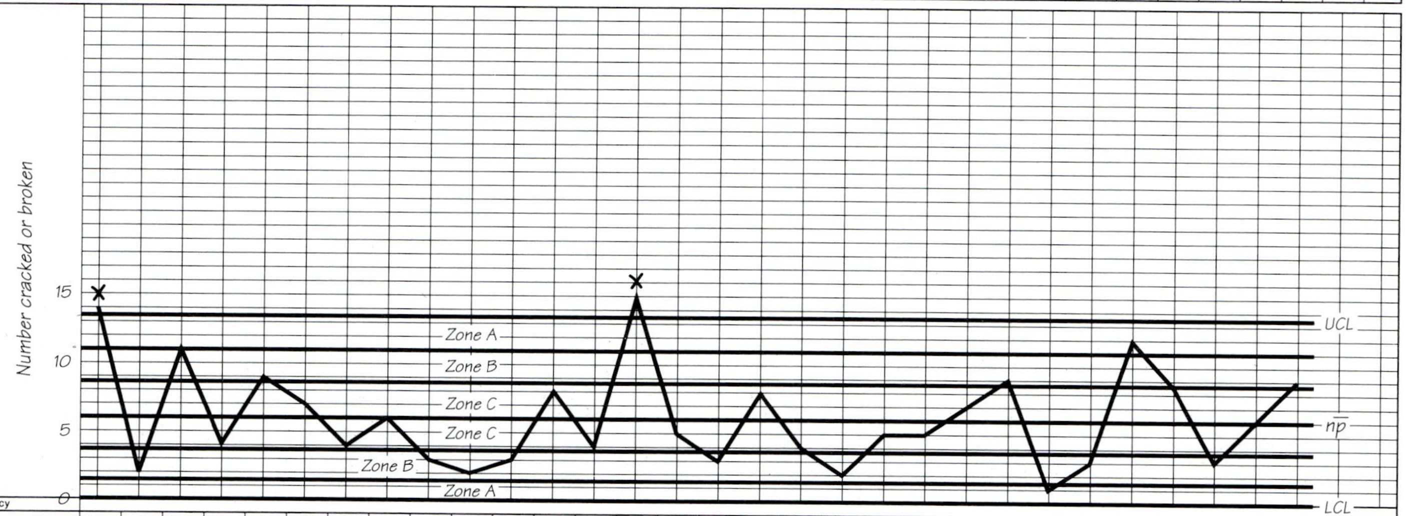

Date	1	2	3	4	5	6	7	8	9	10	11	12	13	14	15	16	17	18	19	20	21	22	23	24	25	26	27	28	29	30
Type of Discrepancy																														
1.																														
2.																														
3.																														
4.																														
5.																														
6.																														
7.																														
8.																														
9.																														
10.																														
Total Discrepancies	14	2	11	4	9	7	4	6	3	2	3	8	4	15	5	3	8	4	2	5	5	7	9	1	3	12	9	3	6	9
Average/% Discrepancies	.14	.02	.11	.04	.09	.07	.04	.06	.03	.02	.03	.08	.04	.15	.05	.03	.08	.04	.02	.05	.05	.07	.09	.01	.03	.12	.09	.03	.06	.09
Sample Size (n)	100	100	100	100	100	100	100	100	100	100	100	100	100	100	100	100	100	100	100	100	100	100	100	100	100	100	100	100	100	100

or plastic film is a continuous area of opportunity. If we're measuring the number of accidents recorded per month, a month represents a continuous area of opportunity. Measurements of the number of errors per hour in data entry or the number of typographical errors made per page have areas of opportunity (an hour or a page) that present enough opportunities for multiple defects to be considered nearly continuous. Imperfections in a complex piece of machinery, such as a computer, have areas of opportunity that aren't strictly continuous; but the large number of individual pieces involved make the areas of opportunity very nearly so, and they're often taken to be continuous.

Both the c chart and u chart are concerned with counts of the number of occurrences of an event over a continuous (or virtually continuous) area of opportunity. Those counts will be whole numbers such as 0, 1, 2, 3, ... The principle behind both types of charts is the same; the primary difference between the two charts is whether the size of the areas of opportunity remains constant from subgroup to subgroup.

If we're to use the c charts or u charts, the events we're studying must be describable as discrete events; these events must occur randomly within some well-defined area of opportunity; they should be relatively rare; and they should be independent of each other. Exact conformance to these conditions isn't always easy to verify. Usually, it's not too difficult to tell whether the events are discrete and whether there's some well-defined area of opportunity. But whether the events are relatively rare is somewhat subjective and requires some process knowledge and experience. The issue of independence is generally revealed by the control chart. That is, if the events aren't random and independent, they'll tend to form identifiable patterns that we introduced in Chapter 5 and will discuss further in Chapter 8. This will produce indications of a lack of control on the control chart.

c Charts

Areas of opportunity that are constant in size are easier to handle than those that vary, in much the same way as constant subgroup sizes in a p chart are easier to handle than those that vary. Constant areas of opportunity might be such things as one unit of a particular model of a TV set, a particular type of hospital room, one printed circuit board, one purchase order, one aircraft canopy, five square feet of paper board, or five linear feet of wire. When all conditions for an area of opportunity chart are met, and when the subgroup sizes remain constant, a c chart is used.

The number of events in an area of opportunity is denoted by c, the count for each area of opportunity. The string of successive c values, taken over time, is used to construct the control chart.

The centerline for the chart is the average number of events observed. It's calculated as

$$\text{Centerline(c)} = \bar{c} = \frac{\text{Total number of events observed}}{\text{Number of areas of opportunity}} \qquad (6.14)$$

The standard error is the square root of the mean, $\sqrt{\bar{c}}$. Upper and lower control limits can be found by adding and subtracting three times the standard error from the centerline, $\bar{c}$. Thus

$$\text{UCL(c)} = \bar{c} + 3\sqrt{\bar{c}} \qquad (6.15)$$

$$\text{LCL(c)} = \bar{c} - 3\sqrt{\bar{c}} \qquad (6.16)$$

Counts, Control Limits, and Zones. As we've already seen with the p charts and np charts, when a process is in a state of control, only very rarely will points fall beyond the control limits. Therefore when a point does fall outside the control limits, we'll consider it an indication of a lack of control and take appropriate action. When the lower control limit is calculated to be negative, we'll use 0 as the lower control limit because, just as with the p chart and np chart, negative numbers of events (such as -3 defects on a radio) are impossible.

Consider a firm that has decided to use a c chart to help keep track of the number of telephone requests received daily for information on a given product. Each day represents an area of opportunity. Over a 30-day period, 1,206 requests are received, an average of 40.2 per day. This value is $\bar{c}$, the centerline. The upper and lower control limits can be found using Equations 6.15 and 6.16:

$$\text{UCL(c)} = 40.2 + 3\sqrt{40.2} = 59.2$$

$$\text{LCL(c)} = 40.2 - 3\sqrt{40.2} = 21.2$$

Actual counts occurring in an area of opportunity will always be whole numbers. Thus a count of 59 is within the control limits, while a count of 60 is beyond the UCL. The A, B, and C zone boundaries are constructed at one and two standard errors from the centerline, respectively. The zone boundaries are

$$\text{Boundary between lower zones B and C} = 40.2 - \sqrt{40.2} = 33.9$$

$$\text{Boundary between lower zones A and B} = 40.2 - 2\sqrt{40.2} = 27.5$$

$$\text{Boundary between upper zones A and B} = 40.2 + 2\sqrt{40.2} = 52.9$$

$$\text{Boundary between upper zones B and C} = 40.2 + \sqrt{40.2} = 46.5$$

Because the actual counts are whole numbers, the observations would fall into zones as follows:

Zone	Counts
Upper A	53, 54, 55, 56, 57, 58, 59
Upper B	47, 48, 49, 50, 51, 52
Upper C	41, 42, 43, 44, 45, 46
Lower C	34, 35, 36, 37, 38, 39, 40
Lower B	28, 29, 30, 31, 32, 33
Lower A	22, 23, 24, 25, 26, 27

The zones each contain a reasonable number of whole numbers and are close enough in size to be workable. But consider the problem that would have been encountered if the process average had been $\bar{c} = 2.4$. Here we'd get

$$UCL(c) = 2.4 + 3\sqrt{2.4} = 7.0$$

$$LCL(c) = 2.4 - 3\sqrt{2.4} = -2.2 \text{ (use 0.0)}$$

$$\begin{array}{l} \text{Boundary between} \\ \text{lower zones B and C} \end{array} = 2.4 - \sqrt{2.4} = 0.9$$

$$\begin{array}{l} \text{Boundary between} \\ \text{lower zones A and B} \end{array} = 2.4 - 2\sqrt{2.4} = -0.7 \text{ (use 0.0)}$$

$$\begin{array}{l} \text{Boundary between} \\ \text{upper zones A and B} \end{array} = 2.4 + 2\sqrt{2.4} = 5.5$$

$$\begin{array}{l} \text{Boundary between} \\ \text{upper zones B and C} \end{array} = 2.4 + \sqrt{2.4} = 3.9$$

As before, because the counts are whole numbers, the observations will fall into zones as follows:

Zone	Counts
Upper A	6, 7
Upper B	4, 5
Upper C	3
Lower C	1, 2
Lower B	0
Lower A	0

These zones are so small that they're practically meaningless. The upper zone C, for example, only has one possible count, 3. When the average count is small, we generally don't make use of the zones in seeking indications of a lack of control. Rather, we focus on points beyond the control limits, runs of points above or below the centerline, and runs upward or downward in the data as indicators of a lack of stability. The exact value of the centerline below which the use of A, B, and C zones becomes impractical requires knowledge of, and experience with, the particular process involved. As a rule of thumb, the zone boundaries shouldn't be used for c charts with average counts of less than 20.0. Once again, there's no substitute for process knowledge and experience; but as the observable count shrinks, the use of variables control charts must be instituted for continued process improvement.

Furthermore, keep in mind that when the average count is small, larger and larger areas of opportunity will be needed to detect imperfections. This will occur as a natural consequence of improved quality through the use of control charts. When the areas of opportunity needed to find imperfections grow unacceptably large, attribute control charts must be abandoned in favor of variables control charts. This usually solves the problem of small values for $\bar{c}$ and is another step on the ladder of quality consciousness.

An Example. Let's consider the output of a paper mill as an example. The product appears at the end of a web and is rolled onto a spool called a reel. Every reel is examined for blemishes, which are imperfections. Each reel is an area of opportunity. Results of these inspections produce the data in Figure 6.19.

The assumptions necessary for using the c chart are well met here, as the reels are large enough to be considered continuous areas of opportunity; imperfections are discrete events and seem to be independent of one another, and they're relatively rare. Even if these conditions aren't precisely met, the c chart is fairly robust, or insensitive to small departures from the assumptions, so we may still safely use it.

In this example the average number of imperfections per reel is

$$\bar{c} = 150/25 = 6.00$$

and the standard error is $\sqrt{6.00} = 2.45$.

Equations 6.15 and 6.16 yield upper and lower control limits:

$$UCL(c) = 6.00 + 3(2.45) = 13.35$$

$$LCL(c) = 6.00 - 3(2.45) = -1.35 \quad \text{(use 0.00)}$$

Figure 6.20 is a control chart using these values.

Small Average Counts. Even though the control chart in Figure 6.20 is useful, frequently, when average counts are small, data appearing as counts will tend to be asymmetric. This may lead to overadjustment (false alarms) or underadjustment (too little sensitivity).

FIGURE 6.19   Number of Blemishes Found in 25 Reels of Paper

Reel	Number of Blemishes	Reel	Number of Blemishes
1	4	14	9
2	5	15	1
3	5	16	1
4	10	17	6
5	6	18	10
6	4	19	3
7	5	20	7
8	6	21	4
9	3	22	8
10	6	23	7
11	6	24	9
12	7	25	7
13	11	Total	150

FIGURE 6.20   c Chart for Number of Blemishes on Reels of Paper

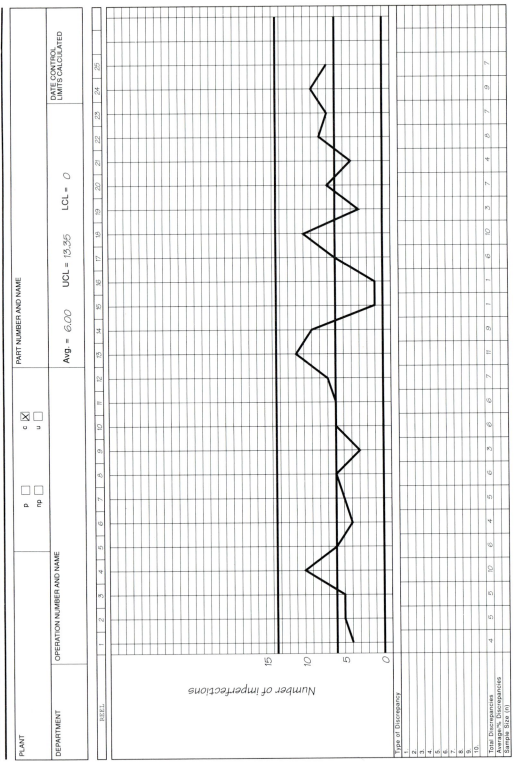

False alarms are indications that the process is exhibiting special variation when no special variation can be found. Most often, these indications will be points on the control chart that are just beyond the upper control limit. False alarms, in and of themselves, can destabilize a stable process. Employees searching for special sources of variation will generally fix something that doesn't need fixing. That is, they'll adjust the process to compensate for nonexistent special sources of variation. This may send the system into a complete state of chaos. Also, false alarms may demoralize employees who may begin to feel that many of their efforts don't result in process improvements.

In some cases, control limits calculated using equations 6.8 and 6.9 may not provide sufficient sensitivity to an indication of a special source of variation. This can result in a loss of opportunity for process improvement.

To avoid both of these problems, we may use a set of fixed control limits for the c chart. These fixed control limits are sometimes called *probability control limits* as they were originally developed using a Poisson probability model. Even though we may be unwilling to make the claim that our data follow this probability model, these fixed control limits provide an excellent and economical rule for separating special and common variation when average counts are less than 20. Figure 6.21 gives values for upper and lower control limits that can be used when average counts are less than 20.

FIGURE 6.21   c Chart Fixed Control Limits

Process Average	LCL	UCL	Process Average	LCL	UCL
0 to 0.10	0	1.5	9.28 to 9.64	2.5	18.5
0.11 to 0.33	0	2.5	9.65 to 10.35	2.5	19.5
0.34 to 0.67	0	3.5	10.36 to 10.97	2.5	20.5
0.68 to 1.07	0	4.5	10.98 to 11.06	3.5	20.5
1.08 to 1.53	0	5.5	11.07 to 11.79	3.5	21.5
1.54 to 2.03	0	6.5	11.80 to 12.52	3.5	22.5
2.04 to 2.57	0	7.5	12.53 to 12.59	3.5	23.5
2.58 to 3.13	0	8.5	12.60 to 13.25	4.5	23.5
3.14 to 3.71	0	9.5	13.26 to 13.99	4.5	24.5
3.72 to 4.32	0	10.5	14.00 to 14.14	4.5	25.5
4.33 to 4.94	0	11.5	14.15 to 14.74	5.5	25.5
4.95 to 5.29	0	12.5	14.75 to 15.49	5.5	26.5
5.30 to 5.58	0.5	12.5	15.50 to 15.65	5.5	27.5
5.59 to 6.23	0.5	13.5	15.66 to 16.24	6.5	27.5
6.24 to 6.89	0.5	14.5	16.25 to 17.00	6.5	28.5
6.90 to 7.43	0.5	15.5	17.01 to 17.13	6.5	29.5
7.44 to 7.56	1.5	15.5	17.14 to 17.76	7.5	29.5
7.57 to 8.25	1.5	16.5	17.77 to 18.53	7.5	30.5
8.26 to 8.94	1.5	17.5	18.54 to 18.57	7.5	31.5
8.95 to 9.27	1.5	18.5	18.58 to 19.36	8.5	31.5
			19.37 to 20.00	8.5	32.5

In the example of the reels of paper, the centerline is 6.00. Therefore, for this application of the c chart, the control limits should properly have come from Figure 6.21. As 6.00 is in the 5.59 to 6.23 range, the values for the lower and upper control limits respectively are 0.5 and 13.5. These values have been used to draw the c chart in Figure 6.22.

Notice that these control limits are almost the same as those created using Equations 6.15 (13.35) and 6.16 (0.00). In general, because the number of events is a whole number, both control charts may show the same indications of a lack of control. In this particular case the resulting control charts are similar, but a count of 0 will indicate a lack of control using the fixed limits, and won't indicate a lack of control using the three-sigma limits.

It's not too unusual to find that the control limits resulting from computations using Equations 6.15 and 6.16 and those resulting from Figure 6.21 are similar, and some users merely ignore these table values. The danger in ignoring the table values is that when average counts are small, the three-sigma limits may generate false indications of a lack of control or fail to signal a lack of control. This can lead to overadjustment or underadjustment of a process, which in and of itself may cause the process to become out of control or, as we've said, may lead to some frustration on the part of those employees trying to search for special causes of variation where none exist.

A note of caution when dealing with c charts: People charged with determining the number of imperfections must be clear and consistent in the definition of an imperfection. Operational definitions, as discussed in Chapter 3, are extremely important, and the individuals identifying imperfections must be properly trained so that they understand the nature of the process. If the individuals identifying imperfections aren't properly trained, some identified imperfections may not actually be imperfections, while some actual imperfections may go undetected—hence the independence of observations of the occurrence of imperfections may suffer. This, in turn, may result in violations of the underlying assumptions used for the c chart, resulting in either the generation of many false alarms, or in undetected out-of-control behavior.

Stabilizing a Process: An Example.   An industrial washing machine manufacturer inspects completed units for defects. Figure 6.23 lists counts of defects found on 24 machines.

The centerline for the control chart is

$$\bar{c} = 1100/24 = 45.8$$

and the control limits can be found using Equations 6.15 and 6.16:

$$LCL(c) = 45.8 - 3\sqrt{45.8} = 25.5$$

and

$$UCL(c) = 45.8 + 3\sqrt{45.8} = 66.1$$

Hence counts of 67 or more and 25 or fewer indicate a lack of control. Figure 6.24 is a control chart for this process.

FIGURE 6.22 c Chart Using Fixed Limits for Number of Blemishes on Reels of Paper

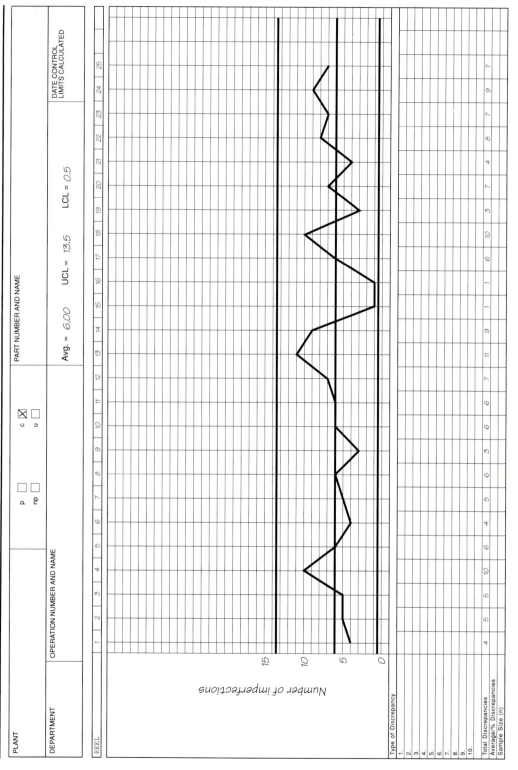

**FIGURE 6.23   Defects Found
on 24 Machines**

Machine Number	Count	Machine Number	Count
1	62	13	51
2	60	14	75
3	36	15	49
4	39	16	52
5	36	17	62
6	47	18	43
7	33	19	70
8	32	20	18
9	74	21	44
10	71	22	20
11	43	23	18
12	39	24	26
		Total	1,100

The process is not in control. Special causes of variation are present. The local operators responsible for the final inspection act so that the special causes of variation for points 9, 10, 14, 19, 20, 22, and 23 are identified and the appropriate corrections are made. The data for points affected by known special causes that have been eliminated are deleted from the data set, and the centerline and control limits are recomputed:

$$\bar{c} = 754/17 = 44.4$$

$$LCL(c) = 44.4 - 3 \sqrt{44.4} = 24.4$$

$$UCL(c) = 44.4 + 3 \sqrt{44.4} = 64.4$$

The new limits are so close to the old limits that the old limits are used for the next 24 machines produced; Figure 6.25 presents the data.

The first five data points for these next 24 machines are well below the lower control limit. Investigation by the local operators reveals that a substitute for the regular inspector counted the defects on those five machines. The substitute wasn't properly trained and didn't identify all the defects correctly. The operators informed management, and management made appropriate changes in policy so that this situation wouldn't recur. These points can now be eliminated from the data. Beginning with machine number 30, all counts are below the process average. Local operators decided that the process has been changed, so a revised control chart is constructed beginning with point number 30.

$$\bar{c} = 674/19 = 35.5$$

$$LCL(c) = 35.5 - 3 \sqrt{35.5} = 17.6$$

$$UCL(c) = 35.5 + 3 \sqrt{35.5} = 53.4$$

FIGURE 6.24 **Control Chart for 24 Washing Machines**

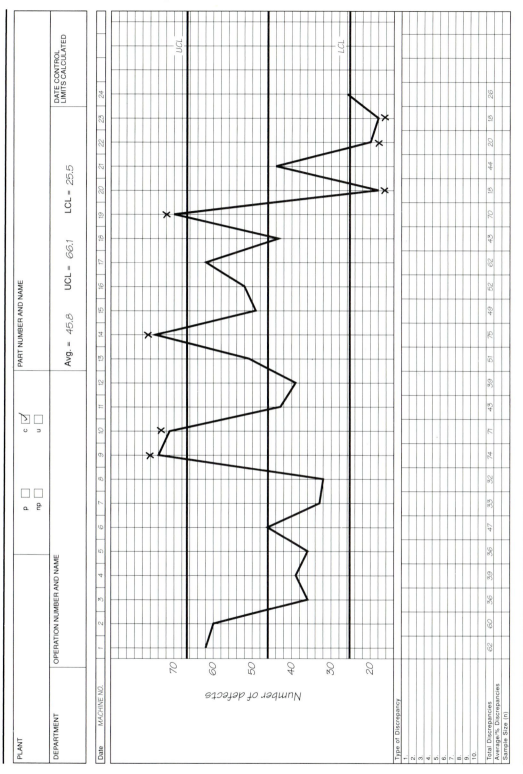

PLANT		PART NUMBER AND NAME		
DEPARTMENT	OPERATION NUMBER AND NAME			DATE CONTROL LIMITS CALCULATED
		Avg. = 45.8	UCL = 66.1	LCL = 25.5

Date	MACHINE NO.	1	2	3	4	5	6	7	8	9	10	11	12	13	14	15	16	17	18	19	20	21	22	23	24

p ☐ c ☑
np ☐ u ☐

Number of defects

Type of Discrepancy

1.																										
2.																										
3.																										
4.																										
5.																										
6.																										
7.																										
8.																										
9.																										
10.																										
Total Discrepancies	62	60	36	39	36	47	33	32	74	71	43	39	51	75	49	52	62	43	70	18	44	20	18	26
Average/% Discrepancies																								
Sample Size (n)																								

FIGURE 6.25 Defects Found on Next 24 Machines

Machine Number	Number of Defects	Machine Number	Number of Defects
25	21	37	46
26	18	38	31
27	7	39	42
28	12	40	44
29	18	41	26
30	32	42	37
31	32	43	26
32	37	44	29
33	39	45	31
34	39	46	34
35	34	47	36
36	39	48	40

Figure 6.26 illustrates the revised control chart. The process, as it stands, now appears to be in a state of control.

Another Example. Consider the case of a mill with a constant work force of 450 employees that has a sign posted at the employee entrance reading "SAFETY IS BETTER THAN COMPENSATION." Informal conversations with employees reveal they consider the sign a reminder to be careful. But management has *not* simultaneously made the work environment safer. There still are cluttered aisles, and spills and leaks of liquids on the floor aren't attended to hastily. The workers know this but have long ago stopped their fruitless efforts at getting management to allocate the resources necessary to create a safer workplace.

To examine the problem, a control chart of the number of accidents per month is constructed. Figure 6.27 illustrates the data for the past 26 months.

As the number of labor-hours per month remains constant, the area of opportunity is considered constant month-to-month. The centerline for the c chart is

$$\bar{c} = 26/26 = 1.0$$

From Figure 6.21, the fixed control limits are

$$LCL(c) = 0 \text{ and } UCL(c) = 4.5$$

Figure 6.28 displays the control chart for the past 26 months. As there are no indications of any special variation, we can conclude that the process is stable and in a state of statistical control. The company may not know it, but it's in the business of producing accidents at the stable rate of one per month. It will continue to do so until some effort is made to change the underlying process. If no change in the process is made, accidents will continue to be produced at this rate. Consequently, the "SAFETY IS BETTER THAN COMPENSATION" sign is unfair. Employees aren't empowered to make system changes that would lower the average number of accidents per month; the sign unjustly and subtly shifts the burden of responsibility for safety from management to the employees.

FIGURE 6.26 Control Chart for Next 24 Washing Machines

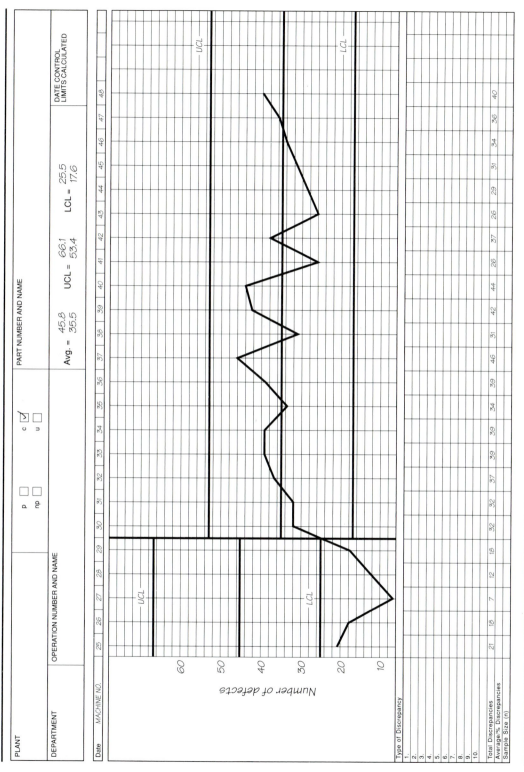

FIGURE 6.27   Accidents per Month

Month	Number of Accidents	Month	Number of Accidents
Jan.	3	Feb.	2
Feb.	2	Mar.	0
Mar.	0	Apr.	0
Apr.	2	May	3
May	1	June	2
June	1	July	0
July	1	Aug.	1
Aug.	0	Sept.	0
Sept.	0	Oct.	1
Oct.	1	Nov.	0
Nov.	1	Dec.	0
Dec.	3	Jan.	1
Jan.	0	Feb.	1
		Total	26

u Charts

In some applications the areas of opportunity vary in size. Generally the construction and interpretation of control charts are easier when the area of opportunity remains constant, but from time to time changes in that area may be unavoidable. For example, samples taken from a roll of paper may need to be manually torn from rolls, so that the areas sampled—the areas of opportunity—will vary; continuous welds in heat exchangers will have varying areas of opportunity depending on the total number and lengths of the welds present in different units; and the number of typing errors in a document will have areas of opportunity that will vary with the lengths of the documents. When the areas vary, the control chart used is a u chart.

The u chart is similar to the c chart in that it's a control chart for the count of the number of events, such as the number of nonconformities over a given area of opportunity. The fundamental difference lies in the fact that during construction of a c chart, the area of opportunity remains constant from observation to observation, while this isn't a requirement for the u chart. Instead, the u chart considers the number of events (such as blemishes or other defects) as a fraction of the total size of the area of opportunity in which these events were possible, thus circumventing the problem of having different areas of opportunity for different observations.

The characteristic used for the control chart, u, is the ratio of the number of events to the area of opportunity in which the events may occur. For observation i, we'll call the number of events (such as imperfections) the observed c_i, and call the area of opportunity a_i. Thus, u_i is the ratio

$$u_i = c_i/a_i \qquad (6.17)$$

for each point.

FIGURE 6.28 Control Chart for Accidents per Month

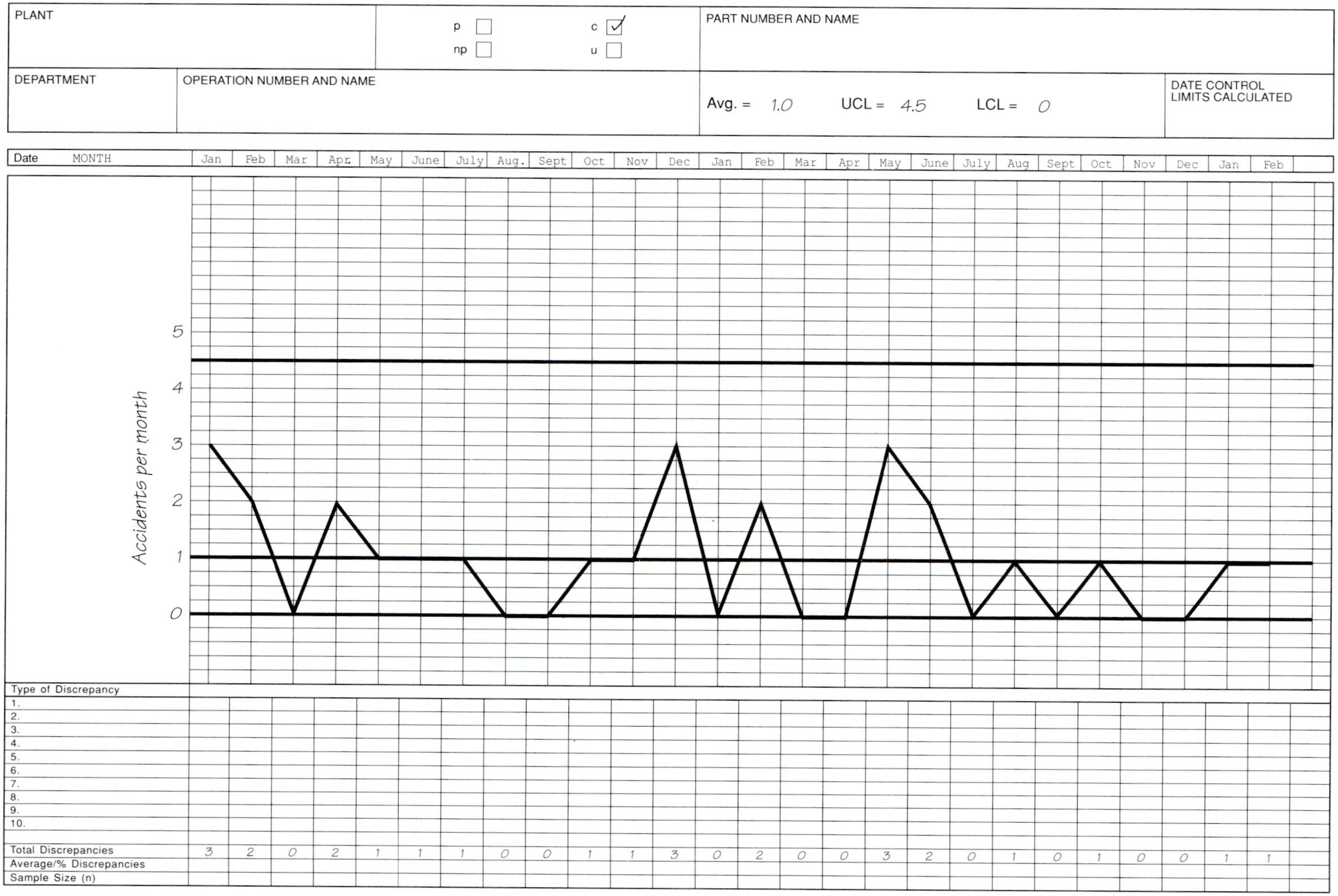
PLANT
p
np
c ✓
u
PART NUMBER AND NAME
DEPARTMENT
OPERATION NUMBER AND NAME
Avg. = 1.0
UCL = 4.5
LCL = 0
DATE CONTROL LIMITS CALCULATED
Date MONTH
Jan Feb Mar Apr. May June July Aug. Sept Oct Nov Dec Jan Feb Mar Apr May June July Aug Sept Oct Nov Dec Jan Feb
Accidents per month
5
4
3
2
1
0
Type of Discrepancy
1.
2.
3.
4.
5.
6.
7.
8.
9.
10.
Total Discrepancies 3 2 0 2 1 1 1 0 0 1 1 3 0 2 0 0 3 2 0 1 0 1 0 0 1 1
Average/% Discrepancies
Sample Size (n)

The average of all the u_i values, $\bar{u}$, provides a centerline for the control chart:

$$\text{Centerline}(u) = \bar{u} = \Sigma c_i / \Sigma a_i \qquad (6.18)$$

Control limits are usually placed at three standard errors on either side of the centerline for each individual subgroup. The standard error is given by the square root of the average u value divided by the subgroup's area of opportunity, $\sqrt{\bar{u}/a_i}$. Since the area of opportunity varies from subgroup to subgroup, so does the standard error. This results in control limits that vary from subgroup to subgroup:

$$\text{LCL}(u) = \bar{u} - 3\sqrt{\bar{u}/a_i} \qquad (6.19)$$

$$\text{UCL}(u) = \bar{u} + 3\sqrt{\bar{u}/a_i} \qquad (6.20)$$

When the lower control limit is negative, a value of 0.0 is used instead.

FIGURE 6.29   **Defects in Rolls of Plastic**

Inspection Lot (i)	Square Feet of Plastic	Area of Opportunity (in 100 square feet) a_i	Number of Defects in Lot c_i	Defects per 100 Square Feet u_i
1	200	2.00	5	2.50
2	250	2.50	7	2.80
3	100	1.00	3	3.00
4	90	0.90	2	2.22
5	120	1.20	4	3.33
6	80	0.80	1	1.25
7	200	2.00	10	5.00
8	220	2.20	5	2.27
9	140	1.40	4	2.86
10	80	0.80	2	2.50
11	170	1.70	1	0.59
12	90	0.90	2	2.22
13	200	2.00	5	2.50
14	250	2.50	12	4.80
15	230	2.30	4	1.74
16	180	1.80	4	2.22
17	80	0.80	1	1.25
18	100	1.00	2	2.00
19	140	1.40	3	2.14
20	120	1.20	4	3.33
21	250	2.50	2	0.80
22	130	1.30	3	2.31
23	220	2.20	1	0.45
24	200	2.00	5	2.50
25	100	1.00	2	2.00
26	160	1.60	4	2.50
27	250	2.50	12	4.80
28	80	0.80	1	1.25
29	150	1.50	5	3.33
30	210	2.10	4	1.90
Totals	4,790	$\Sigma a_i = 47.90$	$\Sigma c_i = 120$	

An Example.   Consider the case of the manufacture of a certain grade of plastic. The plastic is produced in rolls, with samples taken five times daily. Because of the nature of the process, the square footage of each sample varies from inspection lot to inspection lot. Hence the u chart should be used here. Figure 6.29 shows the data on the number of defects, c_i, for the past 30 inspection lots. The number of defects per 100 square feet calculated from Equation 6.17.

Using Equation 6.18 we find the centerline to be

$$\text{Average number of defects/100 sq. ft.} = \bar{u} = 120/47.90 = 2.5$$

The control limits are different for each of the subgroups and must be computed individually for each subgroup using Equations 6.19 and 6.20. Figure 6.30 lists the resulting values.

FIGURE 6.30   **Control Limits for Defects in Rolls of Plastic**

Inspection Lot i	Number of Inspection Units a_i	LCL	UCL
1	2.0	0	5.9
2	2.5	0	5.5
3	1.0	0	7.3
4	0.9	0	7.5
5	1.2	0	6.8
6	0.8	0	7.8
7	2.0	0	5.9
8	2.2	0	5.7
9	1.4	0	6.5
10	0.8	0	7.8
11	1.7	0	6.2
12	0.9	0	7.5
13	2.0	0	5.9
14	2.5	0	5.5
15	2.3	0	5.6
16	1.8	0	6.1
17	0.8	0	7.8
18	1.0	0	7.3
19	1.4	0	6.5
20	1.2	0	6.8
21	2.5	0	5.5
22	1.3	0	6.7
23	2.2	0	5.7
24	2.0	0	5.9
25	1.0	0	7.3
26	1.6	0	6.3
27	2.5	0	5.5
28	0.8	0	7.8
29	1.5	0	6.4
30	2.1	0	5.8

Figure 6.31 illustrates the control chart for this process. No points indicate a lack of control, so there's no reason to believe that any special variation is present. If sources of special variation were detected, we would proceed as we did with the c chart—that is, we would identify the source or sources of the special variation, eliminate them from the system if detrimental, or incorporate them into the system if beneficial; drop the data points from the data set; and reconstruct and reanalyze the control chart.

Two Alternatives for Coping with Variable Control Limits.   Variable control limits can be confusing and can lead to difficulties with calculation and record keeping. Additionally, as the control limits are based on the area of opportunity, future control limits can't be projected from past control limits even for a stable process. Therefore, a given data point may not be identifiable as indicating a lack of control until some time after it occurs, when control limits can be calculated. This can lead to costly delays in corrective actions; there are two common remedies for this problem.

One possible solution is to calculate approximate control limits based upon an average value for the area of opportunity. If the areas of opportunity don't vary by more than about ±25 percent, the control limits can be based on

$$\text{Average area of opportunity} = \bar{a} = \Sigma a_i / k \qquad (6.21)$$

where k is the number of subgroups or areas of opportunity.

Then the values for the upper and lower control limits can be approximated using

$$\text{UCL}(u_{approx.}) = \bar{u} + 3\sqrt{\bar{u}/\bar{a}} \qquad (6.22)$$

and

$$\text{LCL}(u_{approx.}) = \bar{u} - 3\sqrt{\bar{u}/\bar{a}} \qquad (6.23)$$

Points well within these control limits may be considered to indicate only common variation (assuming there are no other indications of a lack of control, as discussed in Chapters 5 and 8). Exact values for the UCL and LCL should be calculated for those points falling close to the control limits to determine whether any special causes of variation are indicated.

In our plastic rolls example, the average area of opportunity and the approximate control limits are found using Equations 6.21, 6.22, and 6.23:

$$\bar{a} = \text{Average area of opportunity} = 47.90/30 = 1.6 \text{ hundred sq. ft.}$$

$$\text{UCL}(u_{approx.}) = 2.5 + 3\sqrt{2.5/1.6} = 6.25$$

$$\text{LCL}(u_{approx.}) = 2.5 - 3\sqrt{2.5/1.6} = -1.25 \text{ (use 0.00)}$$

Figure 6.32 presents a control chart using these values and the data for the plastic rolls given in Figure 6.29. None of the entries falls close to the approximate control limits, nor are there any other indications of a lack of control. Further investigation of special variation doesn't appear warranted. One possible drawback when using approximate control limits is that the selection of those points

FIGURE 6.31 u Chart for Number of Defects in Rolls of Plastic

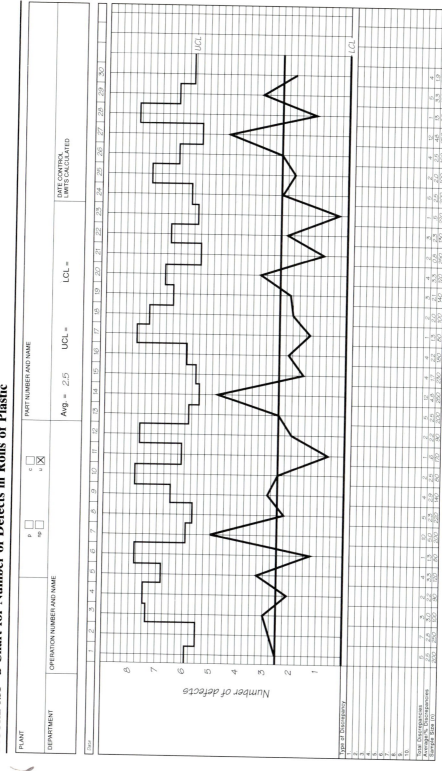

PLANT					PART NUMBER AND NAME			
DEPARTMENT		p ☐	c ☐					
		np ☐	u ☒					
	OPERATION NUMBER AND NAME				Avg. = 2.5	UCL =	LCL =	
							DATE CONTROL LIMITS CALCULATED	

Date	1	2	3	4	5	6	7	8	9	10	11	12	13	14	15	16	17	18	19	20	21	22	23	24	25	26	27	28	29	30

Number of defects

Type of Discrepancy																														
1																														
2																														
3																														
4																														
5																														
6																														
7																														
8																														
9																														
10																														
Total Discrepancies	5	7	2	4	1	10	5	4	2	2	1	2	5	4	4	4	1	3	3	4	3	2	5	5	2	12	13	5	4
Average/% Discrepancies	2.5	2.8	2.2	3.3	1.3	5.0	2.3	2.9	2.5	2.2	6	2.2	2.5	4.8	1.7	2.2	1.3	2.1	3	3.3	0.8	2.3	1	2.5	2.0	2.5	4.8	3.3	1.9
Sample Size (n)	200	250	90	120	80	200	220	140	80	90	170	90	200	250	230	180	80	140	120	250	130	220	200	100	160	250	80	150	210

FIGURE 6.32   u Chart Using Average Area of Opportunity

PLANT

PART NUMBER AND NAME

DEPARTMENT

OPERATION NUMBER AND NAME

p ☐   c ☐
np ☐   u ☒

Avg. = 2.5   UCL = 6.25   LCL = 0

DATE CONTROL
LIMITS CALCULATED

Date LOT

	1	2	3	4	5	6	7	8	9	10	11	12	13	14	15	16	17	18	19	20	21	22	23	24	25	26	27	28	29	30

UCL

Centerline

LCL

Defects per 100 sq. ft.

6
5
4
3
2
1

Type of Discrepancy
1.
2.
3.
4.
5.
6.
7.
8.
9.
10.

Total Discrepancies	2.5	2.8	3.0	2.2	3.3	1.3	5.0	2.3	2.9	2.5	.6	2.2	2.5	4.8	1.7	2.2	1.3	2.0	2.1	3.3	.8	2.3	.5	2.5	2.0	2.5	4.8	1.3	3.3	1.9
Average/% Discrepancies																														
Sample Size (n)																														

needing calculation of exact UCL and LCL values is somewhat arbitrary; this is because the term *close to the approximate control limits* requires a subjective judgment.

To overcome this problem we can use a second remedy for the variable control limits of the u chart. This alternative involves calculating two sets of control limits: an outer set, using for $\bar{a}$ in Equations 6.19 and 6.20 the smallest value for the area of opportunity that will be encountered; and an inner set, using for $\bar{a}$ the largest possible value for the area of opportunity encountered in the data. Any data point outside the outer control limits clearly indicates a lack of control. Any data point lying within the inner set of control limits clearly doesn't indicate any special causes of variation (assuming that there are no other runs or nonrandom patterns indicating a lack of control). Those data points falling between the two sets of control limits require the calculation of exact UCL and LCL values.

Variable Control Limits: An Example.   A manufacturer of chemical process equipment uses an automatic welding machine for long continuous welds. The welds are x-rayed, and the x-rays are examined for imperfections in the welds. Weld lengths are measured in inches and vary with the particular unit or portion of the unit being assembled. Thus the appropriate attribute chart for counts of the number of imperfections is a u chart. The firm has traditionally kept track of imperfections in terms of the rate per 100 linear inches of weld. Figure 6.33 shows data from 25 welds as well as the computed rate per 100 inches.

The average rate is computed using Equation 6.18:

$$\bar{u} = 57/5631 = 1.01 \text{ imperfections per 100 inches}$$

The shortest weld length produced with this equipment is 100 inches, and the longest is 400 inches (areas of opportunity of 1.00 and 4.00 hundred inches, respectively). These values can be used to construct outer and inner control limits, respectively:

$$\text{UCL}(u_{outer}) = 1.01 + 3\sqrt{1.01/1.00} = 4.02$$

$$\text{LCL}(u_{outer}) = 1.01 - 3\sqrt{1.01/1.00} = -2.00 \qquad \text{(use 0.00)}$$

$$\text{UCL}(u_{inner}) = 1.01 + 3\sqrt{1.01/4.00} = 2.52$$

$$\text{LCL}(u_{inner}) = 1.01 - 3\sqrt{1.01/4.00} = -.50 \qquad \text{(use 0.00)}$$

Figure 6.34 illustrates a control chart using these values. Units 16 and 17 are between the inner and outer sets of control limits; hence, exact values for the upper control limits are required. For unit 16, the 254 weld inches correspond to an a_i of 2.54 hundred inches; consequently, from Equation 6.20, the UCL value is

$$\text{UCL}(u) = 1.01 + 3\sqrt{1.01/2.54} = 2.90$$

Unit 16's rate of imperfections, 2.76, doesn't indicate the presence of any special causes of variation. For unit 17, the 144 weld inches correspond to an a_i of 1.44 hundred inches; consequently the UCL value is

$$\text{UCL}(u) = 1.01 + 3\sqrt{1.01/1.44} = 3.52$$

FIGURE 6.33   Imperfections in Welds

Unit No.	Weld Length	Imperfections	Number of Imperfections per 100 Inches (Rate) u
1	187	2	1.07
2	302	1	0.33
3	302	0	0.00
4	172	2	1.16
5	240	5	2.08
6	144	1	0.69
7	120	1	0.83
8	320	2	0.63
9	264	2	0.76
10	180	1	0.56
11	208	1	0.48
12	234	5	2.14
13	180	1	0.56
14	288	3	1.04
15	108	1	0.93
16	254	7	2.76
17	144	6	4.17
18	180	2	1.11
19	288	2	0.69
20	360	3	0.83
21	220	5	2.27
22	156	0	0.00
23	348	1	0.29
24	288	2	0.69
25	144	1	0.69
Totals	5631		26.76

Unit 17's rate of imperfections, 4.17, indicates a special cause of variation.

A careful study of the circumstances surrounding unit 17 reveals that the incorrect welding wire was used in both that unit and unit 16. A change in procedure is instituted to prevent this type of error from recurring—that is, the system has now been changed to eliminate this source of special variation.

As the system has been changed, the data for units 16 and 17 can now be dropped. After this has been done, the process average becomes

$$\bar{u} = 44/5233 = 0.84 \text{ imperfections per 100 inches}$$

The outer and inner sets of control limits can now be recomputed as

$$\text{UCL}(u_{outer}) = 0.84 + 3\sqrt{0.84/1.00} = 3.59$$

$$\text{LCL}(u_{outer}) = 0.84 - 3\sqrt{0.84/1.00} = -1.91 \quad \text{(use 0.00)}$$

$$\text{UCL}(u_{inner}) = 0.84 + 3\sqrt{0.84/4.00} = 2.21$$

$$\text{LCL}(u_{inner}) = 0.84 - 3\sqrt{0.84/4.00} = -.53 \quad \text{(use 0.00)}$$

FIGURE 6.34 Control Chart for Rate of Weld Imperfections

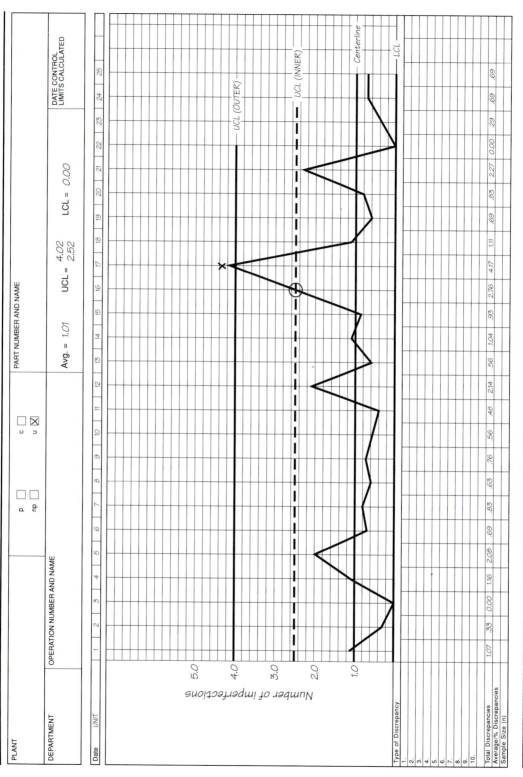

FIGURE 6.35 **Revised Control Chart for Rate of Weld Imperfections**

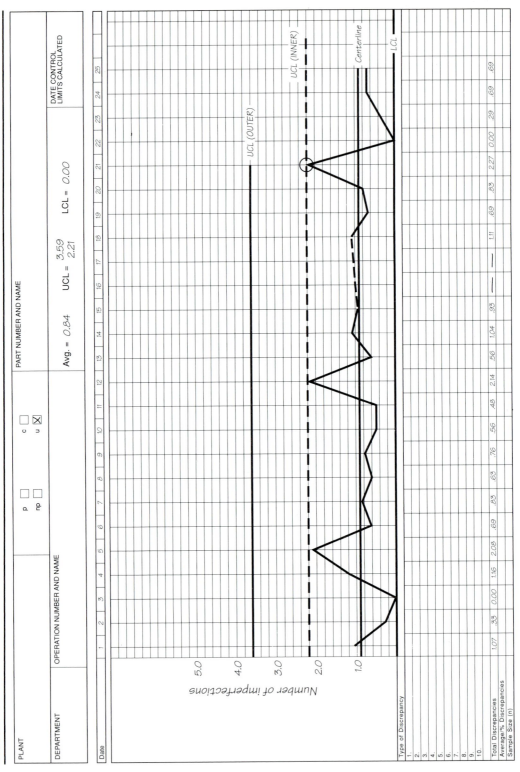

Figure 6.35 shows the revised control chart. Point 21 is beyond the inner upper control limits, so the exact value for the upper control limit must be calculated using Equation 6.20:

$$UCL = 0.84 + 3 \sqrt{0.84/2.20} = 2.69$$

Hence the u value 2.27 at point 21 doesn't indicate the presence of any special source of variation, and the process now appears stable and in a state of statistical control.

Limitations of Attribute Control Charts

As we've mentioned, as processes improve and defects or defectives become rarer, the number of units that must be examined to find one or more of these events increases. If we consider a p chart where the average fraction of nonconforming items is 0.005, then on average we'd need to examine 200 units to have an average count of just 1.00. In the extreme, to maintain a reasonable average count as the area of opportunity grows, 100 percent inspection becomes the rule. This implies inspecting all of the items and sorting those that conform to some specification from those that don't. Not only is this inspection costly, but it's equivalent to accepting the fact that the process is producing a constant fraction of its output as defective and will continue to do so. Hence, attribute control charts are limited in terms of the level of process improvement they make possible. Additional process improvement is possible with variables control charts, to be discussed in Chapter 7.

Another disadvantage of using attribute control charts is that if special variation from several different sources is present, it's hard to identify and isolate the special sources individually. One or more of these special sources might mask another, resulting in a process that appears to be stable but is really operating under the influence of several special sources of variation. As the number of special sources of variation increases, their tendency to mask one another can grow, resulting in further difficulties in the future.

On the other hand, variables control charts use numerical measurements, which make them more revealing and powerful than attribute control charts. Attribute data fail to reveal by how much a unit is beyond an upper or lower specification limit; they only indicate whether a given unit conforms. Therefore the attribute data won't provide as clear a direction for process improvement as will variables data.

Summary

Attribute control charts can be broadly classified into two groups: (1) p charts and np charts based on binomial counts and (2) u charts and c charts based on counts over areas of opportunity.

The p chart can help stabilize a process by indicating a lack of statistical control in some characteristic measured as a proportion of output. Subgroup sizes may be constant or may vary from subgroup to subgroup.

The np chart is mathematically equivalent to a p chart; but the number rather than the proportion of items with the characteristic of interest is charted. Subgroup sizes are generally held constant for each subgroup for the np chart.

A c chart is used when a given single unit of output may have multiple events, such as the number of defects in an appliance or in a roll of paper. The c chart helps stabilize the number of events when the area of opportunity in which the events may occur remains the same for each unit of output from the system.

The u chart is used when counts of the number of events are to be control charted and the area of opportunity varies from unit to unit.

While attribute control charts help identify special process variation, they are only a milestone on the road to never-ending improvement. To continue on the journey, variables control charts must be instituted and used.

Exercises

6.1 A manufacturer of wood screws periodically examines screw heads for burrs. Subgroups of 300 screws are selected and examined using a carefully designed procedure. For the following observations:

Obs. No.	Freq.	Proportion with Burrs	Obs. No.	Freq.	Proportion with Burrs
1	19	0.063	11	7	0.023
2	16	0.053	12	13	0.043
3	11	0.037	13	17	0.057
4	6	0.020	14	29	0.097
5	22	0.073	15	1	0.003
6	2	0.007	16	2	0.007
7	4	0.013	17	19	0.063
8	7	0.023	18	28	0.093
9	5	0.017	19	24	0.080
10	27	0.090	20	23	0.077

a. Find the centerline and standard error.
b. Find the control limits and zone boundaries.
c. Identify any indications of a lack of statistical control and state the reason you believe that a lack of control is indicated.

6.2 A large metropolitan hospital processes many samples of blood daily. Some occasionally get mislabeled or lost, so new samples are required (rework). Subgroups of 50 samples are tracked each day for a 30-day period. Construct a control chart to search for special sources of variation.

Sample Number	n	Missing or Lost	Sample Number	n	Missing or Lost
1	50	4	16	50	3
2	50	4	17	50	5
3	50	6	18	50	1
4	50	5	19	50	6
5	50	2	20	50	5
6	50	0	21	50	3
7	50	6	22	50	1
8	50	1	23	50	0
9	50	2	24	50	4
10	50	0	25	50	3
11	50	4	26	50	5
12	50	2	27	50	6
13	50	3	28	50	2
14	50	0	29	50	3
15	50	2	30	50	1

6.3 A firm with 248 vehicles at one location keeps track of the number out of service each day. *Out-of-service* is defined as unavailable for normal use for four or more hours.

Day	Number Out of Service	Proportion	n
1	7	0.028	248
2	3	0.012	248
3	6	0.024	248
4	2	0.008	248
5	1	0.004	248
6	0	0.000	248
7	12	0.048	248
8	3	0.012	248
9	6	0.024	248
10	4	0.016	248
11	3	0.012	248
12	2	0.008	248
13	7	0.028	248
14	6	0.024	248
15	2	0.008	248
16	5	0.020	248
17	8	0.032	248
18	9	0.036	248
19	3	0.012	248
20	1	0.004	248
21	0	0.000	248
22	2	0.008	248
23	4	0.016	248
24	6	0.024	248
25	4	0.016	248

  a. Find the appropriate centerline for the control chart.
  b. Determine the control limits and zone boundaries.
  c. Determine whether there are any indications of a lack of control.

6.4 A bank is studying the proportion of transactions made using an automated teller machine (ATM). The following represents a series of days and the number of transactions using the ATM:

Day	Total Number of Transactions	Number of ATM Transactions
1	320	87
2	356	92
3	280	75
4	325	109
5	344	69
6	410	99
7	385	120
8	324	86
9	367	111
10	312	90
11	276	86
12	342	106
13	387	65
14	312	91
15	390	131
16	354	78
17	322	89
18	353	98
19	317	81
20	374	58
21	409	104
22	366	81
23	298	87
24	311	72
25	339	84

  a. Find the appropriate centerline for the control chart.
  b. Determine the control limits and zone boundaries.
  c. Are there any indications of a lack of control? What are the indications, and why do they indicate a lack of control?

6.5 A company manufactures 2,000 lawn mowers per day. Every day 40 lawn mowers are randomly selected from the production line. If a mower fails to start on the first pull, it's labeled ''nonconforming.'' Twenty-two days' results are:

Day	n	Number Nonconforming	Fraction Nonconforming
1	40	2	0.050
2	40	3	0.075
3	40	1	0.025
4	40	4	0.100
5	40	3	0.075
6	40	2	0.050
7	40	1	0.025
8	40	1	0.025
9	40	0	0.000
10	40	3	0.075
11	40	2	0.050
12	40	4	0.100
13	40	7	0.175
14	40	2	0.050
15	40	3	0.075
16	40	3	0.075
17	40	2	0.050
18	40	8	0.200
19	40	0	0.000
20	40	1	0.250
21	40	3	0.075
22	40	2	0.050

a. Find the appropriate centerline for the control chart.
b. Determine the control limits and zone boundaries.
c. Find any indications of a lack of control.

6.6 For the following data points, with n = 300,
   a. Find the centerline, control limits, and zone boundaries.
   b. Draw the control chart and indicate any points that demonstrate special variation.

Observation Number	Number of Defectives	Fraction Defective
1	19	.063
2	16	.053
3	11	.037
4	6	.020
5	22	.073
6	2	.007
7	4	.013
8	7	.023
9	5	.017
10	27	.090
11	7	.023
12	13	.043
13	17	.057
14	29	.097
15	1	.003
16	2	.007
17	19	.063
18	28	.093
19	24	.080
20	23	.077

6.7 Samples of 90 retainer rings are examined for the fraction nonconforming. The results for 30 consecutive days are:

Day	Fraction Nonconforming	Day	Fraction Nonconforming	Day	Fraction Nonconforming
1	0.12	11	0.15	21	0.12
2	0.09	12	0.14	22	0.12
3	0.03	13	0.13	23	0.09
4	0.08	14	0.02	24	0.08
5	0.14	15	0.09	25	0.02
6	0.06	16	0.15	26	0.03
7	0.17	17	0.17	27	0.15
8	0.14	18	0.07	28	0.13
9	0.15	19	0.03	29	0.16
10	0.17	20	0.12	30	0.13

Does the process appear to be in a state of statistical control with respect to the fraction nonconforming?

6.8 A given model of a large radar dish represents an area of opportunity in which nonconformities may occur. Results for 25 such assemblies are:

Assembly Number	Number of Nonconformities	Assembly Number	Number of Nonconformities
1	25	14	75
2	60	15	24
3	28	16	50
4	65	17	70
5	91	18	56
6	56	19	21
7	40	20	88
8	54	21	34
9	90	22	82
10	44	23	53
11	62	24	102
12	81	25	64
13	70		

a. Determine the centerline and control limits for the c chart.
b. Are there any indications of a lack of control? What are the indications, and why do they indicate a lack of control?

6.9 A manufacturer of a particular grade of copper tubing experiences flaws in 500-foot coils. The counts of the number of flaws are:

Coil	Count	Coil	Count	Coil	Count
1	7	11	5	21	13
2	2	12	1	22	8
3	6	13	7	23	0
4	4	14	3	24	11
5	5	15	0	25	19
6	15	16	18	26	0
7	4	17	11	27	0
8	0	18	5	28	1
9	0	19	7	29	6
10	2	20	0	30	0

a. Determine the centerline, control limits, and zone boundaries for the c chart.
b. Are there any indications of a lack of control? What are the indications, and why do they indicate a lack of control?

6.10 A large publisher counts the number of keyboard errors that make their way into finished books. The number of errors and the number of pages in the past 26 publications are:

Book Number	Number of Errors	Number of Pages	Book Number	Number of Errors	Number of Pages
1	49	202	14	48	612
2	63	232	15	50	432
3	57	332	16	41	538
4	33	429	17	45	383
5	54	512	18	51	302
6	37	347	19	49	285
7	38	401	20	38	591
8	45	412	21	70	310
9	65	481	22	55	547
10	62	770	23	63	469
11	40	577	24	33	652
12	21	734	25	14	343
13	35	455	26	44	401

a. Determine the centerline and control limits for the u chart.

b. Are there any indications of a lack of control? What are the indications, and why do they indicate a lack of control?

6.11 Lots of cloth produced by a manufacturer are inspected for defects. Because of the nature of the inspection process, the size of the inspection sample varies from lot to lot, making a u chart the appropriate control chart.

Lot Number	100's of Square Yards	Number of Defects
1	2.0	5
2	2.5	7
3	1.0	3
4	0.9	2
5	1.2	4
6	0.8	1
7	1.4	0
8	1.6	2
9	1.9	3
10	1.5	0
11	1.7	2
12	1.7	3
13	2.0	1
14	1.6	2
15	1.9	4

Calculate the centerline and upper and lower control limits:
(a) Using average area of opportunity.
(b) Using varying control limits for each point.
(c) Using two sets of control limits.
Are any special causes of variation present in the data?

Endnotes

1. Write American Society for Quality Control Inc., 310 West Wisconsin Ave., Milwaukee, Wis. 53203.

CHAPTER 7 Variables Control Charts

Introduction

Variables data consist of measurements such as weight, length, width, height, time, temperature, and electrical resistance. Variables data contain more information than attribute data, which merely classify a process's output as conforming or nonconforming, or else count the number of imperfections. Furthermore, because variables control charts deal with measurements themselves, they don't mask valuable information and therefore are more powerful than attribute charts. They use all information contained in the data; this alone makes variables charts preferable when a choice is possible.

There are four principal types of variables control charts: the x-bar and R chart, the x-bar and s chart, the median chart, and the individuals chart. All are used in the never-ending spiral of process improvement.

Quality consciousness must increase for a firm to continue to reduce the difference between customer needs and process performance. Early efforts to control the quality of output often find firms segregating nonconforming production from conforming production. Nonconforming production is then reworked, discarded, downgraded, or otherwise removed from the mainstream of the process output. Each item so removed incurs a greater overall cost to the firm than the production of a conforming item. This is because special attention is required that is, more often than not, labor-intensive and costly. Furthermore, the firm can't measure the harm done to its reputation and self-esteem by even occasionally shipping a defective item.

As a greater awareness of the need to improve quality grows, organizations will use attribute control charts to control and stabilize factors such as the proportion of defectives or the number of defects. However, as that proportion becomes smaller as a result of these efforts, organizations seeking continued process improvement will eventually begin to use variables measurement and variables control charts. As discussed in Chapter 6, a controlled process producing a relatively low fraction of defective units will require relatively large subgroups to detect

228

those defectives. The only way to overcome the need for larger and larger subgroups is to continue upward on the spiral of quality consciousness through the use of variables control charts.

Variables Charts and the PDSA Cycle

As with the attributes control charts, the PDSA cycle provides both an important guideline for proceeding with variables control charts and the mechanism for improved quality through continued process improvement.

Plan

The object or purpose of the control chart must be carefully delineated to effectively use the control chart as a vehicle to reduce the difference between customer needs and process performance. A plan must be established that clearly shows what will be control charted, why it will be control charted, where it will be control charted, when it will be control charted, who will do the control charting, and how it will be control charted.

Consequently, the personnel in an organization must decide which variables to measure. These decisions require the cooperation and input of all those directly or indirectly involved with the process, such as operators, foremen, supervisors, and engineers. Some of the techniques that Chapter 9 discusses (such as brainstorming, cause-and-effect diagrams, check sheets, or Pareto diagrams) may be useful in selecting the process variables that will decrease the difference between customer needs and process performance. Often some feature of the process that has been a source of trouble (resulting in extra cost in scrap or rework) and has failed to yield to corrective efforts is a good place to start. Starting *far upstream,* or near the beginning of a process, will often produce the most dramatic results and may offer the greatest opportunity to alter the factors at the root of downstream special causes of variation.

Subgroup selection is crucial to the proper usage of the control chart. Subgroups should be chosen rationally, as discussed in Chapter 5. The choice should minimize the variation within the subgroups. This will allow isolation of special variation between the subgroups while capturing the inherent process variation within the subgroups. The frequency with which subgroups are selected will help minimize the amount of variation between the subgroups by isolating batches, shifts, production runs, machines, or people. This enables the special variation to be captured, isolated, observed, and analyzed.

Decisions concerning rational subgroups often require the combined knowledge of those directly involved with the process and an experienced statistician. Rational subgroup selection may require a trial-and-error solution, may produce several false starts, and may very well require a great deal of patience.

The method by which the measurements are to be made must be studied carefully. Operational definitions, as discussed in Chapter 3, must be constructed and communicated to people involved with the data collection. At this point any forms to be used for data collection should be selected or designed and the responsibility for construction of the charts assigned.

Do

Data collection and the calculation of control chart statistics constitute the Do stage for constructing variables control charts. It is usually best to collect at least 20 subgroups before beginning to construct a control chart. On rare occasions fewer than 20 subgroups may be used, but a control chart should almost never be attempted with fewer than 10 subgroups.

As Chapter 5 explained, variables charts are made up of two parts: one charts the process variability, and one charts the process location. For instance, in the x-bar and R chart, the R values, or subgroup ranges, are used to track variability. The x-bar values, or subgroup averages, are used for the process location. As the control limits of the portion of the control chart measuring location are based on the average variability ($\overline{R}$ in the x-bar and R chart), the variability-measuring portion of the control chart must be constructed and evaluated first. Only if there is stability in the variability portion can the location portion of the control chart be constructed. For each of the variables control charts, the estimate of the process standard deviation is based on the average value of the measure of variability used for the subgroups. If the process isn't stable, its variability won't be predictable, leading to unreliable estimates of the process standard deviation. If these estimates are used to construct control limits, those limits will also be unreliable and won't reveal special sources of variation when they exist.

After the initial data set has been collected, the centerline, control limits, and zone boundaries (if applicable) should be computed for both portions of the control chart. These should be entered onto the control charts along with the collected set of data points. First, the variability portion of the control chart should be examined for indications of special variation. Any special causes of variation must be studied and the process must be stabilized before the location portion of the control chart is analyzed. This means skipping ahead to the Study stage of the PDSA cycle before completing the Do stage.

If there are no indications of a lack of control in the control chart's variability portion, the location portion of the chart can be analyzed in the Study stage of the PDSA cycle. This completes the Do stage.

Study

Indications of a lack of control, such as patterns of the type introduced in Chapter 5 and discussed in detail in Chapter 8, are studied in the Study stage of the PDSA cycle. Indications of special sources of variation may be found in the chart dealing with variation, in the chart dealing with location, or in both. Whether the variation is common process variation or variation resulting from special causes, once we've found and identified them we proceed to the Act stage to set policy to formalize process improvements resulting from analysis of the control chart.

We must periodically review all aspects of the control chart and make changes where appropriate. If the process itself has been changed in some way, the control limits should certainly be recomputed and the analysis of the process begun anew.

Act

If the variation found in the Study stage results only from common causes, then efforts to reduce that variation must focus on changes in the process itself. When indications of special causes of variation are present, the cause or causes of that special variation should be removed if the variation is detrimental or incorporated into the process if the variation is beneficial.

The focus of the Act stage is on formalizing policy that results directly from the prior study of the causes of process variation. This will lead to a reduction in the difference between customer needs and process performance.

Last in the Act stage, the purpose of the control chart must be reconsidered by returning to the Plan stage. This will help to maintain a focus on improvement and point out those areas that can be most beneficial in reducing the difference between customer needs and process performance.

Subgroup Size and Frequency

The selection of an appropriate control chart depends, in part, on the subgroup size. In turn, many factors affect the decision of ideal subgroup size. Large subgroups are, of course, more expensive than small ones; nevertheless, large subgroups lead to tighter control limits so long as the subgroups are selected rationally to minimize the within-group variation.

Traditionally, subgroup sizes of four or five have been used ever since they were first suggested by W. A. Shewhart.[1] Subgroups should be large enough to detect points or patterns indicating a lack of control when lack of control exists. This requires statistical expertise and an understanding of the process under study. Subgroups of two or three are used when the cost of sampling is relatively high. Individuals charts may be used when only one measurement is available or appropriate as a subgroup. Subgroups of 6 to 14 are occasionally used when we want the control chart to be very sensitive to changes in the process average. When subgroup sizes are 10 or more, we'll see that x-bar and s charts are generally used instead of x-bar and R charts.

The frequency with which subgroups are selected—how often they're selected—depends on the particular application. If quick action is required, the frequency should be greater. A stable process merely being monitored will require less frequent subgroup selection than one being studied or being brought into a state of statistical control for the first time. There are no hard and fast rules for determining the frequency of subgroup selection, and decisions are generally made on the basis of knowledge about the process under study and knowledge of statistics.

x-Bar and R Charts

As the name implies, the x-bar and R chart uses the subgroup range, R, to chart the process variability, and the subgroup average, $\bar{x}$, to chart the process location. Chapter 5 provided an initial look at x-bar and R charts. There we saw that the periodic selection of small subgroups of process output could be very useful in process stabilization and improvement.

Stable processes yield subgroups that will behave predictably, enabling us to construct an x-bar and R chart. The two characteristics, $\bar{x}$ and R, can be estimated by relatively simple procedures. Estimates of the standard errors of both R and $\bar{x}$ are based on the average subgroup range, $\bar{R}$. This not only simplifies the estimation procedure but directly impacts how the control charts must be constructed and analyzed.

The Range Portion

The subgroup ranges are used as a measure of dispersion. When the range for each subgroup is calculated, the result is a sequence of values used to construct both the range portion and the x-bar portion of the control chart.

The process output has a mean and standard deviation that are estimated using the average of the subgroup averages, $\bar{\bar{x}}$, and the average value of the subgroup ranges, $\bar{R}$. Recall from Chapter 5 that for normally distributed processes,

$$\bar{R} = d_2\, \sigma$$

so that the process standard deviation, σ, can be estimated as $\sigma = \bar{R}/d_2$, where values for d_2 are a function of subgroup size and can be found in Table 1. Also recall that the assumption of normality is not required to interpret the x-bar and R chart.

It would not be desirable to use all of the data points together, rather than as subgroups, to estimate the process standard deviation because that would include the variation both within the subgroups and between the subgroups in our measure of dispersion. Hence, our efforts would fail to isolate the between-subgroup variation, and the control chart would be useless.

The sampling distribution of $\bar{R}$, which was discussed in Chapter 5, has a standard error given by

$$\sigma_R = d_3\sigma$$

where the value of d_3 depends on subgroup size, and can be found in Table 1. By substitution, the standard error of $\bar{R}$ can be written as

$$\sigma_R = d_3(\bar{R}/d_2) \tag{7.1}$$

In Chapter 5, these results were used to develop Equations 5.8 and 5.9 for the upper and lower control limits for the range portion of the control chart. The relationships are

$$\text{Centerline(R)} = \overline{R} = \Sigma \, R/k \tag{7.2}$$

$$\text{UCL(R)} = D_4 \, \overline{R} \tag{7.3}$$

$$\text{LCL(R)} = D_3 \, \overline{R} \tag{7.4}$$

where k is the number of subgroups and where $D_3 = [1 - 3(d_3/d_2)]$ and $D_4 = [1 + 3(d_3/d_2)]$ are given in Table 1.

In the development of the upper and lower control limits, three times the estimated standard error of the subgroup ranges, σ_R, is added to and subtracted from the average range value, $\overline{R}$, yielding the factors D_4 and D_3. Recall from Chapter 5 that the zone boundaries for zones A, B, and C are positioned at one and two standard errors on either side of the control chart centerline. Using the estimated standard error of $\overline{R}$ from Equation 7.1, the zone boundaries are given by

$$\begin{array}{ll}\text{Boundary between} & = \overline{R} - 2d_3 \, (\overline{R}/d_2) \\ \text{lower zones A and B} & = \overline{R}(1 - 2d_3/d_2)\end{array} \tag{7.5}$$

$$\begin{array}{ll}\text{Boundary between} & = \overline{R} - d_3 \, (\overline{R}/d_2) \\ \text{lower zones B and C} & = \overline{R}(1 - d_3/d_2)\end{array} \tag{7.6}$$

When the result of Equations 7.5 or 7.6 is a negative number, 0.00 is used instead, as negative ranges are meaningless. Also,

$$\begin{array}{ll}\text{Boundary between} & = \overline{R} + d_3 \, (\overline{R}/d_2) \\ \text{upper zones B and C} & = \overline{R}(1 + d_3/d_2)\end{array} \tag{7.7}$$

$$\begin{array}{ll}\text{Boundary between} & = \overline{R} + 2d_3 \, (\overline{R}/d_2) \\ \text{upper zones A and B} & = \overline{R}(1 + 2d_3/d_2)\end{array} \tag{7.8}$$

An easy way to position the zone boundaries is to divide the distance between the upper control limit and the centerline by 3. The resulting quantity is then added to and subtracted from the centerline to form the boundaries between the C and B zones. The boundaries between the B and A zones are formed by adding and subtracting twice the result of dividing the distance between the upper control limit and the centerline by 3.

For example, if the average range were 50 and the control limits were 20 and 80, the distance between the upper control limit and the centerline would be 30. The resulting zone boundaries are shown in Figure 7.1.

Assuming that the range portion of the control chart is stable, the x-bar portion may be developed.

The x-Bar Portion

The string of subgroup averages computed from the data is used to construct the x-bar portion of the control chart. As discussed in Chapter 5, the sampling distribution of the mean has mean and standard error given by

$$\overline{\overline{x}} = \frac{\Sigma \, \overline{x}}{\text{Number of subgroups}}$$

$$\sigma_{\overline{x}} = (\overline{R}/d_2)/\sqrt{n}$$

FIGURE 7.1   **A, B, and C Zones for a Control Chart**

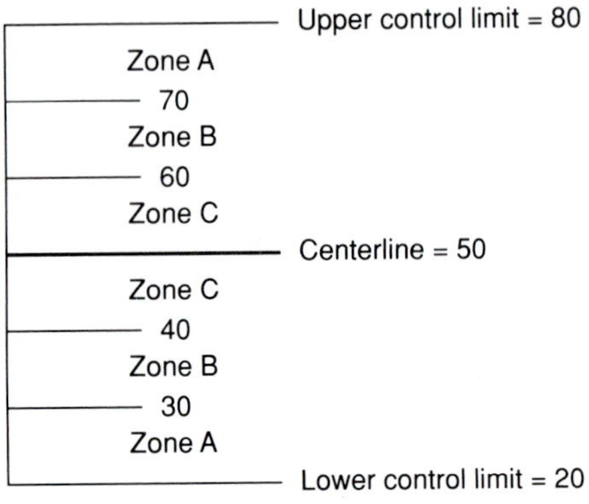

```
                                        ─── Upper control limit = 80
                   Zone A
              ─── 70
                   Zone B
              ─── 60
                   Zone C
         ─────────────────────────────── Centerline = 50
                   Zone C
              ─── 40
                   Zone B
              ─── 30
                   Zone A
              ─────────────────────────── Lower control limit = 20
```

Thus, the centerline of the x-bar control chart is found using Equation 5.10 by taking the average of the subgroup averages, $\bar{\bar{x}}$:

$$\text{Centerline}(\bar{x}) = \bar{\bar{x}} = \Sigma\ \bar{x}/k \qquad (7.9)$$

Using a multiple of three standard errors to construct the control limits, we have

$$\text{UCL}(\bar{x}) = \bar{\bar{x}} + 3\bar{R}/(d_2\sqrt{n})$$

$$\text{LCL}(\bar{x}) = \bar{\bar{x}} - 3\bar{R}/(d_2\sqrt{n})$$

or, more conveniently,

$$\text{UCL}(\bar{x}) = \bar{\bar{x}} + A_2\bar{R} \qquad (7.10)$$

$$\text{LCL}(\bar{x}) = \bar{\bar{x}} - A_2\bar{R} \qquad (7.11)$$

where $A_2 = 3/(d_2\sqrt{n})$ is tabulated as a function of subgroup size in Table 1.

The product $A_2\bar{R}$ represents three times the standard error of the subgroup means. This is useful in forming the A, B, and C zones used in this chart as well in helping to detect patterns indicating a lack of statistical control. Zone boundaries are placed on both sides of the centerline at a distance of one and two times the standard error, respectively. Equivalently, a simple way to position the zone boundaries is to divide the distance between the upper control limit and the centerline by 3. Just as with the range portion, the resulting quantity is then added to and subtracted from the centerline to form the boundaries between the C and B zones. As before, the boundaries between the B and A zones are formed by adding (or subtracting) twice the result of dividing the distance between the upper control limit and the centerline by 3.

$$\begin{array}{l}\text{Boundary between} \\ \text{lower zones A and B}\end{array} = \bar{\bar{x}} - (2/3)A_2\bar{R} \qquad (7.12)$$

$$\begin{array}{ll}\text{Boundary between lower zones B and C} & = \overline{\overline{x}} - (1/3)A_2\overline{R} \qquad (7.13)\end{array}$$

$$\begin{array}{ll}\text{Boundary between upper zones B and C} & = \overline{\overline{x}} + (1/3)A_2\overline{R} \qquad (7.14)\end{array}$$

$$\begin{array}{ll}\text{Boundary between upper zones A and B} & = \overline{\overline{x}} + (2/3)A_2\overline{R} \qquad (7.15)\end{array}$$

x-Bar and R Charts: An Example

Consider the case of a manufacturer of circuit boards for personal computers. Various components are to be mounted on each board and the boards eventually slipped into slots in a chassis. The boards' overall length is crucial to assure a proper fit, and this dimension has been targeted as an important item to be stabilized. (Note: The width, thickness, hardness, or any other characteristic may also be targeted either simultaneously or at another point in time.) Boards are cut from large sheets of material by a single rotary cutter continuously fed from a hopper. At a customer's request, it was decided to create a control chart for the length of circuit boards produced by the process.

After input from many individuals involved with the process, it is decided to select the first five units every hour from the production output. Each group of five items represents a subgroup. This manner of subgroup selection is most likely to isolate the variation over time between the subgroups and, therefore, capture only common process variation within the subgroups.

Boards are measured using an operationally defined method; Figure 7.2 lists the resulting lengths. The average and the range for each subgroup of five elements in Figure 7.2 have been computed and are shown in the last two columns on the right. This arrangement of the data will be used to determine whether special sources of variation between subgroups are evident in measurement-to-measurement changes over time.

Construction of the control chart begins with the range portion. The centerline is found by taking the average of the subgroup ranges using Equation 7.2:

$$\overline{R} = 0.569/25 = 0.023$$

Not only does this value form the centerline for the range portion of the control chart; it forms the basis for estimating the standard error and hence the control limits and zone boundaries as well. Values for D_3 and D_4, from Table 1, are 0.00 and 2.114, respectively. Equations 7.3 and 7.4 yield the control limits:

$$UCL(R) = (2.114)(0.023) = 0.049$$

and

$$LCL(R) = (0)(0.023) = 0.000$$

Equations 7.5, 7.6, 7.7, and 7.8 yield values for the zone boundaries:

$$\begin{array}{ll}\text{Boundary between lower zones A and B} & = 0.023[1 - 2(0.864)/2.326] = 0.006\end{array}$$

FIGURE 7.2   **Cut Circuit Board Lengths**

Time	Sample Number	1	2	3	4	5	Average $\bar{x}$	Range R
9 AM	1	5.030	5.002	5.019	4.992	5.008	5.010	0.038
10	2	4.995	4.992	5.001	5.011	5.004	5.001	0.019
11	3	4.988	5.024	5.021	5.005	5.002	5.008	0.036
12	4	5.002	4.996	4.993	5.015	5.009	5.003	0.022
1 PM	5	4.992	5.007	5.015	4.989	5.014	5.003	0.026
2	6	5.009	4.994	4.997	4.985	4.993	4.996	0.024
3	7	4.995	5.006	4.994	5.000	5.005	5.000	0.012
4	8	4.985	5.003	4.993	5.015	4.988	4.997	0.030
5	9	5.008	4.995	5.009	5.009	5.005	5.005	0.014
6	10	4.998	5.000	4.990	5.007	4.995	4.998	0.017
7	11	4.994	4.998	4.994	4.995	4.990	4.994	0.008
8	12	5.004	5.000	5.007	5.000	4.996	5.001	0.011
9	13	4.983	5.002	4.998	4.997	5.012	4.998	0.029
10	14	5.006	4.967	4.994	5.000	4.984	4.990	0.039
11	15	5.012	5.014	4.998	4.999	5.007	5.006	0.016
12	16	5.000	4.984	5.005	4.998	4.996	4.997	0.021
1 AM	17	4.994	5.012	4.986	5.005	5.007	5.001	0.026
2	18	5.006	5.010	5.018	5.003	5.000	5.007	0.018
3	19	4.984	5.002	5.003	5.005	4.997	4.998	0.021
4	20	5.000	5.010	5.013	5.020	5.003	5.009	0.020
5	21	4.988	5.001	5.009	5.005	4.996	5.000	0.021
6	22	5.004	4.999	4.990	5.006	5.009	5.002	0.019
7	23	5.010	4.989	4.990	5.009	5.014	5.002	0.025
8	24	5.015	5.008	4.993	5.000	5.010	5.005	0.022
9	25	4.982	4.984	4.995	5.017	5.013	4.998	0.035
						Totals	125.029	0.569

Boundary between lower zones B and C = $0.023(1 - 0.864/2.326) = 0.014$

Boundary between upper zones B and C = $0.023(1 + 0.864/2.326) = 0.032$

Boundary between upper zones A and B = $0.023[1 + 2(0.864)/2.326] = 0.040$

Figure 7.3(a) displays the control chart. There are no indications of any special sources of variation, and the process appears stable with regard to its range.

After the stability of the range has been established, the x-bar portion of the control chart may be constructed. If the range is stable, $\bar{R}$ can be used as a basis for the estimate of the standard error for the x-bar portion of the chart. Equation 7.9 yields a centerline value of

$$\bar{\bar{x}} = 125.029/25 = 5.001$$

FIGURE 7.3 x-Bar and R Chart for Cut Circuit Board Lengths

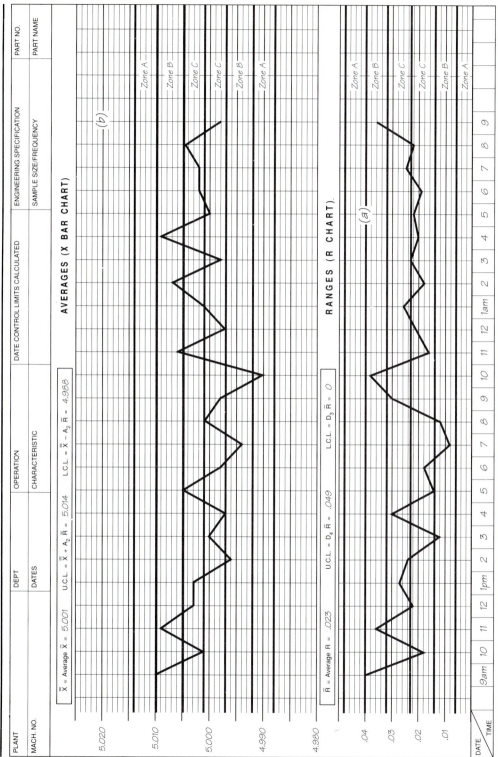

The upper and lower control limits can be found using Equations 7.10 and 7.11. The value for A_2 comes from Table 1 for a subgroup of size 5.

$$\text{UCL}(\bar{x}) = 5.001 + (0.577)(0.023) = 5.014$$

$$\text{LCL}(\bar{x}) = 5.001 - (0.577)(0.023) = 4.988$$

Zone boundaries are found using Equations 7.12 through 7.15:

$$\text{Boundary between lower zones A and B} = 5.001 - (2/3)(0.577)(0.023) = 4.992$$

$$\text{Boundary between lower zones B and C} = 5.001 - (1/3)(0.577)(0.023) = 4.997$$

$$\text{Boundary between upper zones B and C} = 5.001 + (1/3)(0.577)(0.023) = 5.005$$

$$\text{Boundary between upper zones A and B} = 5.001 + (2/3)(0.577)(0.023) = 5.010$$

These values are shown in Figure 7.3(b). There are no indications of any special sources of variation present between the subgroups with respect to time in the x-bar portion, so we can conclude that this process is stable and its output is predictable in the near future, assuming continued stability.

x-Bar and R Charts: Another Example

A manufacturer of high-end audio components buys metal tuning knobs to use in assembling its products. Knobs are produced automatically by a subcontractor using a single machine that's supposed to produce them with a constant diameter. Nevertheless, because of persistent final assembly problems with the knobs, management has decided to examine this process output by requesting that the subcontractor keep an x-bar and R chart for knob diameter. Beginning at 8:30 AM on a Tuesday, the first four knobs are selected every half hour. The diameter of each is carefully measured using an operationally defined technique. The average and range for each subgroup are computed; the data, along with these statistics, are shown in Figure 7.4.

The data are arranged this way to help us determine whether the differences in the subgroups result from special causes over time. The measurement-to-measurement differences here are arranged to trap special variations over time between the subgroups and to confine the common process variation within the subgroups.

Using the subgroup range values and Equation 7.2, the average range can be computed as

$$\bar{R} = 129/25 = 5.16$$

From this, the control limits can be calculated using Equations 7.3 and 7.4. Using a subgroup size of 4, Table 1 gives us values for $D_3 = 0.00$ and $D_4 = 2.282$.

FIGURE 7.4   **Tuning Knob Diameters**

Time	Sample Number	1	2	3	4	Average $\bar{x}$	Range R
8:30 AM	1	836	846	840	839	840.25	10
9:00	2	842	836	839	837	838.50	6
9:30	3	839	841	839	844	840.75	5
10:00	4	840	836	837	839	838.00	4
10:30	5	838	844	838	842	840.50	6
11:00	6	838	842	837	843	840.00	6
11:30	7	842	839	840	842	840.75	3
12:00	8	840	842	844	836	840.50	8
12:30 PM	9	842	841	837	837	839.25	5
1:00	10	846	846	846	845	845.75	1
1:30	11	849	846	848	844	846.75	5
2:00	12	845	844	848	846	845.75	4
2:30	13	847	845	846	846	846.00	2
3:00	14	839	840	841	838	839.50	3
3:30	15	840	839	839	840	839.50	1
4:00	16	842	839	841	837	839.75	5
4:30	17	841	845	839	839	841.00	6
5:00	18	841	841	836	843	840.25	7
5:30	19	845	842	837	840	841.00	8
6:00	20	839	841	842	840	840.50	3
6:30	21	840	840	842	836	839.50	6
7:00	22	844	845	841	843	843.25	4
7:30	23	848	843	844	836	842.75	12
8:00	24	840	844	841	845	842.50	5
8:30	25	843	845	846	842	844.00	4
					Totals	21,036.25	129

Hence

$$UCL(R) = (2.282)(5.16) = 11.78$$

$$LCL(R) = 0(5.16) = 0.00$$

Zone boundaries are computed using Equations 7.5 through 7.8:

Boundary between lower zones A and B $= 5.16[1 - 2(0.880)/2.059] = 0.75$

Boundary between lower zones B and C $= 5.16(1 - 0.880/2.059) = 2.95$

Boundary between upper zones B and C $= 5.16(1 + 0.880/2.059) = 7.37$

Boundary between upper zones A and B $= 5.16[1 + 2(0.880)/2.059] = 9.57$

Figure 7.5 illustrates the control chart. The range at subgroup number 23 is beyond the upper control limit. Furthermore, subgroup number 16 is the eighth

FIGURE 7.5 Initial Range Chart for Tuning Knob Diameters

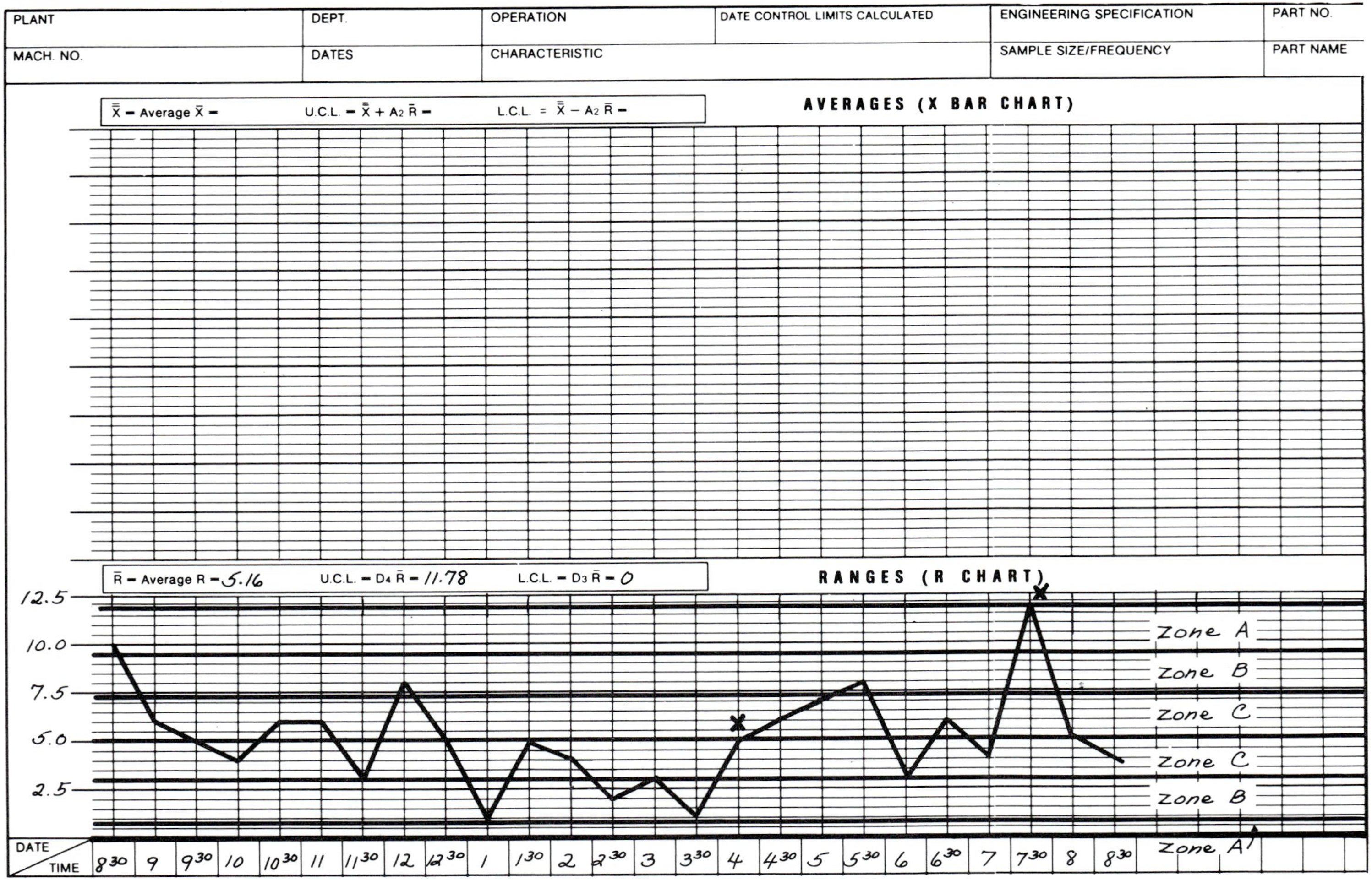

PLANT	DEPT.	OPERATION	DATE CONTROL LIMITS CALCULATED	ENGINEERING SPECIFICATION	PART NO.
MACH. NO.	DATES	CHARACTERISTIC		SAMPLE SIZE/FREQUENCY	PART NAME

AVERAGES (X BAR CHART)

$\bar{X}$ — Average $\bar{\bar{X}}$ —

U.C.L. — $\bar{\bar{X}} + A_2 \bar{R}$ —   L.C.L. = $\bar{\bar{X}} - A_2 \bar{R}$ —

RANGES (R CHART)

$\bar{R}$ — Average R — 5.16

U.C.L. — $D_4 \bar{R}$ — 11.78   L.C.L. — $D_3 \bar{R}$ — 0

Zone A
Zone B
Zone C
Zone C
Zone B
Zone A

| 12.5 | 10.0 | 7.5 | 5.0 | 2.5 |

DATE																								
TIME	8³⁰	9	9³⁰	10	10³⁰	11	11³⁰	12	12³⁰	1	1³⁰	2	2³⁰	3	3³⁰	4	4³⁰	5	5³⁰	6	6³⁰	7	7³⁰	8	8³⁰

consecutive point below the centerline and therefore indicates a lack of control by virtue of the fourth rule presented in Chapter 5. Hence, there are two indications of a lack of control.

An investigation reveals that at 7:25 PM a water pipe had burst in the lunchroom. The episode wasn't serious but caused water to leak from the lunchroom onto the floor beneath the machinery involved in the process. This disruption seems to have caused the lack of control observed at subgroup 23. The operators believe this to be a special cause of variation that shouldn't recur once the plumbing has been repaired. The initial study doesn't reveal any special source of variation for the indication of a lack of control at subgroup 16.

The data for subgroup 23 are then removed from the data set. The repair of the plumbing has permanently removed the conditions leading to this observation. The data for subgroup 16 are left in place, as no special cause of variation can be isolated and removed that would explain its presence. The average value for the subgroup ranges is then recomputed, using Equation 7.2, to reflect the deletion of subgroup 23:

$$\overline{R} = 117/24 = 4.88$$

The revised control limits and zone boundaries are computed using Equations 7.3 through 7.8:

$$UCL(R) = (2.282)(4.88) = 11.14$$

$$LCL(R) = (0)(4.88) = 0.00$$

Boundary between lower zones A and B $= 4.88[1 - 2(0.880)/2.059] = 0.71$

Boundary between lower zones B and C $= 4.88(1 - 0.880/2.059) = 2.79$

Boundary between upper zones B and C $= 4.88(1 + 0.880/2.059) = 6.97$

Boundary between upper zones A and B $= 4.88[1 + 2(0.880)/2.059] = 9.05$

The revised R chart appears in Figure 7.6(a); and because the centerline has been shifted, subgroup number 16 is no longer the eighth consecutive point below the centerline. There are no other indications of a lack of control in the data. Changing the centerline, control limits, and zone boundaries may uncover other indications of a lack of control; but such is not the case here. Hence, the x-bar portion of the chart may be constructed using the average range from the now stable R portion of the chart.

The centerline of the x-bar chart is the average of the subgroup averages, where $\overline{x}$ for subgroup 23 has been subtracted from the total in Figure 7.4. Equation 7.9 yields

$$\overline{\overline{x}} = 20{,}193.50/24 = 841.40$$

FIGURE 7.6 x-Bar and R Chart for Tuning Knob Diameters

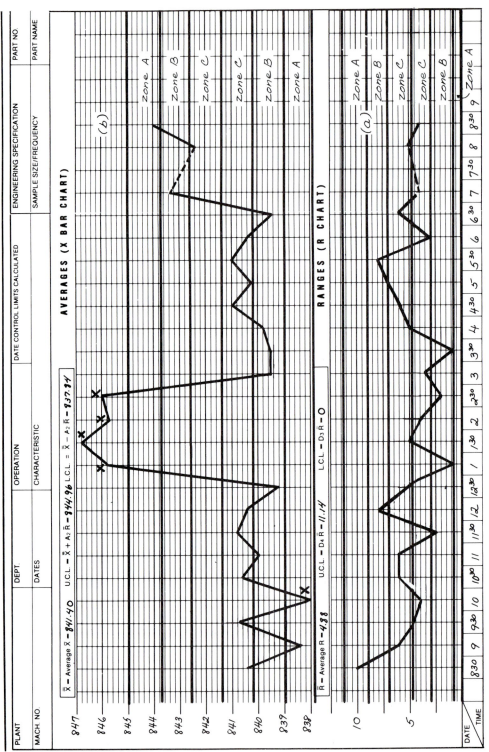

The control limits and the zone boundaries can now be computed using Equations 7.10 through 7.15:

$$\text{UCL}(\bar{x}) = 841.40 + (0.729)(4.88) = 844.96$$

$$\text{LCL}(\bar{x}) = 841.40 - (0.729)(4.88) = 837.84$$

$$\text{Boundary between lower zones A and B} = 841.40 - (2/3)(0.729)(4.88) = 839.03$$

$$\text{Boundary between lower zones B and C} = 841.40 - (1/3)(0.729)(4.88) = 840.21$$

$$\text{Boundary between upper zones B and C} = 841.40 + (1/3)(0.729)(4.88) = 842.59$$

$$\text{Boundary between upper zones A and B} = 841.40 + (2/3)(0.729)(4.88) = 843.77$$

The x-bar portion of the control chart has been completed using these values and appears in Figure 7.6(b). Remember from Chapter 5 that a search for indications of a lack of control should always be made from right to left on the control chart; that is, the search should be made by looking backward in time from the present. There are several indications of a lack of control on the control chart of Figure 7.6(b). Four points are beyond the upper control limit. These occurred sequentially from 1:00 PM to 2:30 PM. Also, at 10:00 AM (fourth subgroup), the average is 838.00, the second of three consecutive points in the lower zone A. This indicates a lack of control by virtue of Rule 2 of Chapter 5.

There's very little doubt that there's at least one source of special variation acting on this process. Finding that source requires some further investigation. The investigation leads to the discovery that at 12:05 PM (just after selection of subgroup number 10), a keyway wedge had cracked and needed to be replaced on the machine. The mechanic who normally makes this repair was out to lunch, so the machine operator made the repair. This individual hadn't been properly trained for the repair, so the wedge wasn't properly aligned in the keyway and subsequent points were out of control. Both the operator and the mechanic agree that the need for this repair wasn't unusual. To correct this problem, management and labor agree to train the machine operator and provide the appropriate tools for making this repair in the mechanic's absence. Furthermore, the maintenance and engineering staffs agree to search for a replacement part for the wedge that's less prone to cracking.

No special source of variation can be found for the indication of a lack of control at 10:00 AM. Indications of special variation will occasionally be found where the special cause won't be identifiable. Failure to identify a special source of variation when one is indicated is no reason to stop analysis; false alarms will happen from time to time.

After the sources of special variation are found and the underlying causes are resolved, the subgroups generated as a result of those sources are dropped from the data. This changes all of the calculated values for centerlines, control limits, and zone boundaries of the control chart.

Accordingly, for the control knobs, we delete subgroups 10, 11, 12, and 13. We must recompute the values for the relevant statistics:

$$\bar{\bar{x}} = 16{,}809.25/20 = 840.46$$

$$\bar{R} = 105/20 = 5.25$$

Using these, the control limits and zone boundaries for the revised R chart become

$$UCL(R) = (2.282)(5.25) = 11.98$$

$$LCL(R) = 0(5.25) = 0.00$$

$$\text{Boundary between lower zones A and B} = 5.25[1 - 2(0.880)/2.059] = 0.76$$

$$\text{Boundary between lower zones B and C} = 5.25(1 - 0.880/2.059) = 3.01$$

$$\text{Boundary between upper zones B and C} = 5.25(1 + 0.880/2.059) = 7.49$$

$$\text{Boundary between upper zones A and B} = 5.25[1 + 2(0.880)/2.059] = 9.74$$

The control chart for the range portion is shown in Figure 7.7(a). There are no indications of a lack of control, so the x-bar chart may be constructed. The control limits and zone boundaries are

$$UCL(\bar{x}) = 840.46 + (0.729)(5.25) = 844.29$$

$$LCL(\bar{x}) = 840.46 - (0.729)(5.25) = 836.63$$

$$\text{Boundary between lower zones A and B} = 840.46 - (2/3)(0.729)(5.25) = 837.91$$

$$\text{Boundary between lower zones B and C} = 840.46 - (1/3)(0.729)(5.25) = 839.18$$

$$\text{Boundary between upper zones B and C} = 840.46 + (1/3)(0.729)(5.25) = 841.74$$

$$\text{Boundary between upper zones A and B} = 840.46 + (2/3)(0.729)(5.25) = 843.01$$

The x-bar portion of the control chart is shown in Figure 7.7(b). The last data point, number 25, now indicates a lack of control. That is, if we ignore the missing entry at data point number 23, data point number 25 is the second of three consecutive points in zone A or beyond, indicating a lack of control by virtue of the second rule of Chapter 5.

At times a conservative approach is warranted—that is, detection of patterns indicating a lack of control will usually be followed by a somewhat costly search for the special cause(s) of variation. A process that has just been altered to eliminate one or more special sources of variation may be allowed to run for a

FIGURE 7.7 Revised x-Bar and R Chart for Tuning Knob Diameters

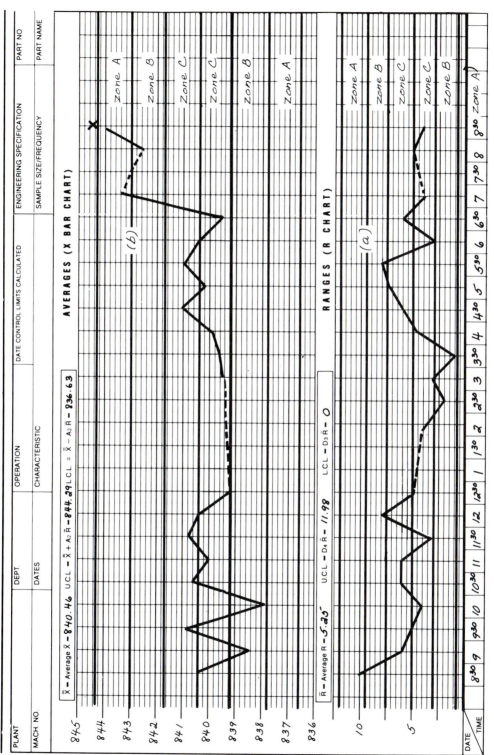

while longer to determine whether any special sources of variation are really present or whether the indication of a lack of control is only a temporary effect that can be attributed to the removal of some of the data. If special sources of variation are present, the data taken from the process output will soon demonstrate evidence of that variation.

This process can be permitted to run as if it were statistically controlled. However, it must be watched closely to ensure that the special sources of variation have been removed and are no longer affecting the process.

x-Bar and s Charts

x-bar and s charts are quite similar to x-bar and R charts. That is, they provide the same sort of information—but x-bar and s charts are used when subgroups consist of 10 or more observations.

In Chapter 4 we saw that the standard deviation of the process output, σ, could be estimated using s, which is computed from Equations 4.9a or 4.9b. s provides an estimate that generally has a smaller standard error than R. The benefits of having a smaller standard error must be weighed against the costs of larger subgroup sizes and more complex calculations. The x-bar and s chart is usually used with larger subgroup sizes, and larger subgroups aren't always desirable.

One reason for this is that s may be viewed as a less robust estimator of the population standard deviation than R, for R is more sensitive to shifts in population shape than s. For subgroup sizes of fewer than 10, the range provides a reasonable statistic with which to estimate the standard error. Also, the range is easier to calculate than s, which gives it an advantage in many situations. So when subgroup sizes are small, the range is used as an estimator for σ.

When subgroup sizes are larger than 10, s is almost always used because as subgroup size increases, s becomes a much more statistically efficient estimator for σ. When the subgroup size is increased, the likelihood of encountering an extreme value increases, so that s, which is less affected than R by extreme values in the data, becomes a better estimator for σ.

Historically, subgroup ranges have been preferred because they're easier to compute than subgroup standard deviations. As the use of electronic calculation has grown, the need to avoid tedious computations has decreased, and the reluctance to use x-bar and s charts has decreased.

The s Portion

The construction of the x-bar and s chart parallels that of the x-bar and R chart in that it begins with an examination of the portion of the chart concerned with the variability of the process. The standard deviation, s, must be calculated for each subgroup. The value for s is the basis for an estimate of the process standard deviation, from which a set of factors for the control limits is developed.

Equation 4.9b is used to compute s for each subgroup:

$$s = \sqrt{\frac{\Sigma x^2 - (\Sigma x)^2/n}{(n-1)}}$$

where n is the number of observations in each subgroup. The sequence of s values is then averaged, yielding $\bar{s}$, which is the centerline for the s chart:

$$\text{Centerline(s)} = \bar{s} = \Sigma\ s/k \qquad (7.16)$$

where k is, once again, the number of subgroups. Additionally, $\bar{s}$ is used to form an estimate of the process standard deviation, σ:

$$\sigma = \bar{s}/c_4$$

where c_4 is a factor that depends on the subgroup size, assumes that the process characteristic is stable and normally distributed, and is given in Table 1.

Control limits for the s chart are constructed by adding and subtracting three times the standard error of $\bar{s}$ from the centerline of the control chart:

$$\text{UCL(s)} = \bar{s} + 3\sigma_\sigma$$

$$\text{LCL(s)} = \bar{s} - 3\sigma_\sigma$$

Assuming that the distribution of the process output was and will be stable, we can derive control limits. As Chapter 4 said, the assumption of normality isn't necessary to interpret the control limits; we can use the Empirical Rule instead.

The sampling distribution of the standard deviation of a stable process has a standard error given by

$$\sigma_\sigma = \sigma\ \sqrt{1 - c_4^2}$$

Hence, the upper control limit for the s chart is

$$\text{UCL(s)} = \bar{s} + 3\sigma\ \sqrt{1 - c_4^2}$$

Using $\bar{s}/c_4$ to estimate σ yields

$$\text{UCL(s)} = \bar{s} + \frac{3\bar{s}\ \sqrt{1 - c_4^2}}{c_4} = \bar{s}\left[1 + \frac{3\ \sqrt{1 - c_4^2}}{c_4}\right]$$

Similarly,

$$\text{LCL(s)} = \bar{s} - \frac{3\bar{s}\ \sqrt{1 - c_4^2}}{c_4} = \bar{s}\left[1 - \frac{3\ \sqrt{1 - c_4^2}}{c_4}\right]$$

A more convenient way to represent these control limits is by defining two new constants, B_3 and B_4:

$$B_3 = 1 - \frac{3\ \sqrt{1 - c_4^2}}{c_4}$$

$$B_4 = 1 + \frac{3\ \sqrt{1 - c_4^2}}{c_4}$$

so that

$$UCL(s) = B_4 \bar{s} \qquad (7.17)$$

$$LCL(s) = B_3 \bar{s} \qquad (7.18)$$

Values for B_3 and B_4 depend on subgroup size and can be found in Table 1.

Boundaries for zones A, B, and C for the s chart are placed at the usual multiples of one and two times the standard error on either side of the centerline. As negative values are meaningless, the zone boundaries cease to exist below the 0.00 line on the control chart. To find the zone boundaries, we divide the difference between the upper control limit and the centerline by 3. This value provides an estimate for the standard error of $\bar{s}$. Adding and subtracting this value from the centerline yields the upper and lower boundaries between zones B and C, respectively, while adding and subtracting two times this value yields the upper and lower boundaries between zones A and B, respectively. These boundaries can be expressed as

$$\begin{matrix} \text{Boundary between} \\ \text{lower zones A and B} \end{matrix} = \bar{s} - (2/3)\bar{s}(B_4 - 1) \qquad (7.19)$$

$$\begin{matrix} \text{Boundary between} \\ \text{lower zones B and C} \end{matrix} = \bar{s} - (1/3)\bar{s}(B_4 - 1) \qquad (7.20)$$

$$\begin{matrix} \text{Boundary between} \\ \text{upper zones B and C} \end{matrix} = \bar{s} + (1/3)\bar{s}(B_4 - 1) \qquad (7.21)$$

$$\begin{matrix} \text{Boundary between} \\ \text{upper zones A and B} \end{matrix} = \bar{s} + (2/3)\bar{s}(B_4 - 1) \qquad (7.22)$$

If the s portion of the control chart is found to be stable, the x-bar portion may be constructed. However, if the s portion indicates of a lack of statistical control, then the x-bar portion can't be safely evaluated until any special sources of variation have been removed and the process stabilized. As we'll see, this is because the estimate of the standard error of the x-bar portion is based on the average value of s. Just as with the x-bar and R chart, if the variability isn't in control, then estimates of the standard error are unreliable, leading to unreliable control limits for $\bar{x}$.

The x-Bar Portion

The centerline for the x-bar chart is the average of the subgroup averages, $\bar{\bar{x}}$, and can be found using Equation 7.9. The control limits are found by adding and subtracting three times the standard error of $\bar{x}$ from the centerline:

$$\bar{\bar{x}} \pm 3 \, \sigma/\sqrt{n}$$

Our estimate of the process standard deviation is

$$\sigma = \bar{s}/c_4$$

so that the control limits become

$$\bar{\bar{x}} \pm 3(\bar{s}/c_4)/\sqrt{n} = \bar{\bar{x}} \pm 3\bar{s}/(c_4\sqrt{n})$$

Letting the constant $A_3 = 3/(c_4\sqrt{n})$, the control limits for $\bar{x}$ can be expressed as

$$UCL(\bar{x}) = \bar{\bar{x}} + A_3\bar{s} \qquad (7.23)$$

and

$$LCL(\bar{x}) = \bar{\bar{x}} - A_3\bar{s} \qquad (7.24)$$

The value for A_3 depends on subgroup size and can be found in Table 1.

The zone boundaries are placed at one and two times the standard error on either side of the centerline. The standard error is expressed as $\bar{s}/(c_4\sqrt{n})$, so that the zone boundaries are

$$\begin{matrix}\text{Boundary between} \\ \text{lower zones A and B}\end{matrix} = \bar{\bar{x}} - 2\,\bar{s}/(c_4\sqrt{n}) \qquad (7.25)$$

$$\begin{matrix}\text{Boundary between} \\ \text{lower zones B and C}\end{matrix} = \bar{\bar{x}} - \bar{s}/(c_4\sqrt{n}) \qquad (7.26)$$

$$\begin{matrix}\text{Boundary between} \\ \text{upper zones B and C}\end{matrix} = \bar{\bar{x}} + \bar{s}/(c_4\sqrt{n}) \qquad (7.27)$$

$$\begin{matrix}\text{Boundary between} \\ \text{upper zones A and B}\end{matrix} = \bar{\bar{x}} + 2\,\bar{s}/(c_4\sqrt{n}) \qquad (7.28)$$

x-Bar and s Charts: An Example

In a converting operation, a plastic film is combined with paper coming off a spooled reel. As the two come together, they form a moving sheet that passes as a web over a series of rollers. The operation runs in a continuous feed, and the thickness of the plastic coating is an important product characteristic. Coating thickness is monitored by a highly automated piece of equipment that uses 10 heads to take 10 measurements across the web at half-hour intervals. Figure 7.8 shows a sequence of measurements taken over 20 time periods.

Arranging the data in this way helps us determine whether there are any special causes of variation from subgroup to subgroup over time.

Values for $\bar{x}$ and s, the subgroup means and standard deviations, have been computed for each subgroup. The values for the averages of these subgroup means and standard deviations are computed using Equations 7.9 and 7.16:

$$\bar{\bar{x}} = 42.43/20 = 2.12$$

and

$$\bar{s} = 2.19/20 = 0.11$$

The s chart must be constructed first, and Equations 7.17 and 7.18 provide the upper and lower control limits:

$$UCL(s) = (1.716)(0.11) = 0.189$$

$$LCL(s) = (0.284)(0.11) = 0.031$$

FIGURE 7.8 **Plastic Coating Thickness**

Head #	8:30	9:00	9:30	10:00	10:30	11:00	11:30
1	2.08	2.14	2.30	2.01	2.06	2.14	2.07
2	2.26	2.02	2.10	2.10	2.12	2.22	2.05
3	2.13	2.14	2.20	2.15	1.98	2.18	1.97
4	1.94	1.94	2.25	1.97	2.12	2.27	2.05
5	2.30	2.30	2.05	2.25	2.20	2.17	2.16
6	2.15	2.08	1.95	2.12	2.02	2.26	2.02
7	2.07	1.94	2.10	2.10	2.19	2.15	2.02
8	2.02	2.12	2.16	1.90	2.03	2.07	2.14
9	2.22	2.15	2.37	2.04	2.02	2.02	2.07
10	2.18	2.36	1.98	2.08	2.09	2.36	2.00
$\bar{x}$	2.14	2.12	2.15	2.07	2.08	2.18	2.06
s	.111	.137	.136	.098	.074	.099	.059

Head #	12:00	12:30	13:00	13:30	14:00	14:30	15:00
1	2.08	2.13	2.13	2.24	2.25	2.03	2.08
2	2.31	1.90	2.16	2.34	1.91	2.10	1.92
3	2.12	2.12	2.12	2.40	1.96	2.24	2.14
4	2.18	2.04	2.22	2.26	2.04	2.20	2.20
5	2.15	2.40	2.12	2.13	1.93	2.25	2.02
6	2.17	2.12	2.07	2.15	2.08	2.03	2.04
7	1.98	2.15	2.04	2.08	2.29	2.06	1.94
8	2.05	2.01	2.28	2.02	2.42	2.19	2.05
9	2.00	2.30	2.12	2.05	2.10	2.13	2.12
10	2.26	2.14	2.10	2.18	2.00	2.20	2.06
$\bar{x}$	2.13	2.13	2.14	2.19	2.10	2.14	2.06
s	.107	.141	.070	.125	.170	.084	.086

Head #	15:30	16:00	16:30	17:00	17:30	18:00	
1	2.04	1.92	2.12	1.98	2.08	2.22	
2	2.14	2.10	2.30	2.30	2.12	2.05	
3	2.18	2.13	2.01	2.31	2.11	1.93	
4	2.12	2.02	2.20	2.12	2.22	2.08	
5	2.00	1.93	2.11	2.08	2.00	2.15	
6	2.02	2.17	1.93	2.10	1.95	2.27	
7	2.05	2.24	2.02	2.15	2.15	1.95	
8	2.34	1.98	2.25	2.35	2.14	2.11	
9	2.12	2.34	2.05	2.12	2.28	2.12	
10	2.05	2.12	2.10	2.26	2.31	2.10	
$\bar{x}$	2.11	2.10	2.11	2.18	2.14	2.10	
s	.101	.136	.115	.121	.113	.106	

The zone boundaries are computed using Equations 7.19 through 7.22:

$$\text{Boundary between lower zones A and B} = 0.11 - (2/3)(0.11)(1.716 - 1) = 0.057$$

$$\text{Boundary between lower zones B and C} = 0.11 - (1/3)(0.11)(1.716 - 1) = 0.084$$

$$\text{Boundary between upper zones B and C} = 0.11 + (1/3)(0.11)(1.716 - 1) = 0.136$$

$$\text{Boundary between upper zones A and B} = 0.11 + (2/3)(0.11)(1.716 - 1) = 0.163$$

The control chart is shown in Figure 7.9(a). As the control chart doesn't indicate a lack of control, the process variability appears stable. The x-bar portion of the chart may now be constructed.

It wouldn't have been correct to analyze the x-bar portion before establishing the stability of the process standard deviation. If the standard deviation weren't stable, the mean value of the subgroup standard deviations would be unreliable; the control limits on the x-bar chart would also be unreliable, as the control limits are based on the value for s̄. The control chart might then fail to reveal patterns indicating a lack of control when they existed. This is why the estimate of the process standard deviation isn't based on the standard deviation taken across all observations.

The average value for x̄, x̄̄, is 2.12. Equations 7.23 and 7.24 are used to form the upper and lower control limits for the x-bar chart:

$$\text{UCL}(\bar{x}) = 2.12 + (0.975)(0.11) = 2.227$$

$$\text{LCL}(\bar{x}) = 2.12 - (0.975)(0.11) = 2.013$$

We then use Equations 7.25 through 7.28 to calculate the zone boundaries:

$$\text{Boundary between lower zones A and B} = 2.12 - 2\{0.11/[(0.9727)(3.16)]\} = 2.048$$

$$\text{Boundary between lower zones B and C} = 2.12 - [0.11/(0.9727)(3.16)] = 2.084$$

$$\text{Boundary between upper zones B and C} = 2.12 + [0.11/(0.9727)(3.16)] = 2.156$$

$$\text{Boundary between upper zones A and B} = 2.12 + 2\{0.11/[(0.9727)(3.16)]\} = 2.192$$

The x-bar chart appears in Figure 7.9(b) and completes the control chart. There are no indications of a lack of control, so the process can be considered to be stable and the output predictable with respect to time as long as conditions remain the same.

It's now appropriate to use some of the methods that will be described in Chapter 9 (such as check sheets, Pareto analysis, or brainstorming) to attempt to reduce the common causes of variation in the never-ending quest to decrease the difference between process performance and customer needs. It's now also appropriate to rearrange the data to look for special sources of variation across the sheet; that is, to attempt to isolate variation on different sections of the sheet passing over the web at the measuring device. A rational subgrouping (as discussed in Chapter 5) to isolate this variation would group the data by measurement head. As measurements are taken every half hour, it might be apt to use all

FIGURE 7.9 x-Bar and s Chart for Plastic Coating Thickness

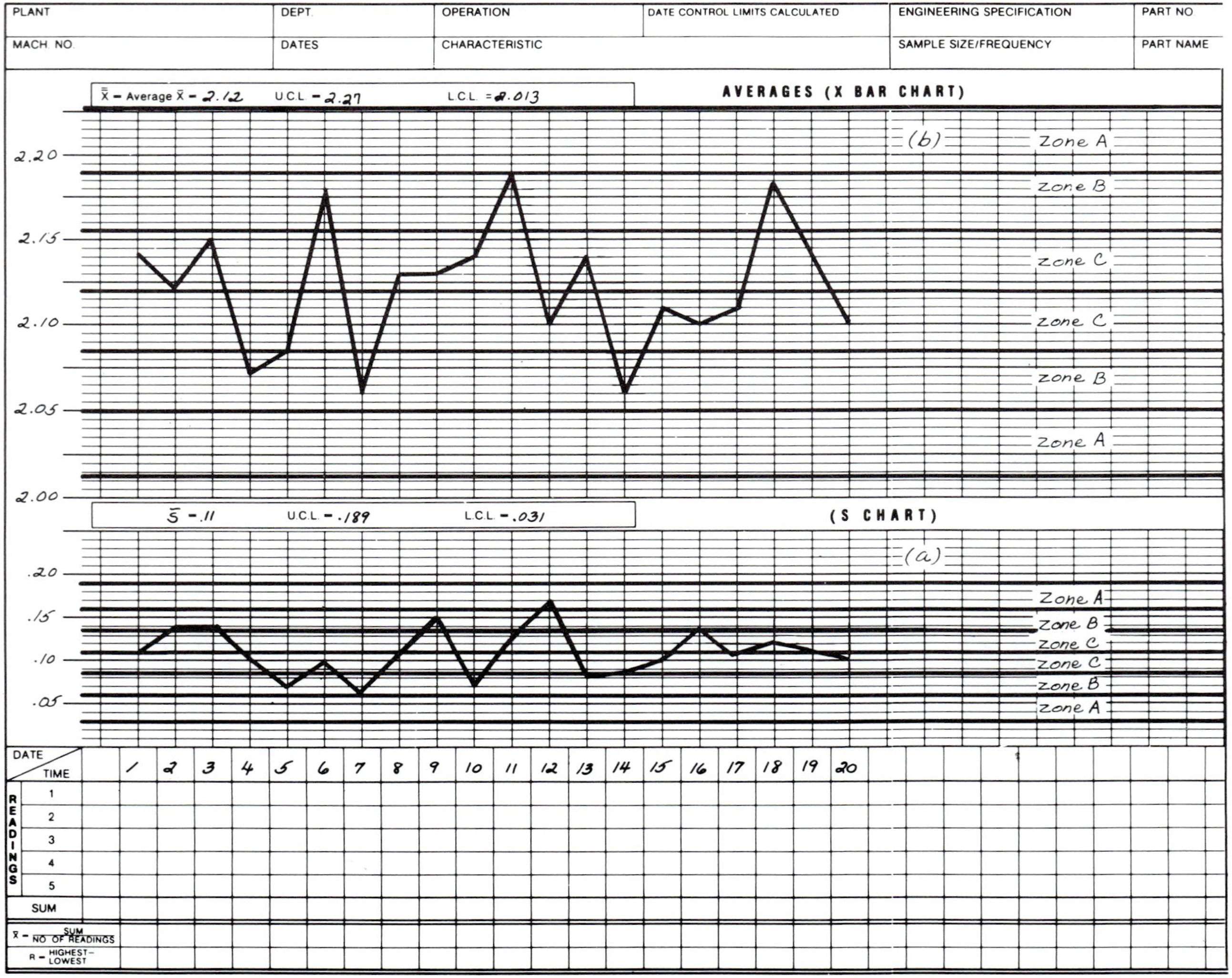

measurements taken on a given shift for each measuring head as a subgroup. This arrangement would yield subgroups with 16 measurements; each subgroup would represent a given head and shift. This subgrouping would address the question of differences between the measurement heads. It would also reveal differences between the shifts. The variation within each subgroup would be variation occurring from that subgroup's measuring head.

Median and Range Charts

Median and range control charts employ the subgroup median, M_e, as a measure of the process location, and the subgroup range, R, as a measure of the process variability. A separate portion of this two-part control chart is constructed with each of these measures in a manner paralleling the construction of the x-bar and R control chart and the x-bar and s control chart.

Conditions for Use

Resistance to control charting may originate in many corners. Difficulty with simple arithmetic is a frequent problem. If proper training hasn't yet been instituted or has been unsuccessful, some employees may be unwilling, unable, or both, to use control charts to do a better job. Median and range charts present an alternative to x-bar and R or x-bar and s charts that may be initially more palatable to some employees. Nevertheless, x-bar and R charts and x-bar and s charts always contain more information from a statistical standpoint and are therefore more powerful than median and range control charts. Median and range control charts should always be replaced by either x-bar and R or x-bar and s control charts as soon as conditions permit.

Control charts for the subgroup medians and ranges involve little arithmetic. Education of those maintaining the charts may be less rigorous, which may permit control charting to be used in circumstances where it might not otherwise be possible. Forms and graphs can be greatly simplified. The control limits to be used can be calculated by individuals other than those lacking the education or experience to do so. The charts can then be used with relatively simple tools for the identification of out-of-control points and/or patterns. Education of the work force is an important commitment on management's part toward never-ending improvement. A logical progression toward more sophisticated methods of quality control can't take place unless all personnel have been properly trained. Once the basic methods have been established with the work force, the transition to x-bar and R charts or x-bar and s charts will be easier and will meet with wider acceptance.

Subgroup Data

Simplicity is often the paramount reason for the use of median and range charts. For this reason, subgroup sizes should generally contain an odd number of elements: three, five, seven, or nine. Even numbers require calculation of the average of the two middle values to form the median, which may defeat the purpose of using this particular chart. Subgroup sizes of more than nine are also usually counterproductive because the size alone may be contrary to the goal of simplicity in the chart.

Because of the circumstances generally encountered, separate data collection forms for each subgroup are recommended for use with median charts. The forms should be clear and simple, with large spaces in which to enter the data. There should be no small lines to confuse the recorder, and the form should be tailored to the specific data being collected. That is, if the subgroup size is five, the form should have exactly five positions for data to be recorded. One form should be used for each subgroup. The form should then have a place for the employee to rank the data and then determine the median and the range for the subgroup. Figure 7.10 illustrates one possible recording chart. After the median and range have been found on the recording chart, they should be entered on the control chart.

Clear instructions and adequate training can't be overstressed. Most commonly, this type of chart is valuable with minimally educated employees. Instructions should be clear; and process improvements resulting from the use of control charts should be highlighted for all those involved to see.

FIGURE 7.10 Median Control Chart Data Recording Sheet

MEDIAN CONTROL CHART DATA RECORDING SHEET

Dept. ____________ Operation: ____________

Date: ________ Time: ________ Recorded by: ____________

Measurements	Ranked measurements	
1: ________	Largest: ________	
2: ________	________	
3: ________	________	← Median
4: ________	________	
5: ________	Smallest: ________	

Largest: ________
(−)
Smallest: ________

Range = ________

Constructing a Median and Range Control Chart

The sampling distribution of the subgroup medians has a mean equal to the average of the subgroup medians, $\overline{M}_e$, and a standard error that can be estimated using a factor, A_6. A_6 is based on the subgroup size, assumes that the process characteristic is stable and normally distributed, and can be found in Table 1.

$$\sigma_{M_e} = (A_6/3)\overline{R} \qquad (7.29)$$

Recall from Chapter 4 that the assumption of normality isn't necessary to interpret the standard error of the subgroup medians; the Empirical Rule may be used instead.

The distribution of the subgroup ranges, discussed earlier in this chapter, has a mean value equal to $\overline{R}$ and a standard error as given in Equation 7.1 of $\sigma_R = d_3(\overline{R}/d_2)$. These values can be used to construct the centerline and control limits for the median and range portions of the control chart.

Median and range control charts should be kept large and relatively easy to read. Here again, the use of small scales and tight lines must be avoided. Chart scales must be large and constructed so that there's no chance that entries will fall outside the range of the entire chart. Scales should appear on both sides of the chart for ease of data entry.

Control limits, in cases where median charts are used, will typically be calculated by someone other than the recorder. We must first compute the average of the subgroup medians and ranges. Just as with the other variables control charts, the range portion must be constructed first so that its stability can be established. The subgroup ranges will, once again, form the basis for the process variability estimate for both the range and median portions. Hence, if the range isn't stable, then control limits for the range and median portions will be unreliable. This may hide points or patterns indicating a lack of control.

Control limits for the range portion are identical to those used in the x-bar and R chart. We use Equation 7.2 for the centerline and Equations 7.3 and 7.4 to determine the upper and lower control limits. If there are no indications of a lack of control, the median portion of the control chart may be constructed. We use Equation 7.30 for the centerline for the control chart and Equations 7.31 and 7.32 to establish the control limits on the subgroup medians:

$$\text{Centerline}(M_e) = \overline{M}_e = \Sigma M_e/k \qquad (7.30)$$

where k is the number of subgroups. Using Equation 7.29 we can create the upper and lower control limits by adding and subtracting three times the standard error from the centerline. Three times the standard error of the subgroup medians is $A_6\overline{R}$ so

$$\text{UCL}(M_e) = \overline{M}_e + A_6\overline{R} \qquad (7.31)$$

$$\text{LCL}(M_e) = \overline{M}_e - A_6\overline{R} \qquad (7.32)$$

Median and Range Charts: An Example

Consider the case of a manufacturer of low-tension ceramic insulators. A mixture of various clays and water is crushed, milled, de-aired, pre-shaped, and then turned on a wheel to achieve the proper final shape. The manufacturer's customers have said that the diameter of the center hole is a critical dimension for the proper ultimate use of the insulator. Consistency in the diameter of the center holes has been identified as an important characteristic. The firm's customers, and hence the manufacturer, want this dimension to be consistently close to nominal with respect to time. To accomplish this goal the manufacturer must identify and remove any special causes of variation and proceed to improve the process by reducing the common variation.

The plant is located in an economically depressed area. The work force is poorly educated but is willing to learn and improve their processes. Median and range control charts are an effective way to introduce control charting in this situation. One particular operator is asked to help construct a control chart for the diameter of the first five ceramic insulator center holes produced in every two-hour period. Figure 7.11 illustrates results of the first 25 of these subgroups.

FIGURE 7.11   Diameters of Insulator Center Holes Measurement (mm)

Sub-group	Time	1	2	3	4	5	Median	Range
1	8:00 AM	9.2	9.6	9.4	9.6	9.0	9.4	0.6
2	10:00	9.8	9.9	9.1	9.5	9.4	9.5	0.8
3	12:00	9.9	9.9	9.4	9.2	9.9	9.9	0.7
4	2:00 PM	9.0	9.5	9.2	9.0	9.8	9.2	0.8
5	4:00	9.1	9.9	9.9	9.7	9.3	9.7	0.8
6	8:00 AM	9.1	9.6	9.1	9.7	9.4	9.4	0.6
7	10:00	9.0	9.8	9.8	9.8	9.3	9.8	0.8
8	12:00	9.9	9.1	9.8	9.2	9.9	9.8	0.8
9	2:00 PM	9.5	9.2	9.6	9.5	9.4	9.5	0.4
10	4:00	9.8	9.9	9.2	9.3	9.8	9.8	0.7
11	8:00 AM	9.7	9.7	9.9	9.4	9.8	9.7	0.5
12	10:00	9.5	9.5	9.4	9.1	9.4	9.4	0.4
13	12:00	9.7	9.2	9.1	9.7	9.1	9.2	0.6
14	2:00 PM	9.8	9.7	9.1	9.1	9.1	9.1	0.7
15	4:00	9.9	9.7	9.5	9.5	9.8	9.7	0.4
16	8:00 AM	9.7	9.4	9.4	9.3	9.1	9.4	0.6
17	10:00	9.0	9.1	9.8	9.8	9.5	9.5	0.8
18	12:00	9.6	9.3	9.0	9.3	9.8	9.3	0.8
19	2:00 PM	9.9	9.9	9.4	9.4	9.1	9.4	0.8
20	4:00	9.8	9.9	9.7	9.3	9.8	9.8	0.6
21	8:00 AM	9.4	9.2	9.8	9.8	9.9	9.8	0.7
22	10:00	9.3	9.3	9.1	9.5	9.6	9.3	0.5
23	12:00	9.0	9.4	9.7	9.5	9.6	9.5	0.7
24	2:00 PM	9.6	9.8	9.5	9.0	9.4	9.5	0.8
25	4:00	9.3	9.2	9.1	9.9	9.9	9.3	0.8
						Totals	237.9	16.7

FIGURE 7.12   Median and Range Chart for Center Hole Diameters

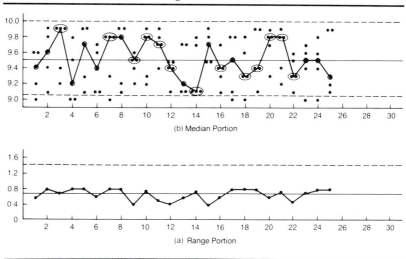

(b) Median Portion

(a) Range Portion

The average of the subgroup medians is found using Equation 7.30:

$$\overline{M}_e = 237.9/25 = 9.52$$

and, using Equation 7.2, we find that the average of the subgroup ranges is

$$\overline{R} = 16.7/25 = 0.67$$

We calculate the control limits for the range portion using Equations 7.3 and 7.4:

$$UCL(R) = (2.114)\ (0.67) = 1.42$$

$$LCL(R) = (0)\ (0.67) = 0.00$$

These are shown in Figure 7.12(a). There are no indications of a lack of control so we may construct the median portion of the control chart.

The control limits for the median portion are calculated using Equations 7.31 and 7.32:

$$UCL(M_e) = 9.52 + (0.691)\ (0.67) = 9.98$$

$$LCL(M_e) = 9.52 - (0.691)\ (0.67) = 9.06$$

These values are shown in Figure 7.12(b).

Most often the median portion of the chart displays all of the individual measurements. Each point is entered as a dot, and the median is circled, as in Figure 7.12(b).

Notice that when more than one data point in a subgroup has the same numerical value, the entry is shown as multiple dots next to each other, enabling the reader to identify all points with the same numerical value in that subgroup.

Additionally, zones A, B, and C haven't been delineated. These should be introduced only when more sophisticated control charts are used.

The subgroups in this example have been selected in sets of five taken every two hours. They will reveal special sources of variation that create patterns in the data between the subgroups, if any exist. Here the control chart will determine whether there are any differences in the subgroups with respect to time.

No points are out of control in the control chart in Figure 7.12, and there are no other indications of special sources of variation. The variation present from measurement to measurement appears to be common process variation. Efforts to reduce this variation should focus on process improvements, using the tools described in Chapter 9.

Individuals Charts

It's not uncommon to encounter a situation where only a single variable value can be periodically observed for control charting. Perhaps measurements must be taken at relatively long intervals, or the measurements are destructive and/or expensive; or perhaps they represent a single batch where only one measurement is appropriate, such as the total yield of a homogeneous chemical batch process. Whatever the case, there are circumstances when data must be taken as individual units that can't conveniently be divided into subgroups.

Individuals charts have two parts: one charting the process variability and the other charting the process average for the single measurements. The two parts are used in tandem much as in the x-bar and R chart. Stability must first be established in the portion charting the variability because the estimate of the process variability provides the basis for the control limits of the portion charting the process average.

Single measurements of variables are considered a subgroup of size one. Hence, there's no variability within the subgroups themselves; and an estimate of the process variability must be made in some other way. An estimate of variability is based on the point-to-point variation in the sequence of single values, measured by the moving range (the absolute value of the difference between each data point and the one that immediately preceded it):

$$R = |x_i - x_{i-1}| \tag{7.33}$$

An average of the moving ranges is used as the centerline for the moving range portion of the chart and as the basis of an estimate of the overall process variation:

$$\text{Centerline(moving range)} = \overline{R} = \Sigma \, R/(k - 1) \tag{7.34}$$

where k is the number of single measurements. As it's impossible to calculate the moving range for the first subgroup because none precedes it, there are only $k - 1$ range measurements; so the sum of the R values is divided by $k - 1$.

As Chapter 5 says, subgroup ranges have an average given by $\overline{R} = d_2\sigma$, and a standard error of $\sigma_R = d_3\sigma$, where σ is the process standard deviation.

The estimate of the process variation is subsequently used to create the 3-sigma control limits for both the moving range portion and the single measurements portion of the control chart.

As we first saw in Chapter 5 and again earlier in this chapter, the control limits for a range are given by

$$\text{UCL(moving range)} = D_4\overline{R} \tag{7.35}$$

$$\text{LCL(moving range)} = D_3\overline{R} \tag{7.36}$$

where D_3 and D_4 depend on subgroup size and are given in Table 1. In this case D_3 is 0.000 and D_4 is 3.267 because the subgroup size for the moving range portion is 2.

For the single measurements portion of the control chart, the centerline is the average of the individual measurements. We find the control limits by adding and subtracting three times the standard deviation of the single measurements, estimated by $\overline{R}/d_2$:

$$\text{Centerline}(x) = \overline{x} = \Sigma\, x/k \tag{7.37}$$

$$\text{UCL}(x) = \overline{x} + 3(\overline{R}/d_2)$$

Using the factor E_2 to represent $3/d_2$, the expression for the upper control limit becomes

$$\text{UCL}(x) = \overline{x} + E_2\overline{R}$$

where E_2 depends on subgroup size and can be found in Table 1. In this case the subgroup size is 2, as we use two observations to calculate each moving range value. Hence, $E_2 = 2.66$, and

$$\text{UCL}(x) = \overline{x} + 2.66\,\overline{R} \tag{7.38}$$

Similarly, the lower control limit is found using

$$\text{LCL}(x) = \overline{x} - 2.66\,\overline{R} \tag{7.39}$$

Constructing an Individuals Chart for Variables Data: An Example

A chemical company produces 2,000-gallon batches of a liquid chemical product, A-744, once every two days. The product is a combination of six raw materials, of which three are liquids and three are powdered solids. Production takes place in a single tank, agitated as the ingredients are added and for several hours thereafter. Shipments of A-744 to the customer are made in bins as single lots when the batches are finished. The chemical company is concerned with the density of the finished product, which it measures in grams per cubic centimeter. As batches are constantly stirred during production, the density is assumed to be relatively uniform throughout each batch. Therefore, management decides that density will be measured by only one reading per batch.

During a 60-day period, 30 batches of A-744 are produced. Figure 7.13 shows the density readings for those batches.

Using Equation 7.33, we calculate the moving range by subtracting the previous observation from each observation, then taking the absolute value. There's

---

FIGURE 7.13   **A-744 Batch Density**

Date	Density	Moving Range	Date	Density	Moving Range
5/6	1.242	—	6/10	1.253	0.018
5/8	1.289	0.047	6/12	1.257	0.004
5/10	1.186	0.103	6/14	1.275	0.018
5/13	1.197	0.011	6/17	1.232	0.043
5/15	1.252	0.055	6/19	1.201	0.031
5/17	1.221	0.031	6/21	1.281	0.080
5/20	1.299	0.078	6/24	1.274	0.007
5/22	1.323	0.024	6/26	1.234	0.040
5/24	1.323	0.000	6/28	1.187	0.047
5/27	1.314	0.009	7/1	1.196	0.009
5/29	1.299	0.015	7/3	1.282	0.086
5/31	1.225	0.074	7/5	1.322	0.040
6/3	1.185	0.040	7/8	1.258	0.064
6/5	1.194	0.009	7/9	1.261	0.003
6/7	1.235	0.041	7/11	1.201	0.060
			Totals	37.498	1.087

---

no moving range for the first observation, so there's one fewer moving range point than data points. The moving ranges are also shown in Figure 7.13.

Using Equation 7.34, the average of the 29 moving range values is

$$\text{Centerline(moving range)} = \overline{R} = 1.087/29 = 0.037$$

The control limits for the moving range portion of the control chart can be found using Equations 7.35 and 7.36:

$$\text{UCL(moving range)} = D_4\,\overline{R} = (3.267)(0.037) = 0.121$$

$$\text{LCL(moving range)} = D_3\,\overline{R} = (0.000)(0.037) = 0.000$$

The control chart for the moving range portion is shown in Figure 7.14(a). The moving range appears to be in a state of statistical control, so it's safe to use the average moving range value to construct the single measurements portion of the chart.

Using Equation 7.37, the centerline for the control chart of the single measurements is

$$\text{Centerline(x)} = \overline{x} = 37.498/30 = 1.250$$

The control limits for the single measurements portion are found using Equations 7.38 and 7.39:

$$\text{UCL(x)} = \overline{x} + 2.66\,\overline{R} = 1.250 + 2.66\,(0.037) = 1.348$$

$$\text{LCL(x)} = \overline{x} - 2.66\,\overline{R} = 1.250 - 2.66\,(0.037) = 1.152$$

Figure 7.14(b) illustrates the control chart for the single measurements portion. The process appears to be in a state of statistical control, for there are no

FIGURE 7.14 Single Measurement and Moving Range Chart for A-744 Density

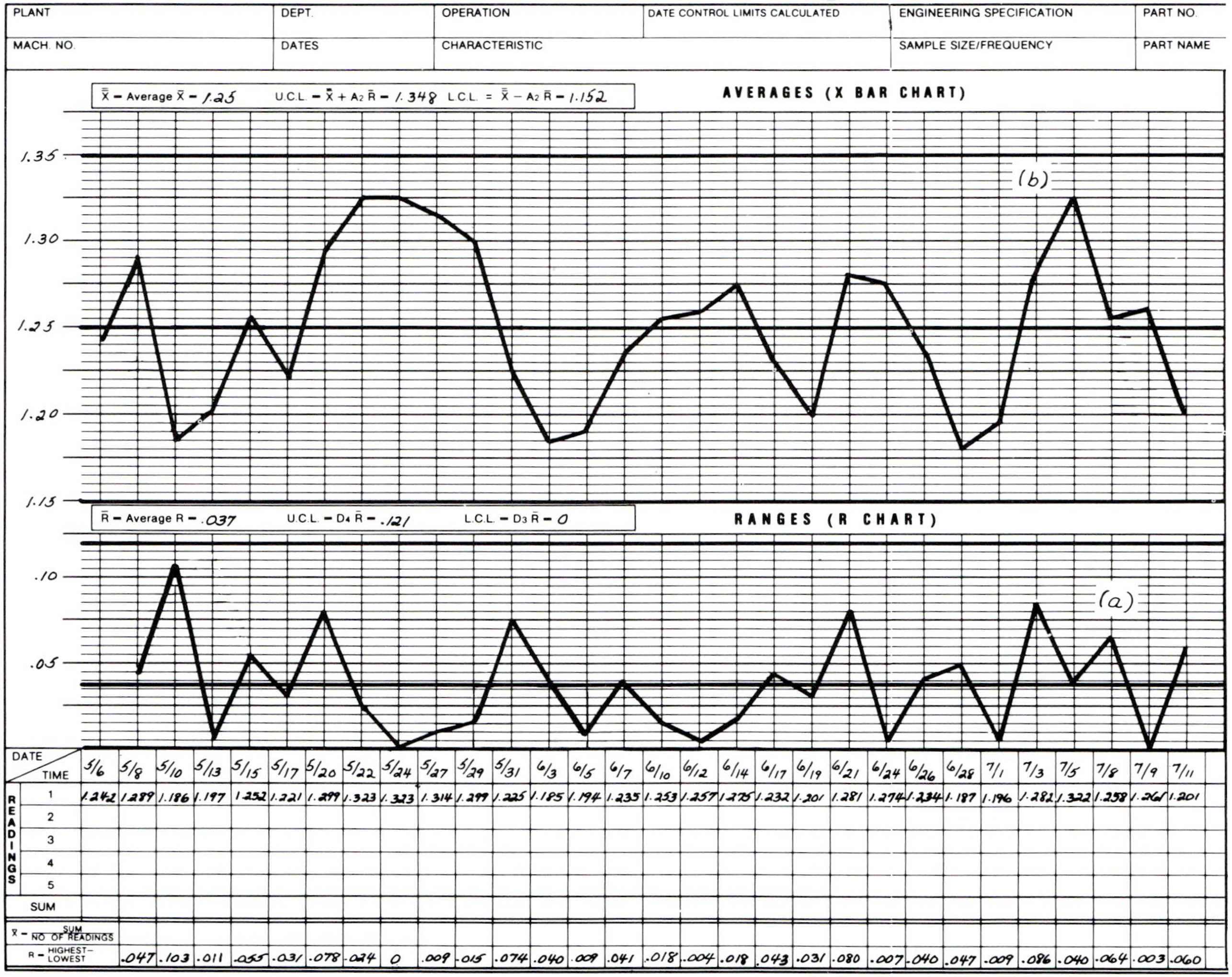
PLANT
DEPT.
OPERATION
DATE CONTROL LIMITS CALCULATED
ENGINEERING SPECIFICATION
PART NO.
MACH. NO.
DATES
CHARACTERISTIC
SAMPLE SIZE/FREQUENCY
PART NAME
X̿ – Average X̄ – 1.25 U.C.L. – X̿ + A2 R̄ – 1.348 L.C.L. = X̿ – A2 R̄ – 1.152
AVERAGES (X BAR CHART)
1.35
1.30
1.25
1.20
1.15
(b)
R̄ – Average R – .037 U.C.L. – D4 R̄ – .121 L.C.L. – D3 R̄ – 0
RANGES (R CHART)
.10
.05
(a)
DATE
TIME
5/6 5/8 5/10 5/13 5/15 5/17 5/20 5/22 5/24 5/27 5/29 5/31 6/3 6/5 6/7 6/10 6/12 6/14 6/17 6/19 6/21 6/24 6/26 6/28 7/1 7/3 7/5 7/8 7/9 7/11
READINGS
1 1.242 1.289 1.186 1.197 1.252 1.221 1.299 1.323 1.323 1.314 1.299 1.225 1.185 1.194 1.235 1.253 1.257 1.275 1.232 1.201 1.281 1.274 1.234 1.187 1.196 1.282 1.322 1.258 1.261 1.201
2
3
4
5
SUM
X̄ – SUM / NO. OF READINGS
R – HIGHEST – LOWEST .047 .103 .011 .055 .031 .078 .024 0 .009 .015 .074 .040 .009 .041 .018 .004 .018 .043 .031 .080 .007 .040 .047 .009 .086 .040 .064 .003 .060

points beyond the control limits and no other signs of any trends or patterns in the data.

Special Characteristics of Individuals Charts

Because each subgroup consists of only one value, and process variation is estimated on the basis of observation-to-observation changes, individuals charts have certain unique characteristics that distinguish them from other control charts. For example, for an individuals chart to be reliable, it's best to have at least 100 subgroups, whereas 25 will suffice for most other control chart forms.

Correlation in the Moving Range.   The moving ranges tend to be correlated. For example, a data point near the centerline followed by one in the upper zone A, followed by one in the lower zone C, will result in two large successive moving range points. As a consequence, large moving range values tend to be followed by other large moving range values, and small moving range values tend to be followed by other small moving range values. Because of this, users must be cautious in applying rules for a lack of control dealing with patterns in the data. For example, using the rule designating the second of two out of three consecutive points in zone A or beyond as indicating a lack of control may generate false alarms. For this reason, it's usually best to be conservative when applying the rules concerning patterns in the data other than points beyond the control limits that indicate a lack of control in moving range charts. For example, instead of 8 consecutive values above or below the centerline indicating a lack of control, we might require 10 or 12. Knowledge and experience are the best guides in establishing policy in this case.

Inflated Control Limits.   The control limits for individuals charts represent process limits because they're measurements on individual items. Hence, they're limits within which the single measurements will fall, provided that the variation in the process over the long run is approximately the same as the variation in the short run. Because the process variation is estimated using the moving range, the short-run (between consecutive subgroups) variation in the process provides the estimate of the process variation. Changes in the short-run or long-run variation may produce unreliable control limits.

One indication that the control limits are unreliable is the occurrence of at least two-thirds of the data points below the centerline of the moving range portion of the control chart.[2] When this happens, control limits should be based on the median of the moving range values, M_{e_R}, rather than those based on the average of the moving range values. If inflated control limits are suspected, control limits based on the median of the moving ranges should be calculated and compared to those based on the average moving range; the narrower of the two sets should be used. Control limits for the moving range portion, based on the median, can be computed using

$$\text{Centerline(median moving range)} = M_{e_R} \qquad (7.40)$$

$$\text{UCL(median moving range)} = D_6 M_{e_R}$$

$$\text{LCL(median moving range)} = D_5 M_{e_R}$$

where values for D_5 and D_6 depend on subgroup size, assume stability and a normal distribution of the process characteristic, and can be found in Table 1. For the individuals chart the subgroup size is two, so the values used are $D_6 = 3.865$ and $D_5 = 0.000$. Hence,

$$\text{UCL(median moving range)} = 3.865 \, M_{e_R} \qquad (7.41)$$

$$\text{LCL(median moving range)} = 0.000 \qquad (7.42)$$

Again, recall that the assumption of normality isn't required to interpret the control limits; the Empirical Rule may be used instead.

The standard error of the single measurements is estimated using M_{e_R}/d_4, where values for d_4 depend on subgroup size and assume stability of the process characteristic. They can be found in Table 1. For subgroup size two, $d_4 = 0.954$. Hence, control limits for the single measurements portion are created by adding and subtracting three times $M_{e_R}/0.954$ from the centerline:

$$\text{UCL(x)} = \bar{x} + 3M_{e_R}/0.954 = \bar{x} + 3.145M_{e_R} \qquad (7.43)$$

Similarly,

$$\text{LCL(x)} = \bar{x} - 3.145M_{e_R} \qquad (7.44)$$

Individuals Charts with Inflated Control Limits: An Example.   Consider the case of the manufacturer of chemicals discussed earlier. The chemical, A-744, is used by the manufacturer's customer as an ingredient in another process sensitive to the quantity of A-744. The manufacturer's customer is seeking to reduce costs by using the A-744 in whole bin lots. As the yield of the batches of A-744 can be expected to vary from its 2,000-gallon target, that yield is a likely candidate for the use of an individuals control chart. The yields of the 30 batches of A-744 are each carefully measured, yielding a sequence of 30 single values. The data and the computed moving ranges are shown in Figure 7.15.

The centerline and control limits for the moving range portion of the control chart are found using Equations 7.34 through 7.36:

$$\bar{R} = 308.5/29 = 10.64$$

$$\text{UCL(moving range)} = D_4 \, \bar{R} = (3.267)(10.64) = 34.76$$

$$\text{LCL(moving range)} = 0.00$$

The control chart for the moving range portion is shown in Figure 7.16(a). The moving range appears stable, so the average moving range value can be used to construct the single measurements portion of the chart.

Using Equation 7.37, the average of the 30 yields is

$$\bar{x} = 60,049.0/30 = 2001.63$$

The control limits for the single measurements portion are found using Equations 7.38 and 7.39:

$$\text{UCL(x)} = \bar{x} + 2.66 \, \bar{R} = 2001.63 + (2.66)(10.64) = 2029.93$$

$$\text{LCL(x)} = \bar{x} - 2.66 \, \bar{R} = 2001.63 - (2.66)(10.64) = 1973.33$$

FIGURE 7.15 **A-744 Batch Yields**

Date	Yield	Moving Range	Date	Yield	Moving Range
5/6	1989.0	—	6/10	2002.3	4.9
5/8	1998.9	9.9	6/12	1999.5	2.8
5/10	2027.4	28.5	6/14	2000.8	1.3
5/13	2001.5	25.9	6/17	2022.4	21.6
5/15	1991.3	10.2	6/19	1998.3	24.1
5/17	2001.3	10.0	6/21	1999.8	1.5
5/20	1997.4	3.9	6/24	2000.9	1.1
5/22	1989.3	8.1	6/26	1994.3	6.6
5/24	1995.5	6.2	6/28	1998.7	4.4
5/27	2014.4	18.9	7/1	2013.5	14.8
5/29	1990.2	24.2	7/3	1998.1	15.4
5/31	1999.6	9.4	7/5	2002.5	4.4
6/3	2008.1	8.5	7/8	2000.2	2.3
6/5	1999.4	8.7	7/9	1996.1	4.1
6/7	1997.4	2.0	7/11	2020.9	24.8
			Totals	60049.0	308.5

The control chart for the single measurements portion is shown in Figure 7.16(b). This portion also appears to be in a state of statistical control. However, an experienced eye detects that more than two-thirds of the data points (20 of the 29 moving ranges) are below the centerline, indicating that the control limits may be artificially inflated and therefore may be hiding indications of special sources of variation.

The median of the 29 moving range values is 8.5. This value can be used to calculate an alternate centerline and set of control limits using Equations 7.40 through 7.42 for the moving range portion of the control chart:

$$\text{Centerline(median moving range)} = 8.5$$

$$\text{UCL(median moving range)} = (3.865)(8.5) = 32.85$$

$$\text{LCL(median moving range)} = 0.00$$

Figure 7.17(a) shows the new control chart for the moving range portion. The process still appears in a state of statistical control, so the median of the moving ranges may be used to construct a new single measurements portion of the chart.

Notice that it is essential to establish stability in the portion of the control chart dealing with variability (the moving range portion) before constructing the portion of the control chart dealing with the single measurements. This is because the control limits for the single measurements portion are based on the estimate of the process variability, generated by the moving range portion. A lack of stability in the moving range portion will produce unreliable estimates of the process variation, resulting in the control chart's failure to properly separate special and common variation.

FIGURE 7.16 Single Measurements and Moving Range Chart for A-744 Batch Yields

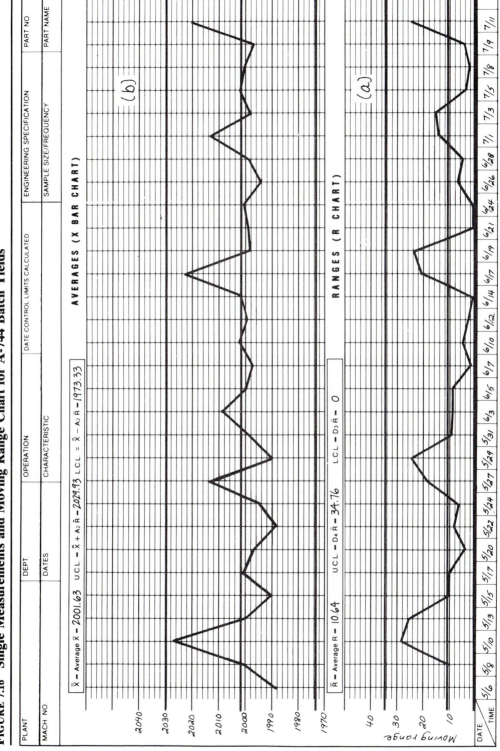

FIGURE 7.17 Single Measurements and Moving Range Chart (Based on Median Moving Range Value) for A-744 Batch Yields

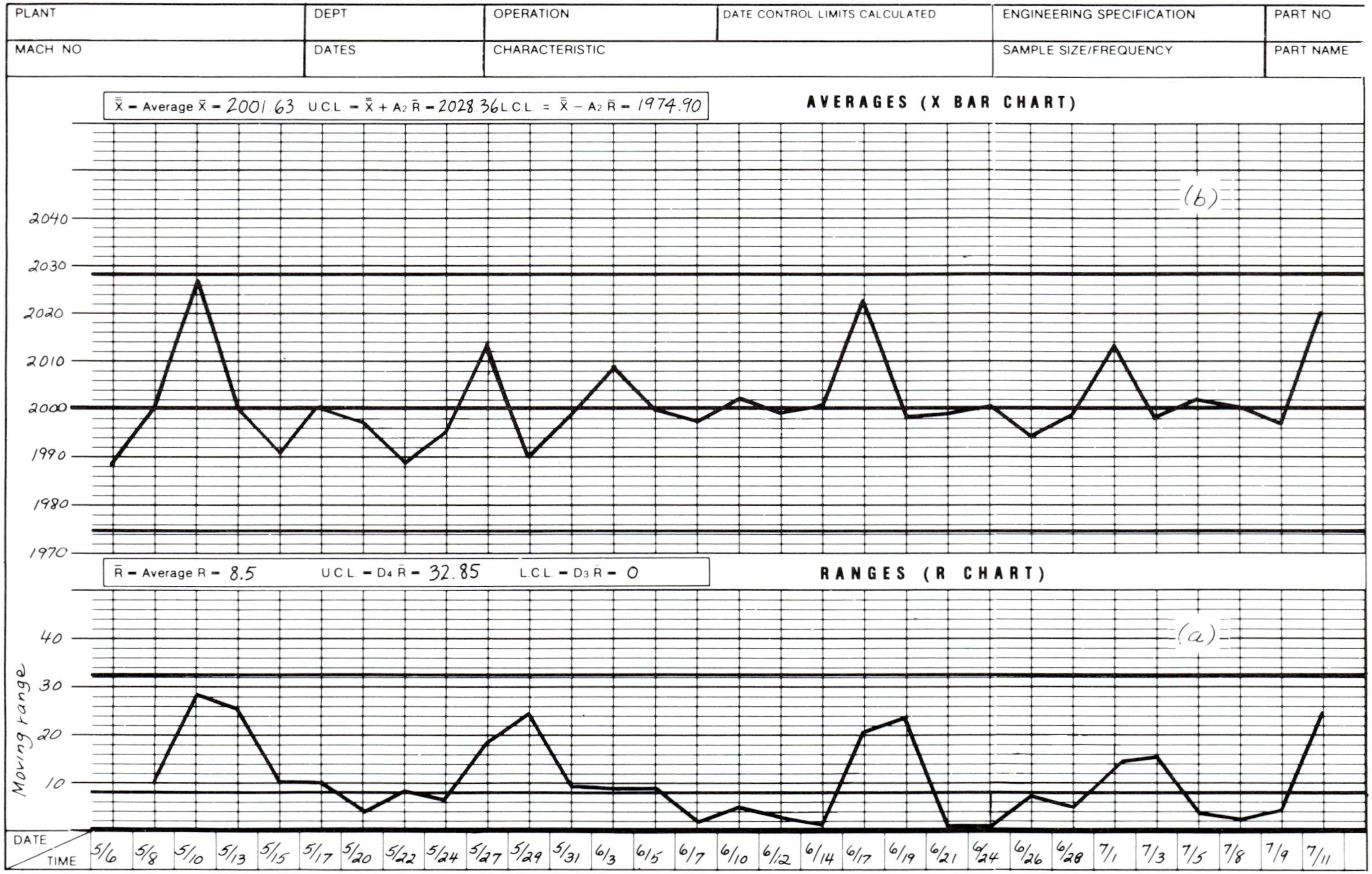

The centerline remains at the average value, 2001.63, and the control limits for the single measurements portion are calculated using Equations 7.43 and 7.44:

$$Centerline(x) = 2001.63$$

$$UCL(x) = 2001.63 + (3.145)(8.5) = 2028.36$$

$$LCL(x) = 2001.63 - (3.145)(8.5) = 1974.90$$

The control chart is shown in Figure 7.17(b). There are no indications of a lack of control. Because the control limits based on the median range are narrower, they are used in this case.

Revising Control Limits for Variables Control Charts

Frequent regular revision of control limits is undesirable and inappropriate. Control limits should be revised only for one of three reasons: a change in the process; when trial control limits have been used and are to be replaced with regular control limits; and when points out of control have been removed from a data set.

Change in Process

Processes change for many reasons. For example, such things as technical improvements, new vendors, new machines, new machine settings, new operational definitions, or new operator instructions may induce process changes. Efforts toward the never-ending reduction of variation may precipitate the change. Whatever the cause, changes in the process itself change the variability and therefore necessitate recalculation of the control limits.

Trial Control Limits

When control charting is initiated for a process (for either a brand new process or an old process that's being charted for the first time), trial control limits are sometimes calculated from the first few subgroups. After about 25 or 30 subgroups become available, these trial limits should be replaced with regular control limits.

Removal of Out-of-Control Points

When out-of-control points used in the calculation of the control limits have been removed from a data set, the control limits must be recalculated. As the removed data values were used to calculate the process mean, range, standard deviation, or other statistics, the removal of these data points will precipitate changes in the centerline, control limits, and zone boundaries. A new centerline and new control limits and zone boundaries must be calculated. As we have seen, this may occasionally reveal other points that are out of control and will help to further identify areas requiring some special action or changes in the process.

Summary

Variables control charts are an important tool for process improvement. Variables control charts not only identify and differentiate between special and common causes of variation but also provide the data essential for process improvement.

x-bar and R charts use subgroups of from 2 to 10 items. When subgroup sizes are larger than 10, x-bar and s charts are used in place of x-bar and R charts. x-bar and R charts and x-bar and s charts are important weapons in the war against variability. Variation of any kind must be continuously reduced if a firm is to battle its way upward on the spiral of never-ending process improvement.

Median and range control charts can serve as a means to introduce control charts to workers with little education. Even though they don't represent a long-term solution approach, these charts can provide a needed introduction when other control charts are too complex.

Individuals charts are useful when, because of high cost or a lack of available data gathering opportunities, only one measurement is possible in a subgroup.

Exercises

7.1 A large hotel in a resort area has a housekeeping staff who clean and prepare all of the hotel's guest rooms daily. In an effort to improve service through reducing variation in the time required to clean and prepare a room, a series of measurements is taken of the times to service rooms in one section of the hotel. Here are the cleaning times for five rooms for 25 consecutive days.

Cleaning and Preparation Time

Room	Day 1	Day 2	Day 3	Day 4	Day 5	Day 6	Day 7	Day 8	Day 9	Day 10
1	15.6	15.0	16.4	14.2	16.4	14.9	17.9	14.0	17.6	14.6
2	14.3	14.8	15.1	14.8	16.3	17.2	17.9	17.7	16.5	14.0
3	17.7	16.8	15.7	17.3	17.6	17.2	14.7	16.9	15.3	14.7
4	14.3	16.9	17.3	15.0	17.9	15.3	17.0	14.0	14.5	16.9
5	15.0	17.4	16.6	16.4	14.9	14.1	14.5	14.9	15.1	14.2

Room	Day 11	Day 12	Day 13	Day 14	Day 15	Day 16	Day 17	Day 18	Day 19	Day 20
1	14.6	15.3	17.4	15.3	14.8	16.1	14.2	14.6	15.9	16.2
2	15.5	15.3	14.9	16.9	15.1	14.6	14.7	17.2	16.5	14.8
3	15.9	15.9	17.7	17.9	16.6	17.5	15.3	16.0	16.1	14.8
4	14.8	15.0	16.6	17.2	16.3	16.9	15.7	16.7	15.0	15.0
5	14.2	17.8	14.7	17.5	14.5	17.7	14.3	16.3	17.8	15.3

Room	Day 21	Day 22	Day 23	Day 24	Day 25
1	16.3	15.0	16.4	16.6	17.0
2	15.3	17.6	15.9	15.1	17.5
3	14.0	14.5	16.7	14.1	17.4
4	17.4	17.5	15.7	17.4	16.2
5	14.5	17.8	16.9	17.8	17.9

a. For each of the 25 subgroups, compute the mean and range.
b. Find the centerline for both the x-bar and range portions of the control chart.
c. Estimate the standard deviation for the time to clean and prepare a room.
d. Find the values for the upper and lower control limits for both portions of the control chart.
e. Find the zone boundaries for all of the zones.
f. Determine whether there are any indications of a lack of control.

7.2 A parcel-sorting facility's management wants to know how much time is needed to sort units in what's termed the primary sort. The variable to be measured is the number of units sorted in a one-minute interval by a given team. Each hour, the first five one-minute intervals are selected. Here are the results.

Number of Units Sorted

Subgroup Number	Observation				
	1	*2*	*3*	*4*	*5*
1	474	386	528	333	465
2	688	367	691	602	569
3	427	500	279	479	721
4	846	506	420	509	409
5	384	365	884	521	562
6	774	889	326	581	270
7	728	902	809	468	318
8	682	614	315	688	594
9	401	418	429	722	781
10	613	394	421	684	675
11	874	546	804	709	469
12	409	591	665	685	540
13	748	818	694	481	290
14	790	796	705	503	710
15	227	409	621	495	891
16	588	789	344	785	724
17	725	802	336	645	782
18	671	691	735	351	853
19	778	795	462	301	549
20	366	691	644	547	705

a. For each of the 20 subgroups, compute the mean and the range.
b. Find the centerline for both the x-bar and range portions of the control chart.
c. Construct the control chart and determine if there are any indications of a lack of control.

7.3 The drive-up window at a local bank is searching for ways to improve service. One teller keeps a control chart for the service time in minutes for the first four customers driving up to her window each hour for a three-day period. Here are the results of her data collection.

Drive-Up Teller Service Times

| | | | | Time | | | | |
Cust.	9 AM	10 AM	11 AM	12	1 PM	2 PM	3 PM	4 PM
1	1.4	3.8	3.6	4.3	4.0	1.3	0.9	4.7
2	2.3	5.2	2.5	1.2	5.2	1.1	4.4	5.1
3	1.9	1.9	0.8	3.0	2.7	4.9	5.1	0.9
4	5.1	4.8	2.9	1.5	0.3	2.3	4.6	4.7

| | | | | Time | | | | |
Cust.	9 AM	10 AM	11 AM	12	1 PM	2 PM	3 PM	4 PM
1	2.8	0.5	4.5	0.6	4.8	2.7	4.2	0.9
2	3.0	2.7	1.9	1.2	2.8	2.0	1.1	4.4
3	4.1	4.7	4.2	2.7	1.1	2.6	4.4	0.6
4	4.8	3.6	0.4	2.5	0.4	2.6	3.1	0.4

| | | | | Time | | | | |
Cust	9 AM	10 AM	11 AM	12	1 PM	2 PM	3 PM	4 PM
1	0.3	3.5	5.2	2.9	3.3	4.0	2.8	0.6
2	2.4	3.4	0.3	1.9	3.7	3.3	0.7	2.1
3	5.0	4.6	2.4	0.8	3.8	5.0	1.6	3.3
4	0.9	3.3	3.9	0.3	2.1	2.8	4.6	2.7

a. For each of the 24 subgroups, compute the mean and the range.
b. Find the centerline for both the x-bar and range portions of the control chart.
c. Estimate the standard deviation for the time to service a customer at the drive-up window.
d. Find the values for the upper and lower control limits for both portions of the control chart.
e. Find the zone boundaries for all zones.
f. Determine whether there are any indications of a lack of control.

7.4 A paper products manufacturer coats one particular paper product with wax. In an effort to control and stabilize the coating process, the employee running the coating machine takes six measurements of coating thickness every 15 minutes during a one-day study period. Here are the results of the data collection.

Coating Thickness

	Time								
	8 AM	*8:15*	*8:30*	*8:45*	*9 AM*	*9:15*	*9:30*	*9:45*	*10 AM*
x̄	2.90	3.30	2.87	0.93	1.43	1.93	2.90	3.03	3.70
R	0.35	0.50	0.32	0.44	0.91	0.42	0.29	0.78	1.01
	Time								
	10:15	*10:30*	*10:45*	*11 AM*	*11:15*	*11:30*	*11:45*	*12 NOON*	*12:15*
x̄	3.70	3.00	3.20	2.80	2.90	3.53	3.90	3.73	3.23
R	1.09	0.95	0.56	0.42	0.55	0.16	1.04	0.90	0.86
	Time								
	12:30	*12:45*	*1 PM*	*1:15*	*1:30*	*1:45*	*2 PM*	*2:15*	*2:30*
x̄	2.73	3.00	3.33	2.50	2.13	1.97	2.70	3.10	2.63
R	0.19	0.84	0.20	0.71	0.90	0.92	1.00	0.56	0.88
	Time								
	2:45	*3 PM*	*3:15*	*3:30*	*3:45*	*4 PM*			
x̄	2.63	3.40	3.20	2.20	2.03	3.23			
R	0.89	1.06	0.66	0.93	0.59	0.91			

a. Find the centerline for both the x-bar and range portions of the control chart.

b. Find the values for the upper and lower control limits for both portions of the control chart.

c. Find the zone boundaries for the upper and lower A, B, and C zones.

d. Determine whether there are any indications of a lack of control.

7.5 Groups of customers arriving at a restaurant must wait to be seated by a hostess. Waiting time is short if a table is available, but it's long if the restaurant is crowded. The hostess records waiting times for the first eight people arriving from 7:00 PM each Friday, Saturday, and Sunday evening for a series of 10 consecutive weeks. She then computes the average and range for each subgroup. Here are the results of the data collection and calculations.

Waiting Times (minutes)

	Date									
	9/2	*9/3*	*9/4*	*9/9*	*9/10*	*9/11*	*9/16*	*9/17*	*9/18*	*9/23*
$\bar{x}$	15.5	23.0	18.0	11.5	12.0	12.3	15.5	13.5	21.5	12.2
R	5.1	6.2	2.6	2.4	4.8	2.6	2.8	4.0	4.8	1.8

	Date									
	9/24	*9/25*	*9/30*	*10/1*	*10/2*	*10/7*	*10/8*	*10/9*	*10/14*	*10/15*
$\bar{x}$	21.5	28.0	16.0	17.0	31.0	21.1	15.0	17.5	17.5	18.0
R	4.0	3.6	3.0	4.4	7.2	2.1	3.6	1.8	1.8	3.6

	Date									
	10/16	*10/21*	*10/22*	*10/23*	*10/28*	*10/29*	*10/30*	*11/4*	*11/5*	*11/6*
$\bar{x}$	17.5	20.0	26.0	17.0	19.0	20.0	16.0	18.0	18.5	19.5
R	3.6	4.8	5.6	3.4	3.4	3.6	3.2	3.8	3.4	3.6

a. Find the centerline for both the x-bar and range portions of the control chart.
b. Find the values for the upper and lower control limits for both portions of the control chart.
c. Find the zone boundaries for the upper and lower A, B, and C zones.
d. Determine whether there are any indications of a lack of control.

7.6 A hospital administrator is studying how long an emergency room patient waits to see a physician during the midnight to 8:00 AM shift. The study is limited to those patients who actually do see a physician, and the amount of waiting time has been carefully, operationally defined. Each day, the first 15 records are studied. Here are the results.

Waiting Time (minutes)

Day 1	Day 2	Day 3	Day 4	Day 5	Day 6	Day 7	Day 8
2	26	19	24	40	31	39	16
32	2	18	33	22	13	41	24
8	40	15	46	23	15	8	27
30	17	18	20	40	4	37	17
38	12	18	32	34	35	17	13
24	14	18	5	7	40	48	30
31	9	44	48	23	41	36	44
46	45	3	20	37	39	32	23
49	13	28	39	31	31	40	50
32	43	47	2	48	17	30	41
32	42	44	16	23	18	38	18
10	33	3	12	45	32	22	14
43	7	13	36	15	8	1	30
27	4	37	47	43	30	33	9
41	34	5	40	5	28	13	44

a. For each of the eight subgroups, compute the mean and the standard deviation.

b. Find the centerline for both the x-bar and s portions of the control chart.

c. Construct the control chart and determine if there are any indications of a lack of control.

7.7 The food services director of an airline wants to measure the weight of food left over on passenger food trays to measure the difference between product performance and customer expectations. Each day, 12 of the trays returned in the main cabin on a particular flight are set aside, placed into a special container, and sent to a lab for weighing. Here are the average and standard deviation of each of these subgroups over a period of 25 days.

Weight of Leftovers (grams)

					Day					
	1	2	3	4	5	6	7	8	9	10
$\bar{x}$	44.5	61.5	66.6	64.9	22.9	12.7	44.2	10.4	82.6	55.2
s	23.5	11.2	28.6	31.9	13.3	12.5	32.8	35.1	39.1	10.9

					Day					
	11	12	13	14	15	16	17	18	19	20
$\bar{x}$	92.9	25.5	92.9	27.5	84.8	82.2	24.7	25.8	79.7	19.5
s	35.5	38.3	27.5	21.3	22.5	35.2	20.7	19.4	14.7	23.5

		Day			
	21	22	23	24	25
$\bar{x}$	10.9	47.5	72.2	28.6	90.0
s	36.7	28.6	25.9	22.7	23.4

    a. Find the centerline for both the x-bar and s portions of the control chart.

    b. Find the values for the upper and lower control limits for both portions of the control chart.

    c. Find the zone boundaries for the upper and lower A, B, and C zones.

    d. Determine whether there are any indications of a lack of control.

7.8 A hospital is studying the amount of time patients spend in their routine admitting procedure. Samples of 12 are selected each day for a 20-day period. Admitting time (which has been carefully operationally defined) is measured in seconds. Here are the subgroup results.

Admitting Time (seconds)

Day 1	Day 2	Day 3	Day 4	Day 5	Day 6	Day 7	Day 8	Day 9	Day 10
362	611	320	621	680	759	372	370	530	494
468	873	944	927	794	665	835	294	881	914
553	768	593	948	650	730	884	480	943	870
390	807	857	817	780	930	930	558	383	272
460	476	710	641	442	369	667	502	316	662
910	816	724	764	372	635	747	595	611	348
707	567	545	986	627	313	390	847	778	447
829	833	526	430	882	843	644	544	531	306
955	521	348	743	756	264	339	853	896	751
705	959	456	451	548	663	664	876	772	415
884	315	576	645	767	991	245	744	719	717
904	414	855	996	745	431	893	816	670	387

Day 11	Day 12	Day 13	Day 14	Day 15	Day 16	Day 17	Day 18	Day 19	Day 20
659	274	797	678	997	242	594	368	806	497
919	754	253	679	205	474	817	850	575	785
603	428	829	351	893	966	381	510	348	806
897	811	857	663	734	823	462	688	298	263
319	916	898	638	474	515	429	201	487	435
499	332	387	928	631	617	786	795	697	337
799	765	918	258	746	894	901	977	249	659
482	961	900	338	642	519	278	715	668	537
615	437	691	446	484	636	472	253	533	786
497	692	600	936	525	547	885	310	985	607
430	380	450	584	685	993	991	412	284	466
765	566	775	535	358	858	557	813	707	564

    a. For each of the 20 subgroups, compute the mean and standard deviation.

    b. Find the centerline for both the x-bar and s portions of the control chart.

    c. Find the values for the upper and lower control limits for both portions of the control chart.

d. Find the zone boundaries for both portions of the control chart.

e. Determine whether there are any indications of a lack of control.

7.9 A tomato farmer has been using an x-bar and R chart to control the yield per plant on his highly mechanized farm. The yield is stable over time in both variability and location. The x-bar portion of the control chart has a centerline value of 10.50 pounds, and the upper control limit is 14.20 pounds.

a. Find the value of the lower control limit.

b. Find the value of the upper and lower A, B, and C zone boundaries.

c. Estimate the value of the standard deviation of the yield of the tomato plants if the subgroup size is six.

7.10 A manufacturer of a special chemical fertilizer is concerned with the pH of finished batches of product. A series of careful measurements reveals:

Batch	pH	Batch	pH	Batch	pH	Batch	pH
1	6.5	26	6.1	51	6.6	76	6.4
2	6.2	27	6.8	52	6.1	77	6.7
3	6.7	28	6.6	53	6.3	78	6.4
4	6.7	29	6.4	54	6.8	79	6.2
5	6.2	30	6.0	55	6.3	80	6.1
6	6.1	31	6.5	56	6.7	81	6.4
7	6.8	32	6.0	57	6.1	82	6.2
8	6.4	33	6.5	58	6.6	83	6.7
9	6.0	34	6.8	59	6.6	84	6.5
10	6.6	35	6.2	60	6.6	85	6.8
11	6.7	36	6.0	61	6.7	86	6.2
12	6.9	37	6.6	62	6.2	87	6.7
13	6.2	38	6.3	63	6.5	88	6.5
14	6.9	39	6.2	64	6.0	89	6.1
15	6.1	40	6.7	65	6.8	90	6.5
16	6.5	41	6.2	66	6.2	91	6.7
17	6.3	42	6.4	67	6.0	92	6.0
18	6.2	43	6.3	68	6.1	93	6.0
19	6.9	44	6.7	69	6.8	94	6.6
20	6.6	45	6.1	70	6.7	95	6.2
21	6.7	46	6.3	71	6.8	96	6.4
22	6.6	47	6.0	72	6.3	97	6.1
23	6.3	48	6.6	73	6.6	98	6.5
24	6.6	49	6.8	74	6.3	99	6.1
25	6.2	50	6.5	75	6.6	100	6.9

a. Find the moving range at each observation.

b. Construct the range portion of the control chart.

c. Are there any indications of a lack of statistical control?

d. Construct the single measurements portion of the control chart.

e. Are there any indications of a lack of control in this portion?

f. Do the control limits appear inflated? Explain.

7.11 A packaging operation is manned by unskilled and relatively uneducated workers. The workers' task is to shovel 30 kilograms of a granular product from a large pile into sacks that are then sealed and placed on pallets for shipping. The scale used by the workers is accurate, and an effort has been made to educate the workers about the need to measure each weight carefully. Given the conditions, initial efforts at process control will utilize a median and range control chart. A sequence of 25 subgroups, each consisting of the weights of five sacks, has been recorded.

Weight (kilograms)

Subgroup Number	Observation				
	1	*2*	*3*	*4*	*5*
1	35.4	35.6	34.8	34.7	34.8
2	36.0	35.6	34.9	34.8	35.9
3	35.2	35.0	35.0	35.4	35.1
4	34.8	35.8	35.2	35.0	34.9
5	34.2	35.0	36.1	34.9	35.1
6	36.0	35.0	35.2	34.8	34.9
7	36.1	34.9	34.5	35.0	35.1
8	35.1	35.0	35.6	34.9	36.2
9	35.0	35.6	36.1	34.8	35.6
10	35.4	35.8	36.0	34.2	36.0
11	35.2	35.3	35.2	35.9	34.8
12	35.9	36.0	35.1	35.1	35.6
13	35.2	35.6	35.0	34.9	35.0
14	35.2	35.6	35.8	35.0	35.1
15	34.9	34.8	35.0	35.2	34.9
16	35.2	35.3	35.2	35.6	35.1
17	35.6	35.8	35.2	35.4	34.9
18	35.2	35.6	35.4	35.6	35.2
19	34.7	34.9	35.6	35.2	35.0
20	35.0	35.1	35.6	35.0	35.1
21	35.6	35.0	35.8	35.2	34.6
22	34.9	35.1	35.6	35.0	35.2
23	35.9	36.0	35.2	36.0	35.2
24	35.2	35.1	35.4	34.9	35.1
25	35.2	35.2	35.1	35.6	34.9

a. Find the median and range for each subgroup.
b. Construct the range portion of the control chart.
c. Are there any indications of a lack of control in the range portion?
d. Construct the median portion of the control chart.
e. Are there any indications of a lack of control in the median portion?

Endnotes

1. W. A. Shewhart, *Economic Control of Quality of Manufactured Product* (New York: D. Van Nostrand, 1931), p. 314.
2. Donald J. Wheeler and David S. Chambers, *Understanding Statistical Process Control* (Knoxville, Tenn.: Statistical Process Controls, 1986), p. 79.

Out-of-Control Patterns

Types of Variation That Create Control Chart Patterns

The purpose of a control chart is to detect special (assignable or exogenous) causes of variation, or special disturbances. Ideally, once detected, special causes can be resolved, leaving a process with only. common (or endogenous) causes of variation. Common sources of variation are endogenous to the system and are not disturbances—they are the system.

Between and Within Group Variation

There are two types of special causes of variation: periodic disturbances and persistent disturbances. Periodic disturbances create special causes of variation that intermittently affect a process. The intermittent nature of these causes tends to affect sampled observations separated in time and, hence, in different subgroups. This is called *between group variation*. The effect of between group variation is to create control chart patterns in which subgroup statistics are beyond the control limits; in other words, its effect is to create control limits too narrow for the subgroup statistics.

Examples of causes of variation that could generate between group variation include

- Chaotic (unstable) functioning of automatic control devices.
- Operator carelessness in setting up machine runs (for example, timing or heat).
- Loose and wobbly braces for holding in place material to be worked on (for example, drilling, cutting, or sanding).
- Overadjustment of a machine.
- Changes in personnel (for example, changes in shift).

Between group variation is demonstrated in the following example. One component part of a certain machine tool is an eccentric cam in which a slot ½ inch deep is milled. The milling machine operator places five cams in the jig, tightens them in with an adjusting screw, and cuts five slots simultaneously.

During a study of this milling operation, it becomes apparent that the slots aren't being held within tolerances. Both foreman and operator complain that the milling cutter has to be changed too often and that the slot goes out of tolerance before the cutter needs resharpening. A control chart analysis is made in an attempt to reduce the downtime required for changing cutters and to increase the operation's productivity.

As five slots are cut at one time, it seems logical to use groups of five simultaneous slots as the subgroup for the control chart. Measurements of slot depth therefore are made on each of five simultaneous slots. About 30 minutes elapse between inspections.

Figure 8.1 shows the x-bar chart for 18 samples of 5 each. The range portion of the chart (not shown) exhibited no indications of a lack of control. Either the process is very erratic or the subgrouping is incorrect. A classification of the possible sources of special variation in the operation reveals

Raw materials: variability in hardness of steel.

Dimensions of cam: variability from preceding operations.

Positioning of the milling jig: variability resulting from operator skill.

Wear in the cutting tool: variability caused by slot's growing shallower as cutter dulls.

As far as the chosen method of subgrouping is concerned, the cams are thoroughly scrambled or randomized before coming to the mills. The first and second sources are therefore included in the subgroups because a cam of any specified hardness or size would be as likely to appear in one subgroup as another. But the third and fourth causes aren't included in the subgroups; their effect is *between* the subgroups. Positioning of the jig affects all five cams in the jig in the same way, but the next group of five might be positioned differently. Tool wear, a

FIGURE 8.1   x-Bar Chart for Milling Slots in Cams

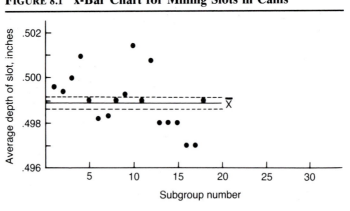

long-term directional effect, should appear as a trend between successive subgroups and wouldn't affect the range within subgroups. Regarding the third cause, positioning of the jig, some variation between successive jig settings is unavoidable, for jig setting depends on the operator's manual skill. Certainly excessive variations from this cause are undesirable.[1]

Persistent disturbances create special causes of variation that continually affect the process. The constant nature of these causes tends to affect all sampled observations and, hence, sampled items both within and between subgroups. Called *within group variation*, this is the most difficult type of variation to identify and interpret. Within group variation creates control chart patterns in which subgroup statistics hug the centerline—in other words, it creates control limits too wide for the subgroup statistics.

Causes of variation that could generate within group variation include

Subcomponents used in final assemblies that come from two or more different sources.

Persistent differences in operators, where their work is mixed further down the line.

Variation in gauges, where measured items are mixed and used in later operations.

Here's an extreme example of within group variation. Shafts are cut to length by two machines: A and B. Each machine cuts 50 percent of all shafts in approximately the same amount of time. Machine A's shafts are all good, but machine B's shafts are all defective. Figure 8.2 presents a schematic of machines A and B.

FIGURE 8.2   Work Flow of Cut-to-Length Operation

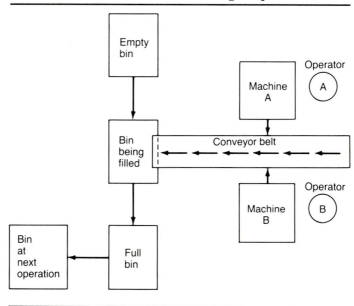

FIGURE 8.3   **Work Flow of Cut-to-Length Operation with an Inspection Station**

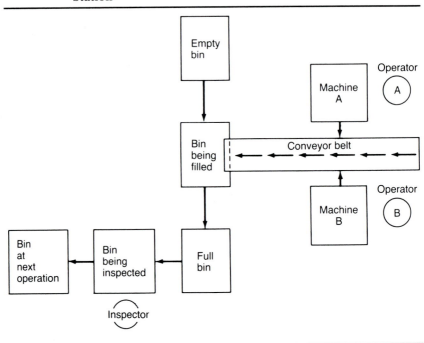

As shafts are finished, they're placed on the conveyor belt and fall into a bin, which can hold 100 shafts. Once a bin is filled, a new one is placed at the end of the conveyor to take its place. Consequently, approximately 50 percent of the shafts in any bin are from machine A; the other 50 percent are from machine B.

Bins are then taken to the next operation. Employees at this next operation have started to complain about defective shafts. An inspection station is set up (Figure 8.3) and a control chart is constructed from 100 percent inspection of every 20th bin (Figure 8.4).

An examination of Figure 8.5's control chart shows that the defective fraction hugs the centerline. Recall that one would expect approximately two-thirds of all subgroup fractions to fall within one standard error of the mean; in this case, 100 percent fall in this region. Stated another way, it is extremely unlikely that a run of 13 or more points (in this case, 15) in a row would fall within a one-sigma band on either side of the mean. The process is unusually quiet.

A novice to control chart interpretation might say that this process exhibits a large degree of stability and predictability, albeit at a very high defect rate; this is totally erroneous. The shaft-cutting process is plagued by within group variation. Each bin is made up of approximately 50 percent defective and 50 percent good shafts, resulting in large within group variation and small between group variation.

FIGURE 8.4   Cut-to-Length Data

Bin	Number of Items in Bin	Defective Items in Bin
1	100	48
2	100	53
3	100	46
4	100	47
5	100	50
6	100	53
7	100	48
8	100	53
9	100	47
10	100	49
11	100	53
12	100	47
13	100	51
14	100	49
15	100	48
Totals	1,500	742

$$\bar{p} = .4947$$

$$UCL = .4947 + 3\sqrt{\frac{.4947(.5053)}{100}} = .6447$$

$$LCL = .4947 - 3\sqrt{\frac{.4947(.5053)}{100}} = .3447$$

As Figure 8.5 shows, this leads to computational procedures that generate control limits too wide for the subgroup statistics.

Three issues must be addressed in this shaft problem. First, the subgrouping should be made on a rational basis; that is, the sample should be made separately from machines A and B. Second, the causes of machine B's defective output must be corrected. Last, both machines should be continually improved using statistical methods.

Distinguishing Within Group Variation from Common Variation

Within group special sources of variation are persistent, as are common sources of variation. However, the critical distinction is that within group special sources of variation are external (exogenous) disturbances to the process, while common sources of variation are internal (endogenous) to the process.

We need to realize that both between and within group special sources of variation must be resolved before the process can be considered stable. As we've discussed, stability is essential to process improvement.

FIGURE 8.5 p Chart for Cut-to-Length Data

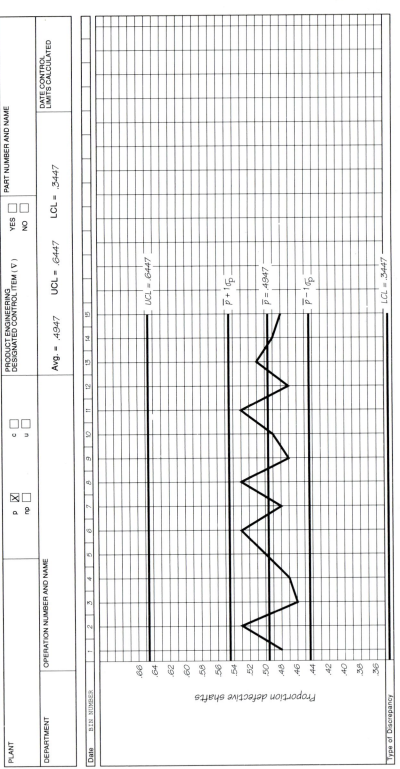

Types of Control Chart Patterns

Identifiable control chart patterns can occur as a consequence of the presence of between and/or within group causes of special variation in a process. Fifteen characteristic patterns have been identified by Western Electric Company engineers[2]: natural patterns; shift in level patterns (sudden shift in level, gradual shift in level, and trends); cycles; wild patterns (freaks and grouping/bunching); multiuniverse patterns (mixtures; stable mixtures associated with systematic variables and with stratification; and unstable mixtures associated with freaks and with grouping/bunching); instability patterns; and relationship patterns (interaction and tendency of one chart to follow another). These patterns are useful because they can be compared to control charts in practice and used as diagnostic tools to detect special sources of variation.

Natural Patterns

A *natural pattern* is one that doesn't exhibit any points beyond the control limits, runs, or other nonrandom patterns and has most of the points near the centerline (approximately two-thirds of the points within a one-sigma band of the centerline). Natural processes aren't disturbed by either between group or within group special causes of variation. The process demonstrates a stable system of variation. Figures 8.6 and 8.7 illustrate a natural process.

It's often necessary to create external disturbances (special sources of variation) to a natural process to create improvements—for example, to move a process's average toward nominal or to reduce the number of defects per unit. The purpose of these external disturbances is to alter the process's basic structure.

Shift in Level Patterns

There are three types of *shift in level patterns:* sudden shift in level, gradual shift in level, and trends.

Sudden Shift in Level Patterns.   A *sudden shift in level pattern* involves a sudden rise or fall in the level of data on a control chart. This is one of the most easily detectable control chart patterns.

Sudden shifts on x-bar charts or on charts for individuals frequently result from a special source of variation, which first shifts the process's average to a new level, but then has no further effect on the process. Sudden shifts on R charts can indicate the presence of some related variable affecting the process variability. For example, the addition of a new untrained worker to a trained and stable work force could increase the variability of output.

Sudden shifts on p charts can indicate such factors as a dramatic change in materials, methods, personnel, or operational definitions. Sudden shifts up indicate process degeneration, while sudden shifts down indicate process improvement, assuming p is the proportion defective.

FIGURE 8.6   Data from a Process Exhibiting a Natural Pattern

---

Day	Date	Subgroup Numbers				
		1	*2*	*3*	*4*	*5*
1	8/30	4.9	5.5	5.3	5.6	5.1
2	31	5.8	5.5	5.6	6.3	5.7
3	9/1	5.9	6.2	5.9	5.8	5.4
4	2	5.9	5.9	6.4	5.3	5.2
5	3	6.2	5.9	5.7	4.9	5.9
6	6	6.0	5.7	5.7	6.3	6.0
7	7	5.2	4.6	5.4	6.1	5.2
8	8	5.1	5.8	6.2	5.9	5.6
9	9	5.8	6.1	5.7	6.5	5.2
10	10	5.2	5.4	5.2	5.8	4.6
11	13	5.2	4.6	5.4	6.1	5.2
12	14	6.2	5.8	5.1	5.2	5.4
13	15	4.9	4.9	4.9	4.9	4.8
14	16	6.1	6.2	5.9	4.5	5.6
15	17	5.3	5.4	5.4	5.4	5.2
16	20	6.3	5.6	5.9	4.7	6.2
17	21	5.4	5.3	5.4	5.2	5.2
18	22	5.6	5.0	5.2	4.9	4.7
19	23	4.7	4.7	5.6	5.0	5.2
20	24	6.0	5.3	5.6	5.0	5.2
21	27	5.1	6.2	5.0	5.2	5.7
22	28	5.6	6.1	5.8	5.9	5.8
23	29	5.5	6.0	5.7	5.0	4.9
24	30	4.9	4.8	6.1	5.3	5.2
25	10/1	4.7	4.9	5.2	6.0	5.7
26	4	6.0	5.1	5.3	4.9	6.1
27	5	5.5	5.0	6.0	5.7	5.0
28	6	5.2	4.9	5.2	5.0	5.3
29	7	5.0	5.2	6.0	4.9	6.1
30	8	5.5	5.1	5.3	4.1	5.0

To illustrate a sudden shift in level on a p chart, daily samples of the first 1,000 medium-sized ratchets produced are taken from a production process and tested for tight levers. The data appear in Figure 8.8. Figure 8.9 illustrates the p chart. Variations are much larger than they should be, as shown by many out-of-control points. Some exogenous factor is apparently preventing consistent quality of work. A study of the assembly process soon reveals that the fixture used in welding the lever is designed so poorly that consistent welds can't be obtained. The fixture is redesigned and more data collected (Figure 8.10). The redesigned fixture has eliminated most of the trouble and reduced the average percentage of tight levers from 4.2 to 0.6 percent. The old and new p charts in Figure 8.11 demonstrate a sudden shift in level.

FIGURE 8.7 Control Chart for a Natural Process

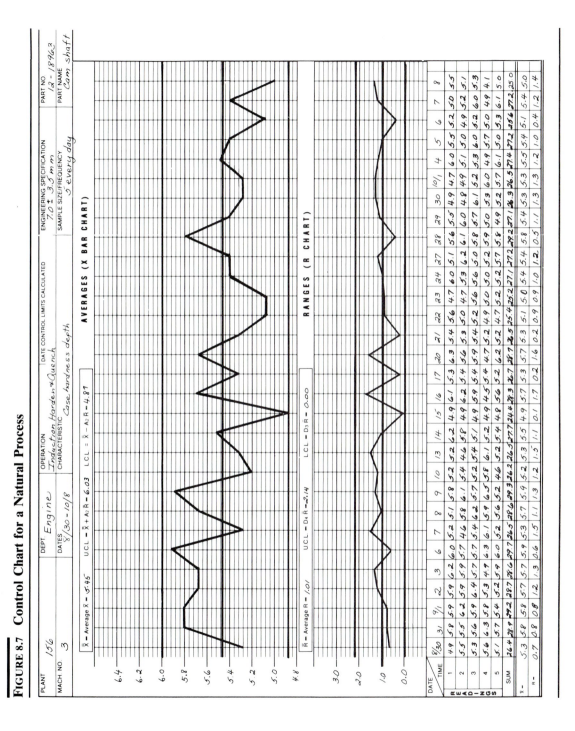

FIGURE 8.8 Ratchet Tight Lever Data

Date	Sample Size	Number Defectives	Fraction Defective
Oct. 3	1,000	25	.025
4	1,000	18	.018
5	1,000	16	.016
6	1,000	20	.020
7	1,000	33	.033
10	1,000	65	.065
11	1,000	30	.030
12	1,000	92	.092
13	1,000	45	.045
14	1,000	26	.026
17	1,000	17	.017
18	1,000	30	.030
19	1,000	8	.008
20	1,000	74	.074
21	1,000	41	.041
24	1,000	29	.029
25	1,000	28	.028
26	1,000	35	.035
27	1,000	90	.090
28	1,000	51	.051
Nov. 2	1,000	53	.053
3	1,000	67	.067
4	1,000	34	.034
5	1,000	55	.055
6	1,000	24	.024
9	1,000	60	.060
10	1,000	81	.081
11	1,000	44	.044
12	1,000	50	.050
13	1,000	46	.046
16	1,000	12	.012
17	1,000	28	.028
18	1,000	40	.040
19	1,000	23	.023
20	1,000	29	.029
23	1,000	27	.027
24	1,000	65	.065
25	1,000	55	.055
26	1,000	69	.069
27	1,000	18	.018
30	1,000	51	.051
Dec. 1	1,000	47	.047
2	1,000	40	.040
3	1,000	52	.052
Totals	44,000	1,843	

Average fraction defective $= \bar{p} = \dfrac{1,843}{44,000} = 0.0419$

$$\bar{p} \pm 3 \sqrt{\frac{\bar{p}(1 - \bar{p})}{n}} = .0419 \pm 3 \sqrt{\frac{.0419(1 - .0419)}{1000}}$$

$$= .061 \text{ and } .023$$

FIGURE 8.9   p Chart for Ratchet Tight Lever Data

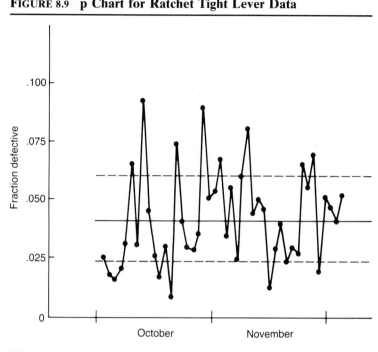

Note in Figure 8.11 that the two points in the latter part of December are out of control as a result of a batch of faulty levers. Nevertheless, the 1.7 percent and 1.5 percent defective pieces found on these two days are fewer than were found most of the time on the old fixture.[3]

Gradual Shift in Level Patterns.   *Gradual shifts in level* generally indicate that some portion of the process has been changed and that the effect of this change is a gradual shift in the average level of output from the process. For example, if new employees are put onto the work floor or new machines or maintenance procedures are set into motion, as they become integrated into the existing process they'll continually and increasingly affect the average level and variation of output. This type of pattern (in the constructive direction) is common in the early stages of quality improvement efforts. Figure 8.12 depicts a gradual shift in level on a p chart.

Trends.   *Trends* are steady changes, increasing or decreasing, in control chart level; they're gradual shifts in level that don't settle down. Trends can result from special sources of variation that gradually affect the process.

Trends on x-bar charts or individuals charts result from disturbances (special sources of variation) that shift the process level up (or down) over time: tool wear; loosening of guide rails or holding devices; operator fatigue; and so on. Figure

---

FIGURE 8.10   Redesigned Ratchet Tight Lever Data

Date	Number Defectives	Fraction Defective
Dec. 4	8	.008
5	10	.010
7	2	.002
8	5	.005
9	10	.010
10	6	.006
11	7	.007
14	6	.006
15	3	.003
16	1	.001
17	2	.002
18	3	.003
21	0	.000
22	4	.004
23	7	.007
24	12	.012
28	9	.009
29	17	.017
30	15	.015
31	10	.010
Jan. 3	8	.008
4	0	.000
7	6	.006
8	0	.000
9	3	.003
10	0	.000
11	1	.001
14	13	.013
15	12	.012
Total	180	

$$\bar{p} = \frac{180}{29,000} = 0.0062$$

$$\bar{p} \pm 3 \sqrt{\frac{\bar{p}(1 - \bar{p})}{n}} = .0062 \pm 3 \sqrt{\frac{.0062(1 - .0062)}{1000}}$$

$$= .0136 \text{ and } .0000$$

8.13 shows trend on an x-bar chart for slot depth caused by tool wear in a cutting jig. Trends are relatively easy to detect, although people who are inexperienced in control chart diagnosis often see trends when they don't exist. Thus caution is advised.

Cycles

Cycles are repeating waves of periodic low and high points on a control chart caused by special disturbances that appear and disappear with some degree of regularity, such as morning start-ups and periodic shifting of operators on x-bar

FIGURE 8.11   p Charts Comparing Ratchet Tight Lever Data Before and After Fixture Redesign

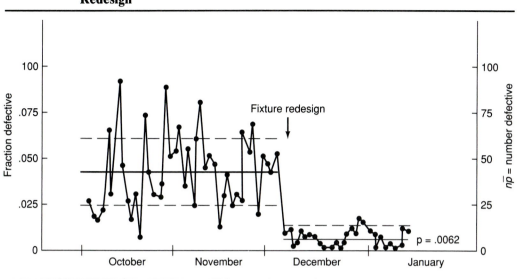

charts; fluctuations in operator fatigue caused by coffee breaks and differences between shifts on R charts; and regular changes in inspectors on a p chart.

If a sampling frequency coincides with the cycle's pattern, sampling may reveal only high or low points; in this case, cycles won't show up on a control chart. The remedy is to sample more frequently to detect the cyclical pattern. Process knowledge is essential to detecting cycles. Figure 8.14 shows a control chart exhibiting cyclical special sources of variation.

Wild Patterns

There are two types of *wild patterns:* freaks and grouping/bunching. Both patterns are characterized by one or more subgroups that are very different from the main body of subgroups.

Freaks.   *Freaks* can be caused by calculation errors or by external disturbances that can dramatically affect one subgroup. They show up on a control chart as points significantly beyond the control limits. Freaks are one of the easiest patterns to recognize. Figure 8.15 shows an example of a freak.

Grouping/Bunching.   *Grouping/bunching* is caused by the introduction into a process of a new system of disturbances that affect a ''group'' or ''bunch'' of points that are close together. Figure 8.16 illustrates grouping/bunching.

FIGURE 8.12 Control Chart for a Gradual Shift in Level

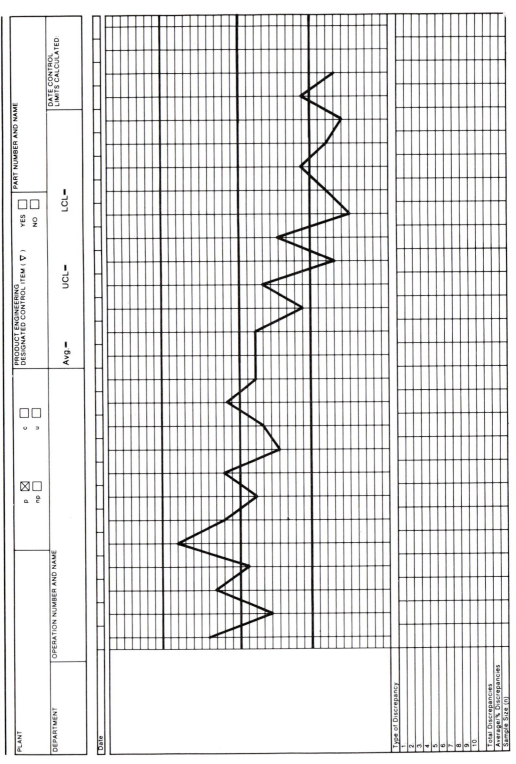

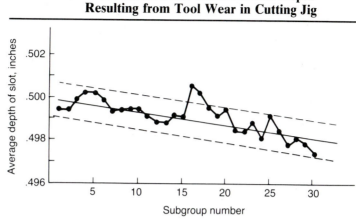

FIGURE 8.13   Trend on an x-Bar Chart for Slot Depth Resulting from Tool Wear in Cutting Jig

Multi-Universe Patterns

There are three *multi-universe patterns* and two groups of associated patterns. The three multi-universe patterns are mixtures, stable mixtures, and unstable mixtures. The first group of associated patterns are systematic variables and stratification related to stable mixtures. The second group of associated patterns are freaks and grouping/bunching related to unstable mixtures.

   All multi-universe patterns are characterized by an absence of points near the centerline (large fluctuations) or by too many points near the centerline (small fluctuations).

Mixtures.   *Mixture patterns* indicate the presence of two or more distributions for a quality characteristic—for example, two distributions of pipe section diameter caused by numerous pipes coming from two different vendors. Mixtures become more apparent the greater the difference between the component distributions. There are two basic forms of mixtures: stable mixtures and unstable mixtures.

Stable Mixtures.   *Stable mixtures* indicate the presence of two or more distributions for a quality characteristic that doesn't change over time with respect to the proportion of items coming from each distribution and/or the average for each distribution. For example, two vendors supply units to a buyer. Seventy percent of the incoming units are purchased from vendor A and weigh 10 pounds on average; the remaining 30 percent of incoming units are purchased from vendor B and weigh 12 pounds on average. As another example, samples are drawn consistently from two shifts or machines. These are stable mixture problems because the proportion of items coming from each distribution is stable over time.

FIGURE 8.14 Control Chart for Cycle Pattern

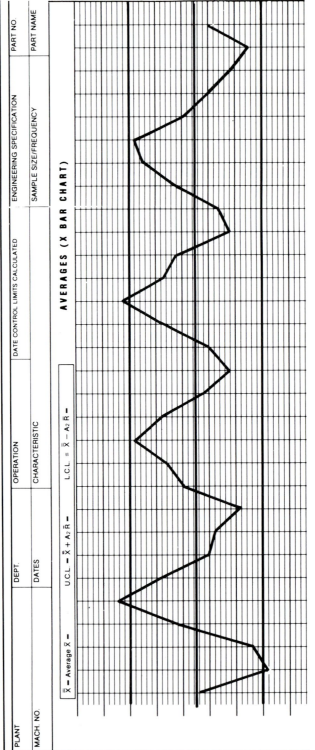

PLANT	DEPT.	OPERATION	DATE CONTROL LIMITS CALCULATED	ENGINEERING SPECIFICATION	PART NO.
MACH. NO.	DATES	CHARACTERISTIC		SAMPLE SIZE/FREQUENCY	PART NAME

$\overline{X}$ – Average $\overline{\overline{X}}$ –   U.C.L. – $\overline{\overline{X}} + A_2 \overline{R}$ –   L.C.L. = $\overline{\overline{X}} - A_2 \overline{R}$ –

AVERAGES (X BAR CHART)

FIGURE 8.15   Control Chart for Freak Pattern

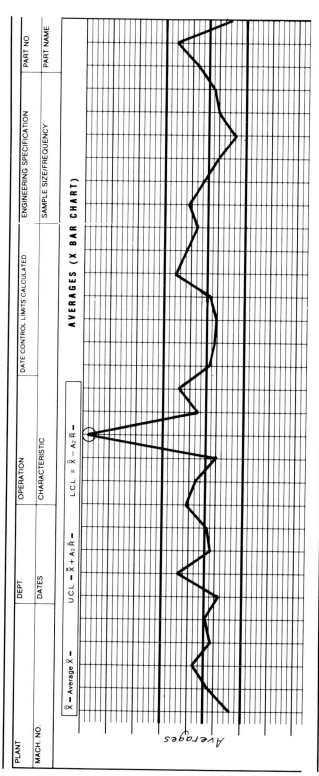

FIGURE 8.16 Control Chart for Grouping/Bunching Pattern

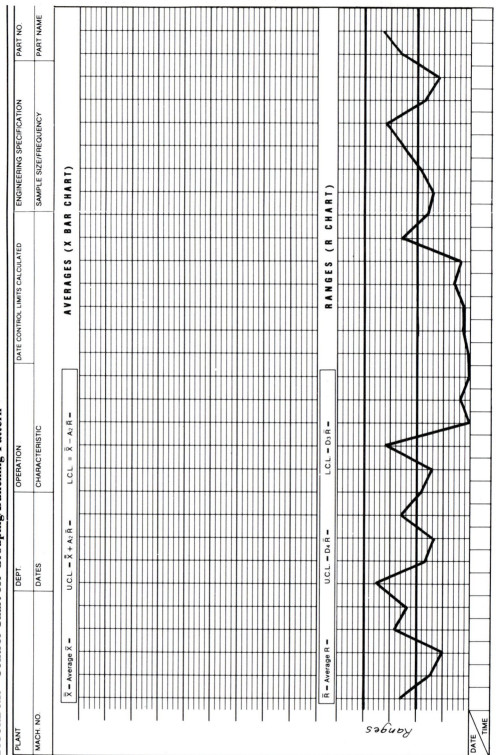

Stable mixture patterns are characterized by an unusually high presence of control chart points near (or beyond) the upper and lower control limits or near the centerline.

Systematic Variables.   If samples are drawn separately from the component distributions, then the stable mixture pattern will appear on the control chart. This is a case of *systematic variables*. For example, if samples are alternately drawn from two shifts that are widely different in output (two distributions), the points on a control chart will sawtooth up and down—shift A high, shift B low, shift A high, shift B low, and so on. This is a systematic variable pattern. An example of the systematic variable form of a stable mixture is differences between tools or differences between shifts, where the data are systematically plotted to bring out the differences between tools or shifts. For example, suppose a box plant produces corrugated boxes. These boxes have many quality characteristics. In this example we'll focus on "glued tab width." The glue tab is what forms the manufacturer's joint of a corrugated box (Figure 8.17). The glued tab width is important to ensure proper strength of the final box.

Ten subgroups of the glued tab widths of four corrugated boxes were drawn from the outputs of shift A and shift B. Figure 8.18 presents the data and x-bar and R chart. The x-bar chart shows a classic sawtooth pattern leading to the unusually high presence of control chart points beyond the control limits. The problem

---

FIGURE 8.17   Glued Tab Width

The glue tab is what forms the manufacturer's joint of the box.
The carrier regulations specify that the overlapped width of the joint must be a minimum of 1¼". The glue tab is made 1⅜" wide so as to meet the minimum 1¼" when the box blank is folded and glued.

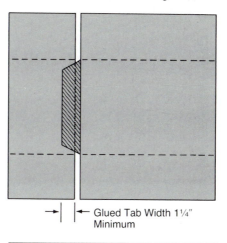

Glued Tab Width 1¼"
Minimum

FIGURE 8.18 Control Chart for Systematic Variable-Stable Mixture Pattern

PLANT		DEPT.	OPERATION	DATE CONTROL LIMITS CALCULATED	ENGINEERING SPECIFICATION	PART NO.
				After shift 10	LSL = 1 - 1/4"	
MACH NO		DATES	CHARACTERISTIC Glued tab width (inches)		SAMPLE SIZE/FREQUENCY 4 every shift	PART NAME Corrugated box

$\bar{\bar{X}}$ = Average $\bar{X}$ = 1.494   UCL = $\bar{\bar{X}} + A_2\bar{R}$ = 1.509   LCL = $\bar{\bar{X}} - A_2\bar{R}$ = 1.479

AVERAGES (X BAR CHART)

Glued width (inches) Averages

1.54
1.53
1.52
1.51
1.50
1.49
1.48
1.47
1.46
1.45

$\bar{R}$ = Average R = .02   UCL = $D_4\bar{R}$ = .046   LCL = $D_3\bar{R}$ = 0

RANGES (R CHART)

Ranges

0.05
0.04
0.03
0.02
0.01
0.00

SAMPLE NO. SHIFT	1		2		3		4		5		6		7		8		9		10	
	A	B	A	B	A	B	A	B	A	B	A	B	A	B	A	B	A	B	A	B
R 1	1.53	1.45	1.51	1.47	1.53	1.54	1.46	1.46	1.52	1.45	1.52	1.45	1.52	1.47	1.52	1.47	1.52	1.53	1.52	1.46
E 2	1.51	1.47	1.52	1.46	1.53	1.46	1.48	1.51	1.54	1.51	1.47	1.45	1.53	1.45	1.53	1.45	1.53	1.46	1.53	1.47
A 3	1.52	1.46	1.51	1.49	1.53	1.47	1.47	1.53	1.51	1.46	1.45	1.51	1.51	1.46	1.51	1.46	1.51	1.51	1.51	1.46
D 4	1.51	1.49	1.54	1.53	1.54	1.47	1.47	1.53	1.51	1.53	1.46	1.53	1.46	1.53	1.49	1.53	1.49	1.51	1.51	1.47
I 5																				
N G S																				
SUM	6.07	5.87	6.10	5.87	6.14	5.93	6.05	5.96	6.07	5.96										
$\bar{X}$ = SUM ÷ NO. OF READINGS	1.52	1.47	1.54	1.47	1.53	1.46	1.51	1.47	1.51	1.47										
R = HIGHEST – LOWEST	.02	.04	.01	.02	.03	.02	.02	.03	.02	.01										

is the presence of a variable that systematically affects the process—in this case, shifts. The control charts in this example should be separated into the x-bar and R charts for shift A and shift B.

Stratification.   If samples are drawn from two or more distributions that have been combined, then the stable mixture pattern can create extremely small differences among statistics on x-bar, R, individuals, or p charts (an unusually high presence of control chart points near the centerline). The small differences on x-bar, R, individuals, or p charts are frequently interpreted by the novice control chart user as representing unusually good control; nothing could be further from the truth.

An excellent example of *stratification* was shown in Figure 8.5. Note the presence of an unusually high number of control chart points near the average proportion defective. This overly quiescent pattern means that the control chart isn't valid for this process. In the presence of stratification, the control chart user must first correct errors in the methods of sampling so that items from component distributions don't get mixed in each sample.

Another illustration of the stratification-stable mixture pattern can be shown via the box plant example discussed in Figures 8.17 and 8.18. Suppose the data on the glued tab width had been collected and recorded by a concerned employee who wanted to make sure he obtained a "representative" sample of output and consequently took two boxes from each shift's output and grouped them together to form one day's sample. The data and x-bar and R chart (Figure 8.19) show unnaturally quiet patterns. The control chart in this example is invalid because the concerned operator sampled simultaneously from both shifts. The proper procedure would have been to sample separately from each shift to create a more rational subgrouping.

Unstable Mixtures.   *Unstable mixtures* indicate the presence of two or more distributions for a quality characteristic that changes over time with respect to the proportion of items coming from each distribution and/or the average for each distribution—for example, a buyer has two vendors for an item, and the proportion coming from each vendor and/or the quality characteristic averages for each vendor change over time. Figure 8.20 depicts an unstable mixture pattern.

Freaks and Grouping/Bunching.   The nature of an unstable mixture pattern implies that the multiple component distributions that make up the distribution of a quality characteristic are sporadically affected by special disturbances. This will cause a systematic variable effect; but this effect will occur unevenly, generating an unusually large number of control chart points near or beyond the control limits. However, those points will be in groups or bunches, depending on the juxtaposition of the various component quality characteristic distributions. This pattern could also result in freaks.

FIGURE 8.19 Control Chart for Stratification-Stable Mixture Pattern

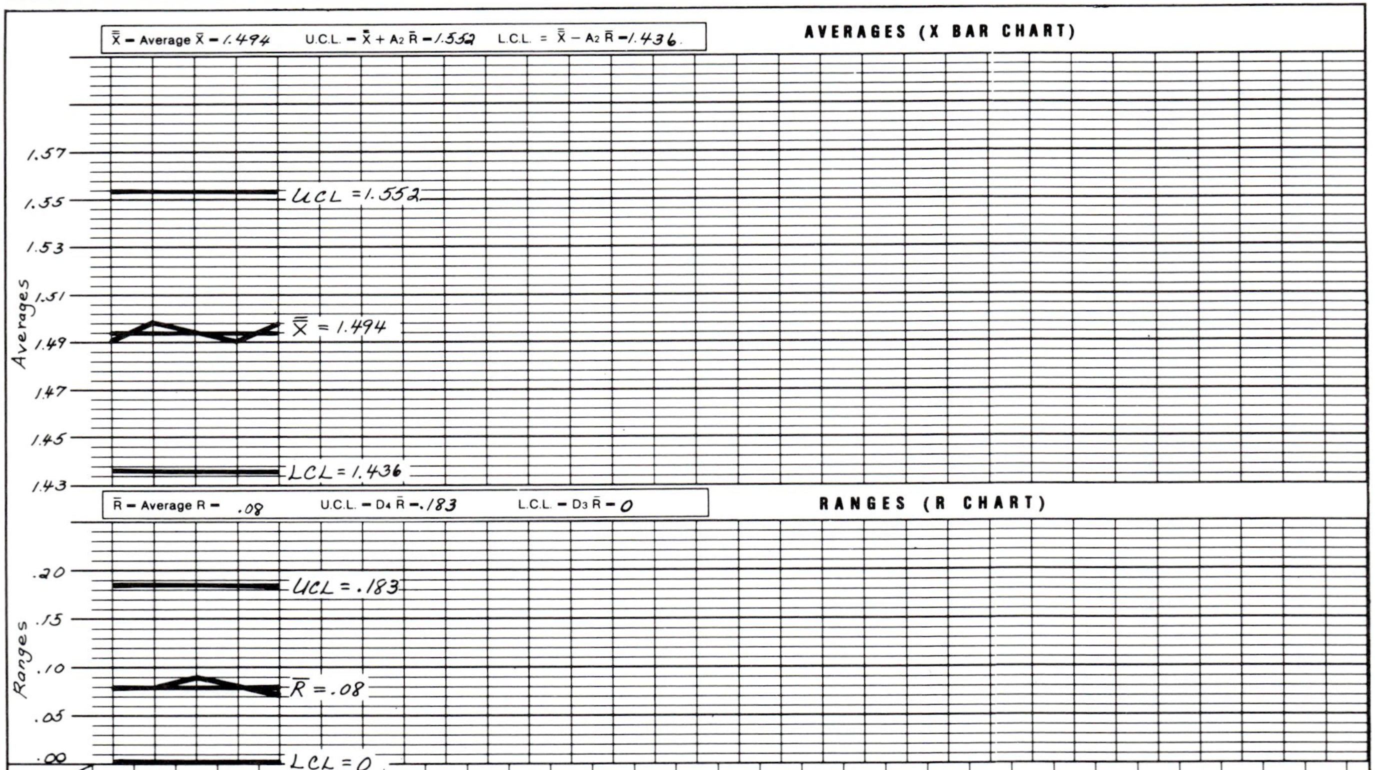

AVERAGES (X BAR CHART)

$\overline{\overline{X}}$ – Average $\overline{X}$ = 1.494   UCL = $\overline{\overline{X}} + A_2 \overline{R}$ = 1.552   LCL = $\overline{\overline{X}} - A_2 \overline{R}$ = 1.436

UCL = 1.552

$\overline{\overline{X}}$ = 1.494

LCL = 1.436

RANGES (R CHART)

$\overline{R}$ – Average R = .08   UCL = $D_4 \overline{R}$ = .183   LCL = $D_3 \overline{R}$ = 0

UCL = .183

$\overline{R}$ = .08

LCL = 0

Averages

1.57
1.55
1.53
1.51
1.49
1.47
1.45
1.43

Ranges

.20
.15
.10
.05
.00

DAY	1	2	3	4	5
R 1	1.53	1.54	1.52	1.52	
E 2	1.51	1.53	1.54	1.53	
A 3	1.45	1.46	1.45	1.47	1.46
D 4	1.47	1.46	1.47	1.45	1.47
I 5					
N G S					
SUM	5.96	5.99	5.98	5.97	5.98
$\overline{X}$ –	1.490	1.498	1.495	1.493	1.495
R –	.08	.08	.09	.08	.07

FIGURE 8.20 Control Chart for Unstable Mixture Pattern

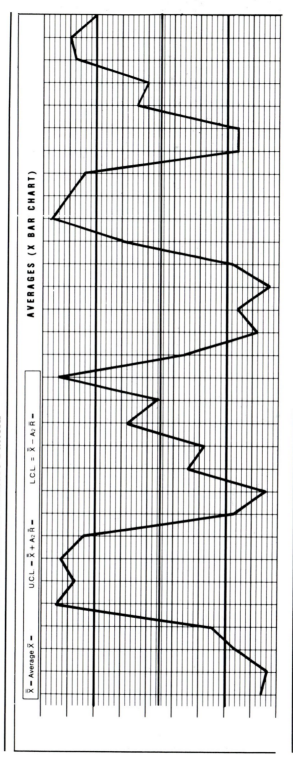

Instability Patterns

Erratic points on a control chart exhibiting large swings up and down characterize a pattern called *instability*. Instability is a possible result of special variation so that the control limits appear too narrow for the control chart. Instability is characterized by large, erratic fluctuations in subgroup statistics and is frequently associated with unstable mixtures.

Instability is caused either by one special disturbance that can sporadically affect the average or variability of a process or by two or more special disturbances, each of which can affect the average and/or variability of a process. These disturbances interact with each other and create complex process disturbances.

Simple instability occurs when one special disturbance creates a wide, bimodal, or multimodal distribution in a quality characteristic or sporadically shifts the process average—for example, occasional lots of material from a supplier that are extremely good or bad, or sporadic adjustment of a machine. Both of these situations create either wide, bimodal, or multimodal quality characteristic distributions or sporadically shift the quality characteristic's average. They create patterns in which control chart limits will appear too narrow for the subgroup statistics. For example, the erratic effect of the bad (or good) lot or the overadjustment of the machine create a situation comparable to the shift-to-shift differences found in the systematic variable pattern. Figure 8.21 depicts a pattern of instability.

Relationship Patterns

There are two types of *relationship patterns:* interaction and tendency of one chart to follow another.

Interaction.   *Interaction patterns* occur when one variable affects the behavior of another variable, or when two or more variables affect each other's behavior and create an effect that would not have been caused by either variable alone.

Interactions between variables are best investigated and understood through a set of statistical tools called experimental design. Interactions can also be investigated and understood through process capability studies.[4]

Interactions can be detected on x-bar, individuals, and p charts by changing the rational subgrouping of the data. For example, if the data in Figure 8.18 are broken into a shift A segment and a shift B segment (i.e., changing the rational subgrouping of the data), the interaction between glued tab width and shifts becomes apparent, as shown in Figure 8.22.

A run of low points on an R or s chart indicates that an interacting variable affecting the variability of the process has been temporarily removed or held at one level. This realization may lead to permanent removal of the interacting variable or continuous maintenance of the interacting variable at one level, consequently reducing the process variability. Figure 8.23 illustrates this. The run of low points indicates the presence of an interacting variable that has temporarily reduced the variability. It's a matter of expertise in the process being studied to identify and manipulate the interacting variable to aid process improvement.

FIGURE 8.21 **Control Chart for Simple Instability**

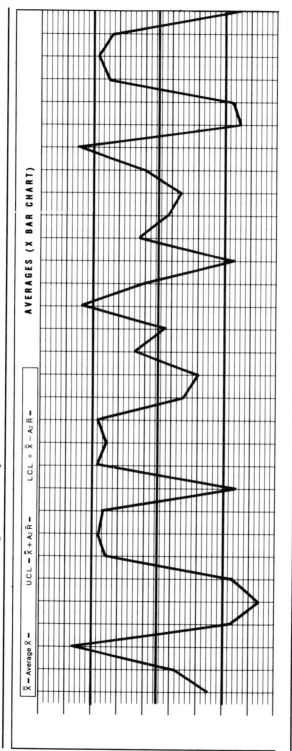

FIGURE 8.22   Control Chart for Interaction Pattern

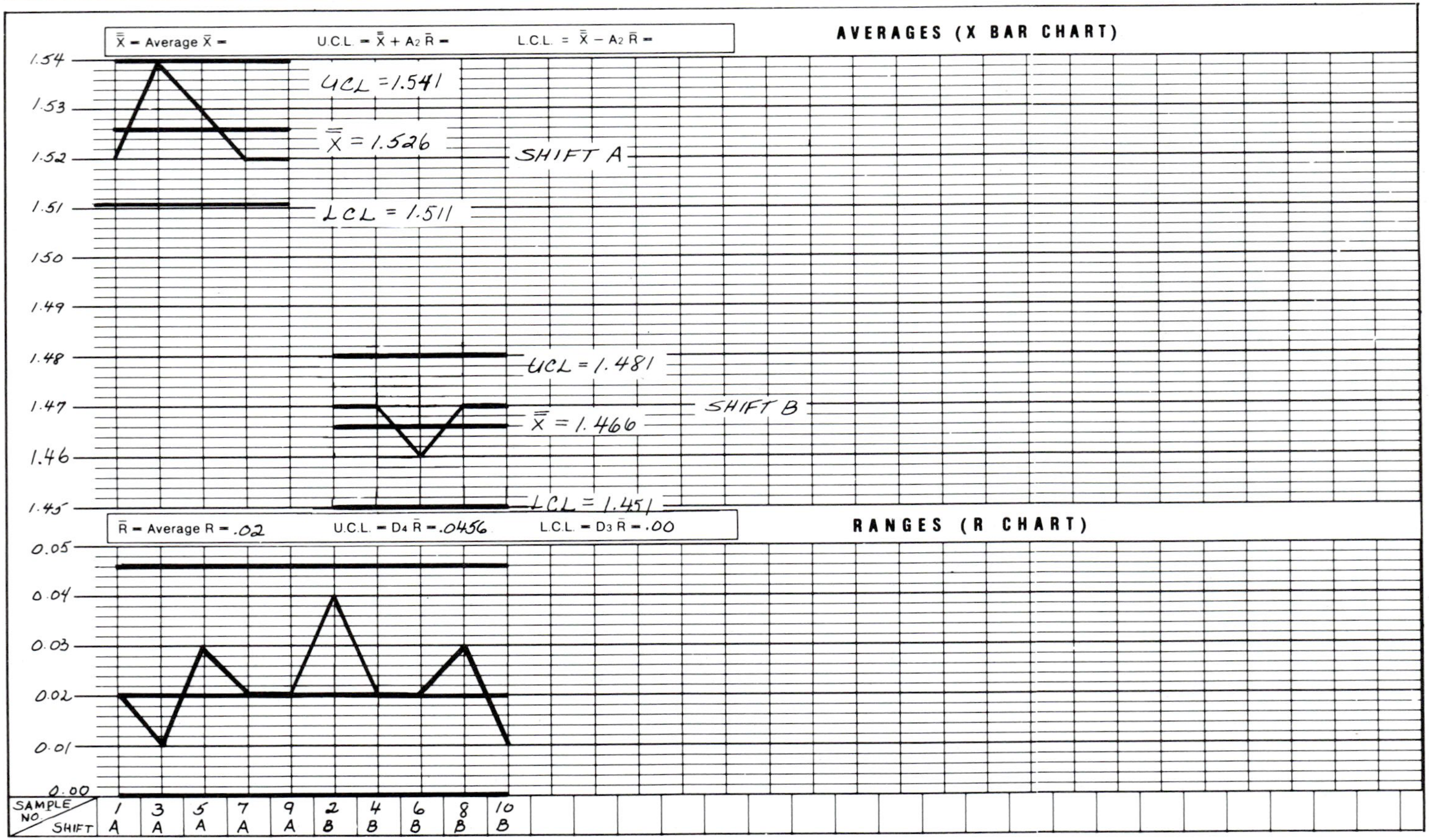

FIGURE 8.23 Control Chart Showing a Run of Low Variation: Interacting Variable

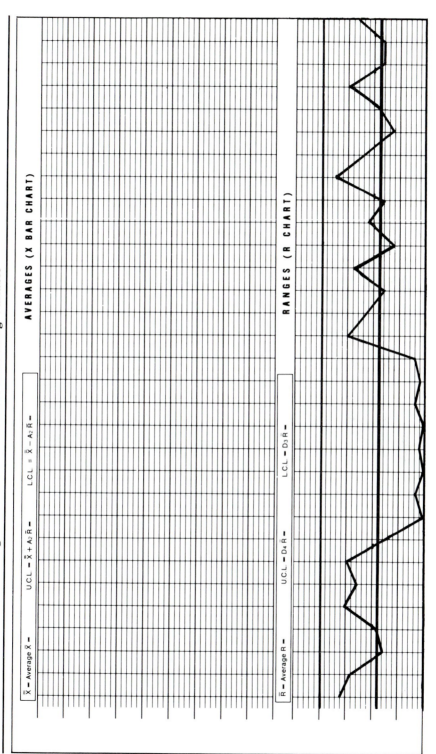

FIGURE 8.24   Control Chart for One Chart Following Another

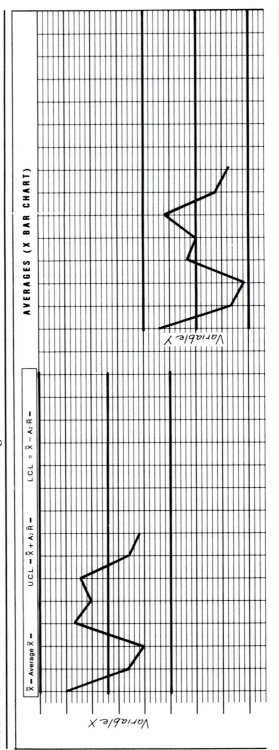

AVERAGES (X BAR CHART)

$\bar{\bar{X}}$ = Average $\bar{X}$ —   U.C.L. = $\bar{\bar{X}}$ + A₂ $\bar{R}$ —   L.C.L. = $\bar{\bar{X}}$ – A₂ $\bar{R}$ —

Variable X

Variable Y

Tendency of One Chart to Follow Another.   These patterns may exist between two or more variables if the control charts for the variables tend to follow each other on a point-to-point basis. This type of pattern most frequently occurs when the control charts in question have been constructed from the same samples. For example, a sample of four items can be measured with respect to three different quality characteristics, and each measurement is plotted on its respective control chart. Figure 8.24 illustrates one chart following another.

Out-of-Control Patterns and the Rules of Thumb

Natural control chart patterns exhibit the following characteristics:

1. Rarely will a point exceed the control limits.
2. Most (but not all) points are near the centerline.
3. A few (but not too many) points are near the control limits.
4. There are no nonrandom patterns among the points.
5. There's neither very high nor very low variability among the points.

If one (or more) of these conditions is missing in a control chart pattern, the pattern will appear unnatural, exhibiting one or more of the following characteristics:

1. Points located beyond the control limits.
2. Absence of points near the centerline.
3. Absence of points near the control limits.
4. Nonrandom patterns among the points.
5. Exceptionally high or low variability among the points.

These characteristics are reflected in Chapter 5's seven rules for detecting out-of-control behavior.

For example, a stratification pattern would exhibit the absence of points near the control limits (too many points near the centerline), indicated by Rule 7; a grouping/bunching pattern would exhibit the absence of points near the centerline (too many points near or beyond the control limits), indicated by Rules 1, 2, 3, and/or 6; a freak pattern would exhibit points beyond the control limits, indicated by Rule 1; a systematic variable pattern would exhibit an unusually small number of runs up and down (a sawtooth pattern, alternating high, low, high, low, high), indicated by Rule 6; and a gradual shift or trend pattern would exhibit an unusually long run of points up or down, indicated by Rules 4 and/or 5.

Summary

We began this chapter by explaining the different types of special variation in an analytic study: between group variation and within group variation. Between group sources of variation are external sources of variation that affect a process periodically, while within group sources of variation are sources of variation that affect a process persistently.

We then presented and described 15 control chart patterns whose detection should be helpful in understanding and eliminating special sources of variation in a process: natural, sudden shift in level, gradual shift in level, trends, cycles, freaks, grouping or bunching, mixtures, stable mixtures with systematic variables, stable mixtures with stratification, unstable mixtures with freaks, unstable mixtures with grouping or bunching, instability, interaction, and tendency of one chart to follow another. These were related to the rules of thumb presented earlier. In general, all rules must be applied cautiously in the context of the process being analyzed.

Exercises

8.1 a. Define special variation.
   b. Define common variation.
   c. Define the two types of special variation: between group variation and within group variation. Explain the difference between them and give examples of each.

8.2 Describe a natural control chart pattern. Give an example of a situation that would lead to a natural control chart pattern.

8.3 Describe a sudden shift in level control chart pattern. Give an example of a situation that would lead to a sudden shift in level control chart pattern.

8.4 Describe a gradual shift in level control chart pattern. Give an example of a situation that would lead to a gradual shift in level control chart pattern.

8.5 Describe a trend control chart pattern. Give an example of a situation that would lead to a trend control chart pattern.

8.6 Describe a cycles control chart pattern. Give an example of a situation that would lead to a cycles control chart pattern.

8.7 Describe a freak control chart pattern. Give an example of a situation that would lead to a freak control chart pattern.

8.8 Describe a grouping or bunching control chart pattern. Give an example of a situation that would lead to a grouping or bunching control chart pattern.

8.9 Describe a mixture control chart pattern. Give an example of a situation that would lead to a mixture control chart pattern.

8.10 Describe a stable mixture control chart pattern. Give an example of a situation that would lead to a stable mixture control chart pattern.

8.11 Describe a systematic variables control chart pattern. Give an example of a situation that would lead to a systematic variables control chart pattern.

8.12 Describe a stratification control chart pattern. Give an example of a situation that would lead to a stratification control chart pattern.

8.13 Describe an unstable mixture control chart pattern. Give an example of a situation that would lead to an unstable mixture control chart pattern.

8.14 Describe an instability control chart pattern. Give an example of a situation that would lead to an instability control chart pattern.

8.15 Describe an interaction control chart pattern. Give an example of a situation that would lead to an interaction control chart pattern.

8.16 Describe a tendency of one chart to follow another control chart pattern. Give an example of a situation that would lead to a tendency of one chart to follow another control chart pattern.

Endnotes

1. William B Rice, *Control Charts in Factory Management* (New York: John Wiley & Sons, 1947), pp. 102–4.

2. AT&T, *Statistical Quality Control Handbook,* 10th printing, May 1984 (Indianapolis: AT&T, 1956), pp. 161–80.

3. Paraphrased from Rice, *Control Charts for Factory Management,* pp. 70–74.

4. AT&T, *Statistical Quality Control Handbook,* pp. 75–117.

Diagnosing a Process

Introduction

As we've seen, statistical control charts are important aids in stabilizing and improving a process, and help decrease the difference between customer needs and process performance. However, other techniques can be used in conjunction with control charts to aid in process stabilization, improvement, and innovation. The methods we'll discuss are brainstorming, cause-and-effect diagrams, check sheets, Pareto analysis, and stratification.

Brainstorming

Brainstorming is a way to elicit a large number of ideas from a group of people in a short period of time. Members of the group use their collective thinking power to generate ideas and unrestrained thoughts. Brainstorming is used for several purposes: to determine problems to work on; to find possible causes of a problem; to find solutions to a problem; and to find ways to implement solutions.[1] Brainstorming was fully developed and utilized by the ancient Greeks; it was known as *heuristics*.[2] Dr. Alex Osborn revived brainstorming in the 1940s in his work in advertising, after which the technique became popular in industrial applications.[3]

Effective brainstorming should take place in a structured session. The group should be small in number, between 3 and 12 as a rule of thumb; having too large a group deters participation. Composition of the group should depend on the issue being examined. The group should include a variety of people, not all of whom should be technical experts in the particular area. The group leader should be experienced in brainstorming techniques. The leader's task is to keep the group focused, prevent distractions, keep ideas flowing, and record the outputs. The brainstorming session should be a closed-door meeting with no interruptions that might interfere with the group's creative process or cause distractions. Seating should promote the free flow of ideas. A U-shape or circle arrangement is suggested. The leader should record the ideas so everyone can see them, preferably on a flip-chart, blackboard, or illuminated transparency.

Procedure

The following steps are recommended for a brainstorming session:

1. Select the topic or problem to be discussed.
2. Each group member makes a list of ideas on a piece of paper. This should take no longer than 10 minutes.
3. Each person reads one idea at a time from her list of ideas, sequentially, starting at the top of the list. As ideas are read, they should be recorded and displayed by the group leader. Group members continue in this circular reading fashion until all the ideas on everyone's list are read.
4. If a member's next idea is a duplication, that member goes on to the subsequent idea on his list.
5. Members are free to pass on each go-round but should be encouraged to add something.
6. The leader then requests each group member, in turn, to think of any new ideas she hadn't thought of before. Hearing others' ideas will probably result in related ideas. This is called *piggybacking*. The leader continues asking each group member, in turn, for new ideas, until they can't think of any more.
7. If the group reaches an impasse, the leader can ask for everyone's "wildest idea." An unrealistic idea can stimulate a valid one from someone else.

Rules

Certain rules should be observed by the participants to ensure a successful brain-storming session—otherwise, participation may be inhibited.

1. Don't criticize, by word or gesture, anyone's ideas.
2. Don't discuss any ideas during the session, except for clarification.
3. Don't hesitate to suggest an idea because it sounds silly. Many times a "dumb" idea can lead to the problem solution.
4. Only one idea should be suggested at a time by each team member.
5. Don't allow the group to be dominated by one or two people.
6. Don't let brainstorming become a gripe session.[4]

Aids to Better Brainstorming

A relaxed atmosphere in which people feel free to suggest any kind of idea enhances the brainstorming session. Here are five techniques that may improve brainstorming by giving people ways to come up with new ideas.[5]

1. *Modification* is changing some aspect of an existing product or service. An example is lower-priced movie tickets for senior citizens.
2. *Magnification* is enlarging a product or service, such as giant economy-size packages.
3. *Minification* is altering a product or service so it becomes smaller or less complex. Examples are portable radios and televisions, electronic calculators, and no-frills airline travel.
4. *Substitution* is using a certain material or service in place of what has traditionally been employed. Examples are using polyester instead of cotton, plastic in place of metal, and nurse-midwives instead of physicians.
5. *Rearrangement* is altering the configuration of basic elements in a product or service—for example, some housing developments use several floor plans but all homes have the same basic features.

An Example of Brainstorming

Consider a group of six people, one from each department of an organization, who brainstorm about the problem of excessive employee absenteeism. They've already decided on the topic to be discussed, so they can proceed to making their lists of causes. After completing their lists, they read their ideas, sequentially, one at a time. The designated leader records the ideas on a flip chart. The first person's list of possible causes of excessive employee absenteeism is

1. Low morale.
2. No penalties for absence.
3. Boredom with job.
4. Personal problems.

The second person's list is

1. Dislike of supervisor.
2. Drug problems.
3. Performance anxiety.
4. Anger over pay.
5. Work-related accidents.

Other members have similar lists. After all have read their lists and the causes have been recorded, the leader requests any new ideas that have emerged. Piggybacking on one of the first person's causes—"personal problems"—might result in another cause, "family problems." Asking for wild ideas might generate a response such as "addiction to video games" or "rundown bathroom facilities."

After all of the ideas have emerged, each group member gets a copy of the list to study. The group meets again and evaluates the ideas. They rank them in order of importance and decide that low morale, drug problems, and boredom with job are the three most critical causes of absenteeism. They are then in the position to develop an action plan to deal with these causes.

Cause-and-Effect Diagrams

The *cause-and-effect (C&E) diagram* is a structured problem-solving technique developed in 1943 by Professor Kaoru Ishikawa, president of the Musashi Institute of Technology in Tokyo.[6] He found that most plant personnel were overwhelmed by the number of factors that could influence a process. Consequently, Ishikawa developed and applied C&E diagrams to help plant personnel cope with the myriad of factors that affected their processes and to solve problems. Cause-and-effect diagrams (also known as *Ishikawa diagrams* or *fishbone diagrams*) are widely used in manufacturing and service industries.

Cause-and-effect analysis is used after brainstorming to organize the information generated in a brainstorming session. It includes gathering and organizing possible reasons or causes of a problem, selecting the most probable cause, and verifying possible causes until a valid cause-and-effect relationship, which leads to a solution, is established.

The technique consists of defining an occurrence or problem (effect). This is usually a result of a brainstorming session in which the team identifies something it wants to correct or change. Once the effect is defined, the factors that contribute to it (causes) are delineated. These are the possible reasons that the problem exists. While there may be only one or two actual causes of a problem, there are probably many potential causes that could appear on the C&E diagram. The relationships between all the contributing causes are illustrated on a C&E diagram. Figure 9.1 shows a simple generic example of a C&E diagram.

Procedure

As the purpose of using a C&E diagram is to understand the factors affecting a process, the diagram must represent the perspective of many different people involved in the process rather than the views of only one or two individuals. Brainstorming sessions that allow for a broad base of process observation are ideally suited for drawing C&E diagrams. Guidelines for the session are that

FIGURE 9.1   Generic Cause-and-Effect Diagram

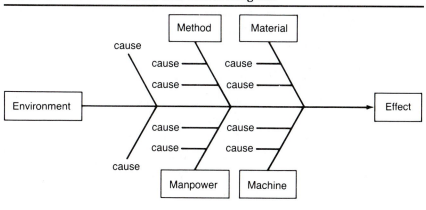

1. Everyone should be encouraged to participate.
2. No criticism should be made of any suggestion.
3. Suggestions don't have to be limited to factors in one's own work area.
4. A period of observation, between the time that the chart is started and the time that it's finished, may be helpful.
5. Concentration on how to eliminate the trouble, rather than getting involved with justifications of why the trouble has occurred, is desirable.

Constructing a C&E Diagram

The following steps are recommended for constructing a C&E diagram.

1. *State the problem.*   Clearly define the problem or effect to avoid confusion and focus the discussion. It's far easier to delineate causes of a well-defined problem than to try to do so for a vaguely defined problem. After a problem has been clearly defined, write it on the right side of a flip chart and draw an arrow or pointer to it.

2. *Identify major causes.*   Identifying major causes is important because it gives the C&E process a structure. But it's often difficult to identify major causes until many potential major causes have been studied. A common method of determining major causes is to use universal major causes (such as machines, methods, material, manpower, and environment) on a C&E diagram, as Figure 9.1 shows.

3. *Brainstorm subcauses.*   The group should brainstorm to uncover all possible subcauses that create the problem. Subcauses should be recorded on the diagram. Remember that only causes of the problem should be suggested at this point, not solutions. Figure 9.2 shows a C&E diagram for errors in producing airline tickets with major causes and subcauses.[7] Four major areas that may cause airline ticket errors are delineated. They're further broken down into subcauses and subsubcauses during subsequent group brainstorming, until all possible causes of the problem have surfaced.

4. *Allow time to ponder the causes before evaluating them.*   Questions to consider at this time include

- Is this cause a variable or an attribute?
- Has the cause been operationally defined?
- Is there a control chart or running record of data for this cause?
- Does this cause interact with other causes?

This information is valuable because it can aid in better understanding a cause's impact on the problem under investigation.

FIGURE 9.2   Cause-and-Effect Diagram for Airline Ticket Errors

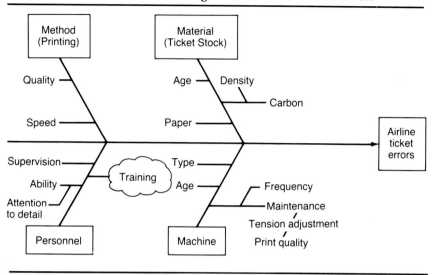

5. *Circle likely causes.*   Causes of the problem should be evaluated. The most likely cause or causes should be circled on the C&E diagram. For example, *training* is circled in Figure 9.2.

6. *Verify the cause.*   Analyze the problem's most likely cause by gathering data to see if it has a significant impact on the problem. If the most likely cause doesn't significantly impact the problem, the group should verify the next most likely choice to determine if it has a significant impact, and so on.

Root Cause Analysis

In some instances, use of a C&E diagram doesn't result in finding the problem's actual underlying causes. The group may become aware of this when discussing the cause or implementing a solution. At either point, the group can examine a cause in more depth; that is, the group can perform cause-and-effect analysis when cause classification doesn't reach the root of the problem. Each cause originally identified can potentially be examined in a much more detailed manner by asking *who, what, where, when, why,* and *how* about each cause. In essence, each cause now becomes an effect (or problem) in a C&E diagram.

Returning to the example of the C&E diagram for airline ticket errors, we can illustrate the use of root cause analysis. Some causes identified were quality of printing, training of personnel, age of the ticket stock, and type of machine. Training of personnel is selected for further analysis. Figure 9.3 illustrates this concept. This procedure could be used to peel back the layers of a problem as we would peel off the layers of an onion—to get to the heart of a problem.

FIGURE 9.3   Root Cause-and-Effect Diagram

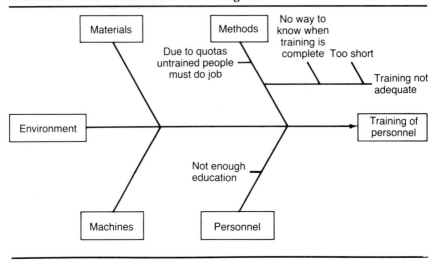

FIGURE 9.4   Cause-and-Effect Diagram of a Process for Making Chocolate Mousse

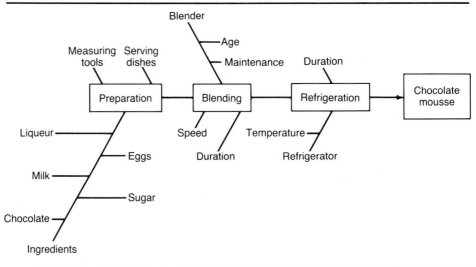

Process Analysis

Another form of a C&E diagram is for *process analysis*. It's used when a series of events (steps in a process) creates a problem and it's not clear which event or step is the major cause of the problem. Figure 9.4 illustrates a process analysis for making a chocolate mousse.[8] Note that each category or subprocess is examined for possible causes. After discovering causes from each step in the process, we select and verify significant causes of the problem.

Summary

The C&E diagram should be used as a framework for collective efforts. If a process is stable, it will help organize efforts to improve the process. If a process is chaotic, the C&E diagram will help uncover areas that can help stabilize the process.

Check Sheets

Check sheets are used for collecting or gathering data in a logical format (called *rational subgrouping,* as discussed in Chapter 5). The data collected can be used in constructing a control chart, a Pareto diagram (to be discussed in the next section), or a histogram. Check sheets have several purposes, the most important being to enable the user(s) to gather and organize data in a format that permits efficient, easy analysis. The check sheet's design should facilitate data gathering.

Process improvement is aided by determining what data or information is needed to reduce the difference between customer needs and process performance. There are unlimited types of data that can be gathered in any organization, including information on the process, products, costs, vendors, inspection, customers, employees, administrative tasks, paperwork, sales, and personnel. Virtually any aspect of an organization can yield facts or data amenable to improving the interdependent system of stakeholders.

Remember, data can be visible or invisible. Visible data are quantifications of product or process characteristics; they can be either attribute or variables in nature. Visible data are collected with check sheets. Invisible data are unknown and/or unknowable; that is, they can't be quantified. Invisible data include the cost of an unhappy customer or the benefit of a prideful employee. They include the most important business data, yet they can't be studied using check sheets and related tools.

Types of Check Sheets

In designing a check sheet, you must determine (1) what the user(s) are attempting to learn by collecting the data and (2) what action the user(s) will take, given the results. This information will facilitate proper design of the check sheet to optimize benefit from the data. We'll discuss three types of check sheets: attribute check sheets, variables check sheets, and defect location check sheets.[9] In reality, there are an unlimited number of formats for a check sheet because the user(s) can develop them based on the data needed to solve a particular problem and can be creative and invent a check sheet if the data aren't amenable to an already established check sheet format. Remember, you must consider invisible data when making decisions.

FIGURE 9.5   Attribute Check Sheet of Defects in Corrugated Board Boxes

Type of Defect	Monday				Tuesday				Total
	8–10 AM	*10–12* AM	*12–2* PM	*2–4* PM	*8–10* AM	*10–12* AM	*12–2* PM	*2–4* PM	*Total*
Smeared print	||		|||	|	|||		|		10
Box not glued properly	|		||	|||	||	|||	|	|||	15
Wrong symbol/ letter	|	||||	||	☰	||		||		16
Symbol/letter in wrong location	|||		☰	||	||	|||	☰	||	22
Total	7	4	12	11	9	6	9	5	63

Attribute Check Sheet.   Gathering data about defects in a process is necessary for stabilization, improvement, and innovation of the process. As there are many possible causes for any given defect, the logical way to collect data is to determine the number or percentage of defects generated by each cause. Based on the information collected, appropriate action can be taken to improve the process. Figure 9.5 shows an *attribute check sheet* for the causes of defects on corrugated board boxes.

This check sheet was created by tallying each type of defect during four two-hour time periods each day; it shows the types of defects and how many of each type occurred during each time period. Keeping track of these data for several days provides management with information on which to base improvements to the process.

Variables Check Sheet.   Gathering data about a process also involves collecting information about variables, such as size, length, weight, and diameter. These data are best represented by organizing the measurements into a frequency distribution on a *variables check sheet*.

Figure 9.6 is a variables check sheet showing the frequency distribution of the length of logs in a sample of 95 trees discussed in Chapter 4. Recall that the raw data in Figure 4.7 could be displayed as a frequency distribution in Figure 4.9. These absolute frequencies appear in Figure 9.6. This type of check sheet is a simple way to examine the distribution of a process characteristic and its relationship to the specification limits (the boundaries of what's considered acceptable log lengths in Figure 9.6); the number and percentage of items outside the specification limit is easy to identify so that appropriate action can be taken to reduce the number of defectives.

Defect Location Check Sheet.   Another way to gather information about defects in a product is to use a *defect location check sheet*. This is a picture of a product

FIGURE 9.6   **Variables Check Sheet for Length of Logs in a Sample of 95 Trees**

400 but under 700	‖‖ ‖‖	9
		Specification = 700
700 but under 1,000	‖‖ ‖‖	8
1,000 but under 1,300	‖‖ ‖‖ ‖‖ ‖‖	20
1,300 but under 1,600	‖‖ ‖‖ ‖‖ ‖‖ ‖‖ ‖‖ ‖‖	35
1,600 but under 1,900	‖‖ ‖‖ ‖‖ ‖‖	18
		Specification = 1,900
1,900 but under 2,200	‖‖	5

FIGURE 9.7   **Defect Location Check Sheet**

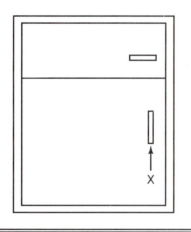

Examiner:	Date:	Serial No.
John May	*8/18/94*	**HO 13988**
Remarks: *Bolt holding button of handle is missing. Handle loose.*		

or a portion of it on which an inspector indicates the location and nature of the defect. Figure 9.7 shows a defect location check sheet for collecting data regarding defects on the fronts of a model of refrigerator door. It shows the location of a defect where the handle is attached to the refrigerator door. If this check sheet were used, and we determined that most defects on the refrigerator doors occurred in the same location, we could perform further analysis. After investigation, we could find the cause and implement a plan to eliminate the problem.

Pareto Analysis

Pareto analysis is a tool used to identify and prioritize problems for solution. It's based on the work of Italian economist Vilfredo Pareto (1848–1923). Pareto focused attention on the concept of "the vital few versus the trivial many." The vital few are the few factors accounting for the largest part (percentage) of a total. The trivial many are the myriad of factors that account for the small remainder. Several other researchers have popularized this approach to prioritizing problem solving, most notably Joseph Juran[10] and Alan Lakelin[11] (a time management specialist). Lakelin formulated the 80–20 rule based on an application of the Pareto principle. This rule says that approximately 80 percent of the value or costs come from 20 percent of the elements. For example, 80 percent of sales originate from 20 percent of customers, or 80 percent of phone calls are made by 20 percent of customers, or 80 percent of dollar inventory is accounted for by 20 percent of the items.

The Pareto diagram is a simple bar chart of the type discussed in Chapter 4, with the bars representing the frequency of each problem, arranged in descending order. That is, the tallest bars are on the left side of the chart descending to the right side. While Pareto analysis is commonly thought of as a problem-solving tool, it really helps determine what problems to solve, rather than showing how to solve them. The process of arranging the data, classifying it, and tabulating it helps determine the most important problem to be worked on.[12]

Constructing a Pareto Diagram

The following steps are recommended for constructing a Pareto diagram.[13] We illustrate the use of a Pareto diagram with an example concerning sources of defective cards for a particular data entry operator.

1. *Establish categories for the data being analyzed.*   Data should be classified according to defects, products, work groups, size, and other appropriate categories; a check sheet listing these categories should be developed. In the example that follows, data on the type of defects for a data entry operator will be organized and tabulated.

2. *Specify the time period during which data will be collected.*   The time period to be studied will depend on the situation being analyzed. Three issues important in setting a time period to study are: (1) selection of a convenient time, such as one week, one month, one quarter, one day, or four hours, (2) selection of a time period that's constant for all related diagrams for purposes of comparison, and (3) selection of a time period that's relevant to the analysis, such as a specific season for a certain seasonal product. In the data entry example, the time period is four months—January through April 1994.

In the example, the types of defects are recorded as they occur during the time period and are totaled, as Figure 9.8 shows.

3. *Construct a frequency table arranging the categories from the one with the largest number of observations to the one with the smallest number of observations.*   The frequency table should contain a category column; a frequency column indicating the number of observations per category with a total at the

FIGURE 9.8   Record of Defects for Data Entry Operator: Checksheet to Determine the Sources of Operator 004's Defective Entries (1/94–4/94)

Major Causes of Defective Entries	Month				*Total*
	1/94	*2/94*	*3/94*	*4/94*	
Transposed numbers	7	10	6	5	28
Out of field	1		2		3
Wrong character	6	8	5	9	28
Data printed too lightly		1	1		2
Torn document	1	1		2	4
Creased document			1	1	2
Illegible source document			1		1
Total	15	20	16	17	68

FIGURE 9.9   Frequency Table of Defects for Data Entry Operator: Pareto Analysis to Determine Major Causes of Defective Entries for Operator 004 (1/94–4/94)

Major Cause of Defective Entries	*Frequency*	*Relative %*	*Cumulative Frequency*	*Cumulative %*
Transposed numbers	28	41.2	28	41.2
Wrong character	28	41.2	56	82.4
Torn document	4	5.9	60	88.3
Out of field	3	4.4	63	92.7
Data printed too lightly	2	2.9	65	95.6
Creased document	2	2.9	67	98.5
Illegible source document	1	1.5	68	100.0
Total	68	100.0		

bottom of the column; a cumulative frequency column indicating the number of observations in a particular category plus all frequencies in categories above it; a relative frequency column indicating the percentage of observations within each category with a total at the bottom of the column; and a relative cumulative frequency column indicating the cumulative percentage of observations in a particular category plus all categories above it in the frequency table.

An "other" category, if there is one, should be placed at the extreme right of the chart. If the "other" category accounts for as much as 50 percent of the total, the breakdown of categories should be reformulated. A rule of thumb is that the "other" bar should be smaller than the category with the largest number of observations.

The frequency table for the data entry example (Figure 9.9) shows that two types of defects (transposed numbers and wrong characters) are causing 82.4 percent of the total number of defective entries.

4. *Construct a Pareto diagram.*   Here are the steps required to construct the Pareto diagram.

a. Draw horizontal and vertical axes on graph paper and mark the vertical axis with the appropriate units, from zero up to the total number of observations in the frequency table.

b. Under the horizontal axis, write the most frequently occurring category on the far left, then the next most frequent to the right, continuing in decreasing order to the right. In the data entry example, "transposed numbers" and "wrong character" are the most frequently occurring defects and are positioned to the far left; "illegible source document" accounts for the fewest defective cards and appears at the far right of the chart.

c. Draw in the bars for each category. For some applications, this may provide enough information on which to base a decision; but often the percentage of change between the columns must be determined. Figure 9.10 displays the bars for the data entry example.

d. Plot a cumulative line (cum line) on the Pareto diagram. Indicate an approximate cumulative percent scale on the right side of the chart and plot a cumulative line ("cum" line) on the Pareto diagram. To plot the

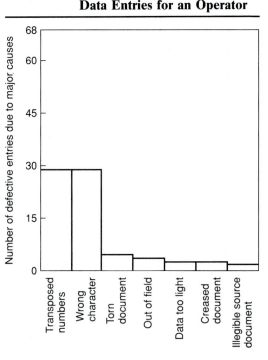

FIGURE 9.10   Pareto Diagram of Defective Data Entries for an Operator

cum line, start at the lower left (zero) corner and move diagonally to the top right corner of the first column. In our example, the top of the line is now at the 28 level, as in Figure 9.11(a). Repeat the process, adding the number of observations in the second column. In our example, the line rests on the 56 level, as in Figure 9.11(b). The process is repeated for each column, until the line reaches the total number of observations level that includes 100 percent of the observations, as in Figure 9.11(c).

e. Title the chart and briefly describe its data sources. Without information on when and under what conditions the data were gathered, the Pareto diagram won't be useful.

Cost Pareto Diagrams

Sometimes Pareto diagrams can have more impact when problems or defects are represented in terms of their dollar costs. We can calculate the dollar cost for a particular type of defect by evaluating the unit cost each time the particular type of defect occurs, and then multiplying that figure by the number of times that particular type of defect occurs. We then construct a Pareto diagram using the dollar cost of a defect (rather than the number of defective units) as the vertical

FIGURE 9.11   Cum Line Plotting

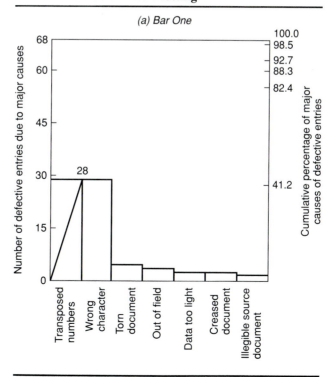

FIGURE 9.11   *(concluded)*

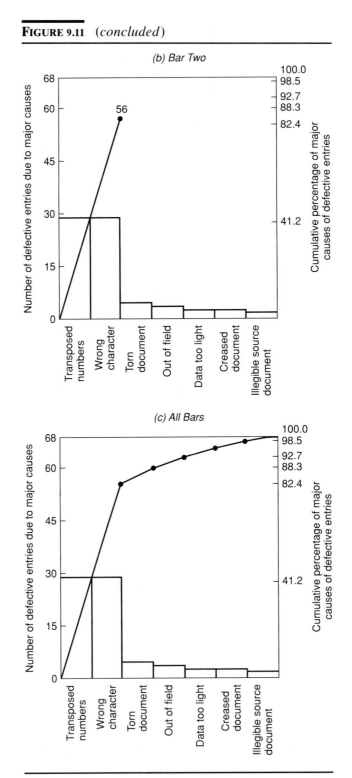

FIGURE 9.12   Frequency Table of Causes of Defective Data Entries with Costs

Defect	Number of Defective Entries	Cost per Defective Entry ($)	Dollar Cost for Type of Defect ($)
Transposed numbers	28	$0.05	$1.40
Wrong character	28	$0.05	$1.40
Torn document	4	$1.00	$4.00
Out of field	3	$0.05	$0.15
Data too light	2	$0.05	$0.10
Creased document	2	$1.00	$2.00
Illegible source document	1	$0.05	$0.05
Total	68		

FIGURE 9.13   Cost Pareto Diagram of Defective Data Entries

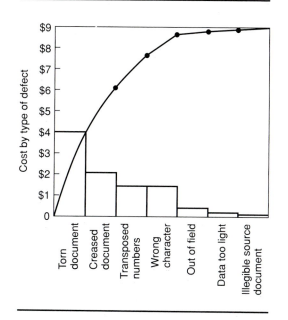

axis. A frequency table with costs per defect and a cost Pareto diagram for the data entry example appear in Figures 9.12 and 9.13. When the dollar cost of each defect is considered, a reordering of defect categories occurs.

This reordering occurs because of the high cost of some types of defects. For example, although there were only four "torn documents," the cost to the company for each torn document is $1 because of machine jams and downtime. Therefore, the dollar cost is $4, making it the most expensive problem. The 28 cards containing "transposed numbers," although high in number, drop down to a lower bar position in terms of dollar cost ($1.40), because the unit cost to correct an entry containing transposed numbers is only $0.05 per entry; only a replacement entry is required to resolve the "transposed numbers" problem.

Note that cost Pareto diagrams consider only visible and known costs; they don't consider invisible and unknown costs such as the cost of an unhappy customer receiving a job containing transposed numbers. This is an important limitation on the effective use of cost Pareto diagrams.

Use of Pareto Diagrams

Pursuing never-ending improvement depends on cooperation between everyone concerned. The Pareto diagram is useful in focusing a group's attention on a common problem and obtaining team cooperation.

As resources, manpower, and time are limited—and the need to improve is critical—it's crucial to concentrate on the most important problems, those represented by the tallest bars on the Pareto diagram. It's often wiser to reduce a tall bar by half than to reduce a short bar to zero. Reducing the tallest bar is a considerable accomplishment, whereas reducing short bars leads to less overall improvement.[14]

Pareto diagrams, as well as the other tools discussed in this book, can be used for improvement in all areas of organizations. They're an important tool for analyzing a problem, whether it be in production, administration, research, maintenance, or office work.

Use of Pareto Diagrams for Root Cause Analysis

Pareto diagrams can be used to determine the root causes (underlying causes) of problems. For example, root cause analysis can be used to improve safety in a mill. First, we construct a Pareto diagram to find out in which department the major injuries occur, as in Figure 9.14(a), assuming all departments are approximately the same size in terms of personnel. Clearly, from our illustration, the maintenance department accounts for the largest number of accidents. Another Pareto diagram, Figure 9.14(b), is constructed to determine how the accidents to maintenance personnel occur. From this chart, we determine that injuries result chiefly from foreign objects entering the eye. We can now take appropriate measures to improve safety—for example, insisting that maintenance personnel wear approved safety goggles.

If the safety effort's focus had been to determine the most serious injuries, perhaps those that disabled employees, the chart would be constructed differently (Figure 9.15). This is much akin to cost Pareto diagrams. Here, although foreign objects in the eye may have accounted for the greatest number of accidents, they aren't the most serious in terms of causing disability. Back strains account for the highest number of disabling injuries. The nature of the analysis we perform can yield different priorities for improvement.

FIGURE 9.14   Accident Pareto Diagram

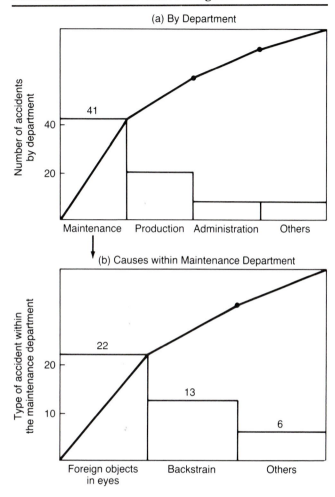

Use of Pareto Diagrams to Improve Chaotic Processes

Pareto diagrams can focus attention on the major problem on which a team concentrates its efforts. This is effective when the ordering of problems remains stable over time—that is, if the process is stable. But if the process is in a state of chaos, Pareto diagrams may not be effective, as the following examples show.[15]

A Maintenance Nightmare.   The foreman of a department considered one of his factory's worst in terms of downtime realizes he must try to improve the situation. The department has 72 machines, each with 36 spindles, for a total of 2,592 spindles that can break down. The foreman, trying to deal with a chaotic situation, constructs a rudimentary Pareto diagram (Figure 9.16). On this chart he lists the

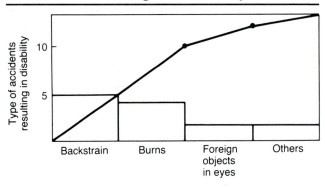

FIGURE 9.15 **FIGURE 9.15   Pareto Diagram of Disability Accidents**

FIGURE 9.16   Foreman's Pareto Diagram

Other Breakdowns	Machine	Spindle Breakdowns
	1	XX
	2	X
X	3	XX
XXXX	4	X
X	5	XXXXXXX
	6	XXXX
X	7	X
	8	
XX	9	X
	10	XXXXXX
	.	
	.	
	.	
	72	

machines by number and creates two categories: spindle breakdowns and other breakdowns. After collecting data, he sees that spindle breakdowns are the major source of problems and also identifies the machines that are experiencing the spindle breakdowns and therefore need an overhaul. Thus, he can plan for the overhauls in order of priority and work them into his maintenance schedule. After three months, the number of breakdowns is reduced by about 70 percent.

Here the Pareto diagram clarified information that had been muddled in the chaotic day-to-day operations. This is an example of a successful application of a Pareto chart to an out-of-control process. It's successful because it tracks deterioration—a one-way process. The machines aren't going to fix themselves—they'll continue to deteriorate unless repairs are made. (In a sense, deterioration is stable instability; its effects persist until a change is made.) The next example demonstrates what happens when Pareto diagrams are applied to an unstable process with changing special causes of variation.

Shifting Causes of Defects.   A quality control specialist is interested in the causes of defects in rejected parts. She divides the causes into eight categories, each representing a different reason for rejection. Data are collected and used to create a Pareto diagram for one month. For this month foreign material is found to be the major cause for rejections, followed by damaged edges, others, stress, damaged mounts, bad trim, bubbles, and sinks. According to the Pareto diagram, foreign material and damaged edges account for 45 percent of the defective units, making these problems the logical starting points for process improvement.

Various distractions keep the quality control specialist from beginning work on process improvement after the first month. At the end of the second month she again prepares a Pareto diagram. This time stress and sinks are the two major problems. Yet in the first month, these had been the fourth and eighth largest reasons for rejection. The causes of defects have shifted over time, indicating process instability.

Pareto diagrams won't be effective if used on a chaotic process because the process isn't ready for improvement. The process must first be stabilized via control charts, as Chapters 5 through 7 related.

Use of Pareto Diagrams to Gauge Improvement

Pareto diagrams can help us determine whether efforts toward process improvement are producing results. If effective actions have been taken, the order of the items on the horizontal axis will change. Figure 9.17 shows Pareto diagrams before and after improvements are implemented on a stable process producing corrugated board boxes. On the basis of the before-improvement chart, the major cause of a defective box is diagnosed as "symbol/letter in wrong location." After improvements are made to the process, "symbol/letter in wrong location" becomes the least frequent source of trouble, demonstrating the improvements' effect through Pareto diagrams. This is a powerful tool when used in this way because it can mobilize support for further process improvement and reinforce continuation of current efforts.

Stratification

Stratification (not to be confused with the stratification control chart pattern discussed in Chapter 8) is a procedure used to describe the systematic subdivision of process data to obtain a detailed understanding of the process's structure. Stratification can be used to break down a problem to discover its root causes and set into motion appropriate corrective actions, called *countermeasures*. Stratification is important to the proper functioning of the PDSA cycle. For example, the number of traffic accidents in Japan peaked in 1970. For each accident, the police officer present at the accident scene was required to complete an accident report

FIGURE 9.17   Pareto Diagram Showing Type and Number of Defects in a Corrugated Box Process Before and After Improvement

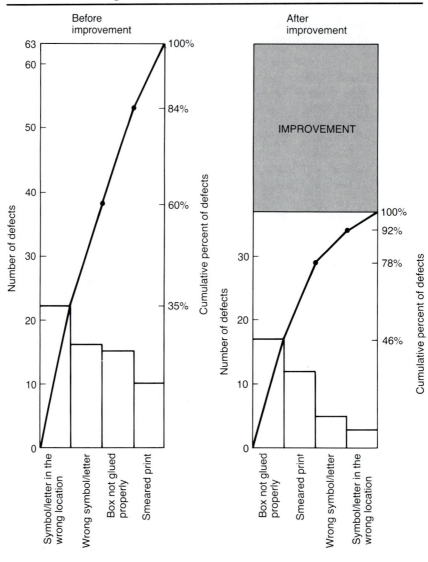

relating the accident's cause. The most common cause of accidents listed on the report, after analysis, was "careless driving." Unfortunately, this stratification didn't yield the root cause of the accidents. Subsequently, a more in-depth stratification of the accident data—particularly the "careless driving" category—yielded specific locations where accidents occurred with high frequency. Determination of these locations gave the police and the Department of Highways information they needed to set appropriate countermeasures to improve road conditions. These actions drastically cut the number of traffic accidents in Japan. Proper stratification showed the root causes of the problem and led the way for establishing proper countermeasures.[16]

Tools for Stratification

Stratification and Pareto Diagrams.   Figure 9.18 shows how stratification is used when performing root cause analysis with Pareto diagrams. Here 110 observations are made. We see that by breaking down a problem into its subcomponents (stratifying the 110 items into appropriate subcomponents A, B, etc.) and by breaking each subcomponent into its subcomponents (stratifying the 50 items in A into A_1 through A_6 and the 40 items in B into B_1 through B_4), we can focus on one or more of the root causes of a process or product problem, from which we can establish a countermeasure for resolving the problem.

FIGURE 9.18   Pareto Diagrams with Stratification

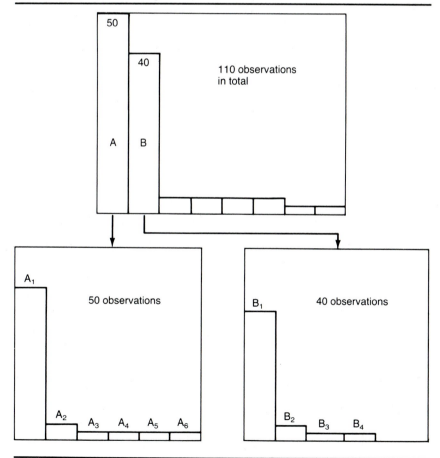

In general, when all categories in a Pareto diagram are approximately the same size, as in Figure 9.19(a), stratifying on another product or process characteristic should be done until a Pareto diagram like the one in Figure 9.19(b) is found. Figure 9.19(a) is called an *old mountain stratification* (the mountain is worn flat) because no one category emerges as the obvious factor on which to take process or product improvement action. Figure 9.19(b) is called a *new mountain stratification* (the mountain is young and has high peaks) because one or two categories emerge as the obvious starting point for process or product improvement action.

Stratification and Cause-and-Effect (C&E) Diagrams.   Figure 9.20 shows how stratification is used when performing root cause analysis with C&E diagrams. We see that second-tier C&E diagrams can be constructed to study in depth any cause

FIGURE 9.19   Pareto Diagram with Same-Size Categories

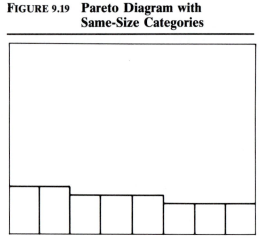

(a) Old Mountain Pareto Diagram

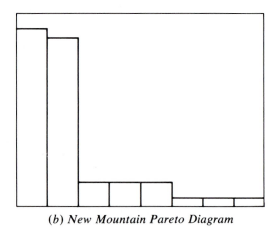

(b) New Mountain Pareto Diagram

FIGURE 9.20 Stratification and Cause-and-Effect Diagrams

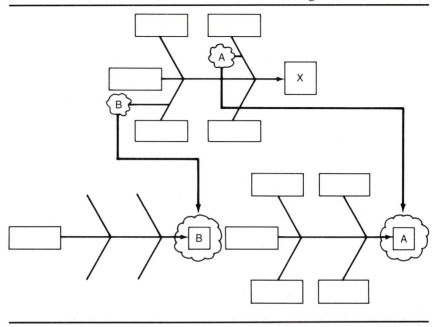

shown on a first-tier C&E diagram, and so on. For example, a C&E diagram used to study "Problem X" generated two major possible causes for Problem X: A and B. Consequently, both A and B are studied through their own C&E diagrams. This stratification could go on indefinitely until one or more of the root causes of Problem X are determined and appropriate countermeasures are set that should lead to process or product improvement.

Stratification with Pareto Diagrams and Cause-and-Effect Diagrams. Figure 9.21(a) shows a Pareto diagram. Figure 9.21(b) shows a C&E diagram focusing exclusively on one of the bars in the Pareto diagram in Figure 9.21(a); this is the *correct* way to stratify a Pareto diagram to study in depth the root causes of a problem. Figure 9.21(c) shows a C&E diagram focusing on all the bars in the Pareto diagram in Figure 9.21(a); this is the *incorrect* way to stratify a Pareto diagram to study in depth a problem's root causes. A C&E diagram should be used to stratify one bar from a Pareto diagram at a time to get an in-depth understanding of the corresponding cause (bar) before any other cause (bar) is studied.

Stratification with Control Charts, Pareto Diagrams, and C&E Diagrams. Figure 9.22 shows how a Pareto diagram can be used to identify common causes of variation from a stable process, and how these common causes of variation can be stratified through Pareto diagrams or C&E diagrams to determine root causes of process or product problems so appropriate countermeasures can be established.

FIGURE 9.21   **Stratification with Pareto Diagrams and Cause-and-Effect Diagrams**

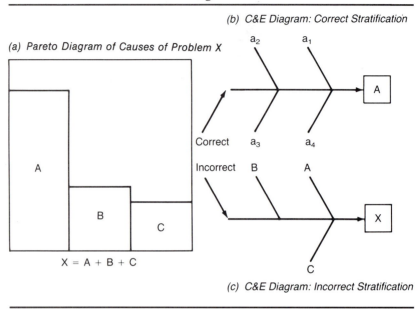

(a) Pareto Diagram of Causes of Problem X

(b) C&E Diagram: Correct Stratification

X = A + B + C

(c) C&E Diagram: Incorrect Stratification

Other Combinations of Tools for Stratification.   Most tools and techniques presented in this text can be used in combination with each other to stratify data about a process or product problem, enabling us to search for a problem's root cause(s). Once we've determined the root cause(s) of a product or process problem, we can establish appropriate countermeasures leading to process or product improvement(s).

Dangers and Pitfalls of Poor Stratification

Failure to perform meaningful stratification can result in establishing inappropriate countermeasures, which can then result in process or product deterioration. For example, analyzing a random sample of 100 records of repairs performed in a factory by maintenance personnel in which accidents occurred could lead to the isolation of alleged root causes of accidents. Countermeasures designed to prevent future accidents would be established based on these alleged root causes. However, this analysis could lead to the determination of causes that aren't unique to repairs resulting in accidents but are typical of both accident and nonaccident repairs. Consequently, the alleged root causes would be ineffective in establishing significant countermeasures to prevent future accidents. Alternatively, if a random sample of 50 records of repairs performed in a factory by maintenance personnel in which accidents occurred were compared with a random sample of 50 records of repairs performed in the same factory (under the

FIGURE 9.22   Stratification with Control Charts, Pareto Diagrams, and Cause-and-Effect Diagrams

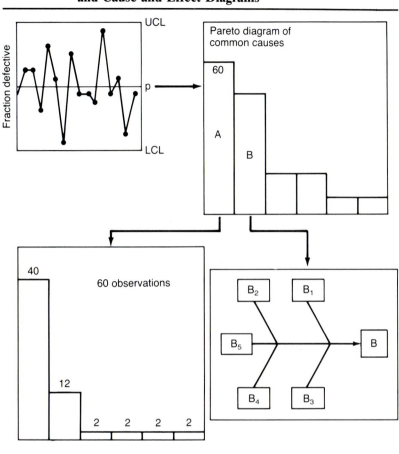

same conditions) in which accidents didn't occur, this comparison could lead to the isolation of the real root causes of accidents. This alternative procedure would isolate real root causes because stratification focused attention on the differences between accident and nonaccident repairs.

Single-Case Boring

Stratification isn't possible when only a few data points are available—for example, if data exist on only four accident cases in a plant. In this situation, meaningful analysis is still possible; it's called *single-case boring*. Single-case boring is a procedure in which a few cases are studied in great depth. This procedure may yield information that could lead to conditional countermeasures or countermeasures appropriate for the situation in which they were discovered—for example, a countermeasure established to prevent a particular type of accident, in a particular situation, by a particular individual.

Summary

This chapter explains tools and methods for stabilizing a process (taking action on special sources of variation) and for improving a process (taking action that will reduce common variation and center the process's average on nominal). These tools and methods are brainstorming, cause-and-effect diagrams, check sheets, Pareto diagrams, and stratification. Brainstorming is a way to bring forth a large number of ideas from a group about a process or product problem. A cause-and-effect diagram can be used to organize ideas about a problem collected during a brainstorming session and focus attention on one idea as a possible solution to the problem. A check sheet is a data collection form used to gather data logically, as in rational subgroups. A Pareto diagram is used to identify and prioritize problems for solution. Finally, stratification describes the systematic subdivision of population or process data to obtain detailed understanding of the structure of the population or process. Stratification can be used to break down a problem to discover its root causes and set into motion appropriate countermeasures. All these tools and methods, in conjunction with control charts, can be used in the PDSA cycle to relentlessly decrease the difference between customer needs and process performance.

Exercises

9.1 a. What is the purpose of brainstorming?
   b. List the seven steps required to conduct a brainstorming session.
   c. List the rules a group should follow to have an effective brainstorming session.

9.2 a. What is the function of a cause-and-effect diagram?
   b. Draw a generic cause-and-effect diagram.
   c. Construct a cause-and-effect diagram to explain the causes of being late to work.
   d. Explain the purpose of root cause analysis.
   e. Construct an example of a root cause analysis.
   f. Explain the purpose of process analysis.
   g. Construct an example of process analysis.

9.3 a. Explain the purpose of check sheets.
   b. Construct an attribute check sheet form to present the reasons for absenteeism in a factory.
   c. Construct a variables check sheet form to present the diameters of bearings produced in a factory.

d. Construct a defect location check sheet to present blemishes on cordless telephones.

9.4 a. Explain the purpose of Pareto analysis.

b. Given the following check sheet of causes of absenteeism in a factory for January 1995, construct a Pareto diagram.

Cause of Absenteeism	Number of Occurrences
Personal illness	28
Child's illness	46
Car broke down	4
Personal emergency	10
Day made long weekend	3
Other	12

9.5 a. Given the following subcauses of "child's illness" from the preceding Pareto diagram, perform a root cause analysis. (Stratify the data to determine the root cause of absenteeism resulting from "child's illness".)

Cause of "Child's Illness" Absenteeism	Number of Occurrences
Had to take child to doctor	8
Couldn't leave child in day care	33
Other reasons	5

b. Suggest a possible countermeasure to rectify the preceding problem.

Endnotes

1. Virgil Rehg, *Quality Circle Manual for Coordinators and Leaders* (Wright-Patterson AFB, OH), p. 19.
2. J. F. Beardsley & Associates, International, *Quality Circles: Member Manual,* 1977, p. 40.
3. Ibid.
4. Rehg, *Quality Circle Manual,* pp. 19–20.
5. Beardsley & Associates, *Quality Circles,* pp. 42–43.
6. Kaoru Ishikawa, *Guide to Quality Control,* 11th printing, 1983 (Tokyo: Asian Produc-

tivity Organization, 1976), pp. 26–28. Available through UNIPUB, Box 433, Murray Hill Station, New York, NY 10157.

7. A Camp, University of Miami School of Business Administration, *Methods for the Improvement of Quality and Productivity,* Class Project, 1986.

8. Ibid.

9. Ishikawa, pp. 29–35.

10. Beardsley & Associates, *Quality Circles,* p. 88.

11. Ibid.

12. Ishikawa, p. 43.

13. Ibid., pp. 43–44.

14. Ibid., pp. 45–46.

15. Adapted from Donald J Wheeler, and David S Chambers, *Understanding Statistical Process Control,* 2d ed. (Knoxville, TN: SPC Press, 1992), pp. 316–17.

16. Speech by Dr. Noriaki Kano, Science University of Tokyo, given at Florida Power and Light Company, March 1987.

PART IV

Process Performance in Analytic Studies

This section covers process performance in analytic studies. It looks at two aspects of process performance: specifications and process capability studies.

Chapter 10 analyzes performance and technical specifications with examples of each type of specification. Technical specifications are subdivided into individual unit specifications, acceptable quality level (AQL) specifications, and distribution specifications. The significance of each type of technical specification to a process's performance is discussed, as is the fallacy of defining quality as conformance to specifications. Chapter 10 also introduces created dimensions (dimensions created in the act of assembly) and discusses statistical laws that govern their behavior.

Chapter 11 concerns process capability studies and quality improvement stories. Process capability studies provide important information about a process's condition to facilitate process improvement actions. Two types of process capability studies are discussed: attribute studies and variables (measurement) studies. In addition, process capability indices are presented as quantitative measures of the status of a process. Finally, Chapter 11 presents quality improvement (QI) stories. The QI story is an efficient format through which employees can present and publicize their quality improvement efforts. QI stories standardize quality control reports, avoid logical errors in analysis, and connect improvement efforts with organizational objectives. QI stories can also be used to enable employees to think in terms of the PDSA cycle in their daily work.

10 Specifications

Introduction

The Deming cycle works to diminish the difference between process performance and customer needs. For this difference to be diminished, specifications for customer needs must be developed and revised, and process performance must be measured against these specifications.

If an organization is operating in a "defect detection" mode, specifications serve as vehicles to sort conforming and defective product or service. Unfortunately, when an organization operates in a defect detection mode, there's no feedback loop in the process to enable management to diminish the difference between customer needs (specifications) and process performance.

If an organization is operating in a "defect prevention" mode, specifications allow management to operationalize, measure, and diminish the proportion of product or service that fails to meet specifications. This is accomplished by creating a process history via attribute control charts. These attribute charts are then used in conjunction with other methods (as discussed in Chapter 9) as a basis for process stabilization and improvement. These improvements can take the form of a decrease in fraction defective, a decrease in number defective, or a decrease in number of defects per unit.

If an organization is operating in a "never-ending improvement" mode, specifications are important signposts to management for continuously reducing process variation within specifications and moving the process's mean to a level that creates customer satisfaction. In this mode of management, the Deming cycle is fully functional. Quality-of-performance studies lead to product or service redesign (revised specifications). Quality-of-conformance studies diminish the difference between revised specifications and process performance. Quality-of-performance studies check to see how the products or services are performing in the hands of the user.

This chapter discusses various types of specifications to better clarify their role in process improvement.

Types of Specifications

Specifications fall into two broad categories: performance specifications and technical specifications.

Performance Specifications

Performance specifications address a need. For example, a vendor agrees to supply an air conditioning system sufficient to maintain a temperature range between 65 and 70 degrees Fahrenheit for a computer facility in Room 325 of the customer's Spring River Plant. Performance is guaranteed for a period of no less than three years. Here the customer's requirements are stated for the vendor.

Technical Specifications

Technical specifications describe performance at delivery. There are three types of technical specifications: individual unit specifications; acceptable quality level (AQL) specifications; and distribution specifications.

Individual Unit Specifications.   *Individual unit specifications* state a boundary, or boundaries, that apply to individual units of a product or service. An individual unit of product or service is considered to conform to a specification if it's on or inside the boundary or boundaries.

Individual unit specifications are made up of two parts, which together form a third part. The first part of an individual unit specification is the *nominal value*. This is the desired value for process performance mandated by the customer's needs. Ideally, if all quality characteristics were at nominal, products and services would perform as expected over their life cycle. The second part of an individual unit specification is a *tolerance*. A tolerance is an allowable departure from a nominal value established by design engineers that's deemed nonharmful to the functioning of the product or service over its life cycle. Tolerances are added and/or subtracted from nominal values. The third part of an individual unit specification is a *specification limit*. Specification limits are the boundaries created by adding and/or subtracting tolerances from a nominal value. It's possible to have two-sided specification limits:

USL = Nominal + Tolerance
LSL = Nominal − Tolerance

where USL is the upper specification limit and LSL is the lower specification limit; or one-sided specification limits (i.e., either USL or LSL).

An example of an individual unit specification and its three parts can be seen in the specification for the "case hardness depth" of a camshaft. A camshaft is considered to be conforming with respect to case hardness depth if each individual unit is between 7.0 mm ± 3.5 mm (or 3.5 to 10.5 mm). The nominal value in that specification is 7.0 mm; the two-sided tolerance is 3.5 mm; the lower specification limit is 3.5 mm (7.0 mm − 3.5 mm); and the upper specification limit is 10.5 mm (7.0 mm + 3.5 mm). As stated earlier, a camshaft is considered to conform with respect to case hardness depth if its case hardness depth is between 3.5 and 10.5 mm, inclusive.

From our earlier discussion of the philosophy of continuous reduction of variation, we saw that the goal of modern management shouldn't be 100 percent conformance to specifications ("Zero Defects") but the never-ending reduction of process variation within specification limits so that all products/services are as close to nominal as possible. Specified tolerances become increasingly irrelevant as process variation is reduced so that the process's output is well within specification limits.

Acceptable Quality Level (AQL) Specifications. *Acceptable quality level (AQL) specifications* state a requirement that must be met by most individual units of product or service, but it allows a certain proportion of the units to exceed the requirements. For example, cam shafts shall be acceptable if no more than 3 percent of the units exceed the specification limits of 3.5 and 10.5 mm. This type of specification limit is frequently referred to as an Acceptable Quality Level. AQL specifications are much like individual unit specifications, except they have one added negative feature: they formally. support the production of a certain percentage of defective product or service. This attitude toward the production of defective product/service is devastating to the interdependent system of stakeholders of an organization.

Distribution Specifications. *Distribution specifications* define an acceptable distribution for each product or service quality characteristic. In an analytic study, a distribution is defined in terms of its mean, standard deviation, and shape. However, from the Empirical Rule discussed in Chapter 4, it's not necessary to make any assumptions about the shape of the distribution. That is, virtually all data from a stable process will fall between the mean plus or minus three standard deviations.

As an example of a distribution specification, the case hardness depth of a camshaft shall be stable with an average depth of 7.0 mm and a standard deviation not to exceed 1.167 mm. In other words, individual units shall be distributed around the average with a dispersion not to exceed 3.50 mm on either side of the average since for a stable process, virtually all of the output will be within three standard deviations on either side of the mean [7.0 mm ± 3(1.167 mm) = 7.0 mm ± 3.50 mm = 3.50 to 10.50 mm].

The mean and standard deviation are simply directional goals for management when using distribution specifications. Management must use statistical methods to move the process average toward the nominal value of 7.0 mm and to decrease the process standard deviation as far below 1.167 mm as possible. Distribution requirements are stated in the language of the process and promote the never-ending improvement of quality.

FIGURE 10.1   **Final Cooked Temperature of Five-Ounce Hospital Steaks**

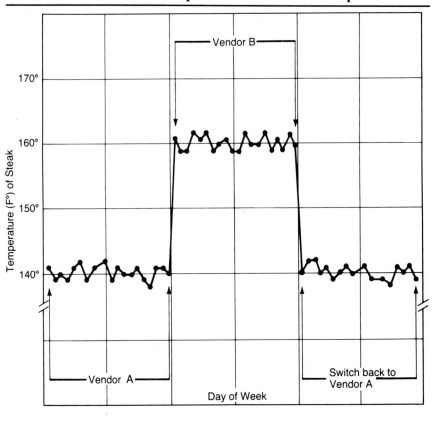

Note: Most patients can detect a 10°F
difference in steak temperature. The
difference in temperature between vendor A's
and vendor B's steak when subjected to the
hospital's cooking regimen is approximately 20°F.

Distinguishing between Performance Specifications and Technical Specifications

Performance specifications aren't commonly used in business; instead, technical specifications are used. Unfortunately, this can cause major problems because technical specifications may not produce the desired performance specifications.

As an example, consider a hospital that serves medium (versus rare or well-done) steak to patients who select steak for dinner.[1] The performance desired is patient satisfaction within nutritional guidelines. But performance specifications aren't used. Instead, a technical specification of five ounces of steak is substituted; it's assumed they're equivalent.

A hospital purchasing agent switches from meat vendor A to meat vendor B to secure a lower price, while still meeting the technical specification of five ounces. He doesn't discuss or inform the hospital nutritionist and kitchen staff of the switch in vendors. The hospital nutritionist begins receiving complaints from patients that the steak is tough and well done. She investigates and finds that vendor A's steaks were thick, while vendor B's are thin (but longer and wider). She realizes via statistical monitoring methods that the thinner steaks get hotter more quickly and hence cook faster, given the usual preparation regimen (Figure 10.1). She concludes, "If I'd known that the steaks had been changed, I could have accommodated the change without creating patient dissatisfaction." The purchasing agent says, "I met the technical specification of five ounces." The problem lies in assuming that technical specifications equal performance specifications. But this isn't necessarily true.

The Fallacy That Conformance to Technical Specifications Defines Quality

Mere conformance to specification limits is insufficient to achieve the quality level required to compete effectively in today's marketplace. Management must constantly try to reduce process variation around a nominal value within specification limits to achieve the degree of uniformity required to produce products or services that function exactly as promised to the customer over their life cycle. The belief that there's no loss from products that are within specification limits regardless of the size of the deviation from the nominal value was discussed in Chapter 5 and will be discussed further in Chapter 12.

Created Dimensions

When parts are assembled, new dimensions are created; these new dimensions have statistical distributions.[2] For example, if two boards are glued together to form a double-thick board, the distribution of the thickness of the double-thick boards is a newly created dimension. Management must be able to control and reduce the variation of these created dimensions so the final assemblies will perform perfectly for the customer over the product's life cycle. Understanding and controlling these created dimensions requires working knowledge of the statistical rules of created dimensions. If management doesn't pay attention to their statistical characteristics, these dimensions will fail to be within specification limits and will cause problems in production or service, will increase costs, and will lead to customer dissatisfaction. The discussion of specifications earlier in this chapter applies to created dimensions as well.

Law of the Addition of Component Dimension Averages

If component parts are assembled so that the individual component dimensions are added to one another, the average dimension of the assembly will equal the sum of the individual component average dimensions. Figure 10.2 illustrates this concept.[3] If three component parts are glued together (assuming the glue takes no measurable dimensions), the average width of the assembled part equals the sum of the average individual part widths:

$$\overline{X}_{assembly} = \overline{X}_1 + \overline{X}_2 + \overline{X}_3 \tag{10.1}$$

where

$$\overline{X}_{assembly} = \text{Average width of the assembly}$$

$$\overline{X}_1 = \text{Average width of part 1,}$$

$$\overline{X}_2 = \text{Average width of part 2, and}$$

$$\overline{X}_3 = \text{Average width of part 3.}$$

Part 1 has an average thickness of 10 mm, part 2 has an average thickness of 20 mm, and part 3 has an average thickness of 30 mm. Consequently, the average thickness of the final assembly is the sum of all three averages, 60 mm (10 mm + 20 mm + 30 mm):

$$\overline{X}_{assembly} = \overline{X}_1 + \overline{X}_2 + \overline{X}_3$$
$$= 10 \text{ mm} + 20 \text{ mm} + 30 \text{ mm}$$
$$= 60 \text{ mm}$$

The preceding law holds only if the processes generating the components are in statistical control.

Law of the Differences of Component Dimension Averages

If component parts are assembled so that the individual component dimensions are subtracted from one another, the average dimension of the assembly will then equal the difference between the individual component average dimensions. Figure 10.3 illustrates this concept.[4] If a bolt is projected through a steel plate, the average length of the bolt projection through the steel plate equals the difference between the bolt's shank length and the width of the steel plate:

$$\overline{X}_{bolt\ projection} = \overline{X}_s - \overline{X}_p \tag{10.2}$$

where

$$\overline{X}_{bolt\ projection} = \text{Average length of the bolt shank projection through the steel plate,}$$

$$\overline{X}_s = \text{Average length of the bolt shank, and}$$

$$\overline{X}_p = \text{Average width of the steel plate.}$$

FIGURE 10.2   Addition of Averages

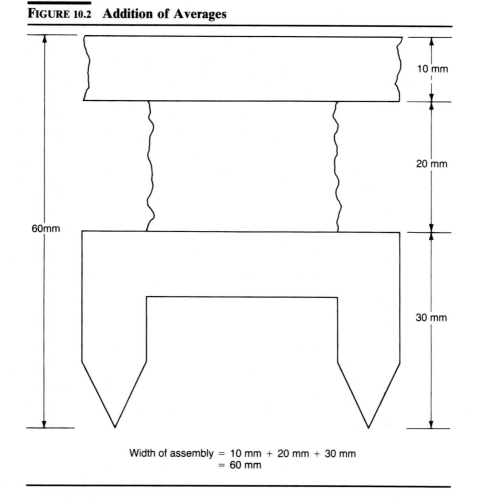

Width of assembly = 10 mm + 20 mm + 30 mm
= 60 mm

The bolt shank has an average length of 12 mm, and the steel plate has an average width of 8 mm. Consequently, the average bolt projection through the steel plate is 4 mm (12 mm − 8 mm):

$$\bar{X}_{\text{bolt projection}} = \bar{X}_s - \bar{X}_p$$

$$= 12 \text{ mm} - 8 \text{ mm}$$

$$= 4 \text{ mm}$$

Again, the preceding law holds only for component processes in statistical control.

Law of the Sums and Differences of Component Dimension Averages

If component parts are assembled so that the individual component parts are added and subtracted from one another, the average dimension of the assembly

FIGURE 10.3   Differences of Averages

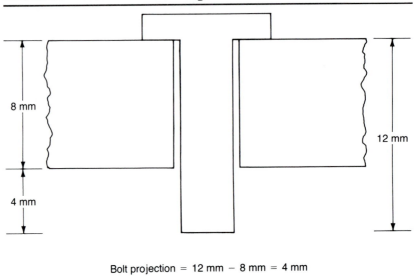

Bolt projection = 12 mm − 8 mm = 4 mm

will then equal the algebraic sum of the individual component average dimensions. Figure 10.4 illustrates this concept.[5] If a bolt is screwed through a steel plate and washers are inserted on either side of the plate, the average length of the bolt projection through the steel plate and washers then equals the difference between the sum of the widths of the two washers and the steel plate, and the length of the bolt shank:

$$\overline{X}_{\text{bolt projection}} = \overline{X}_s - (\overline{X}_{w1} + \overline{X}_p + \overline{X}_{w2}) \qquad (10.3)$$

where

$\overline{X}_{\text{bolt projection}}$ = Average length of the bolt shank projection through both washers and the steel plate,

$\overline{X}_s$ = Average length of the bolt shank,

$\overline{X}_{w1}$ = Average width of the top washer,

$\overline{X}_{w2}$ = Average width of the bottom washer, and

$\overline{X}_p$ = Average width of the steel plate.

The bolt shank has an average length of 40 mm, the steel plate has an average thickness of 27 mm, the top washer has an average thickness of 3 mm, and the bottom washer has an average thickness of 4 mm. Hence, the average bolt projection through the steel plate and both washers is 6 mm [40 mm − (3 mm + 27 mm + 4 mm)]:

FIGURE 10.4   Sums and Differences of Averages

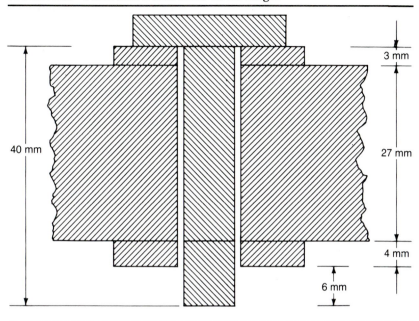

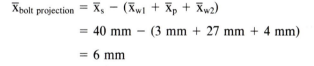

$$\overline{x}_{\text{bolt projection}} = \overline{x}_s - (\overline{x}_{w1} + \overline{x}_p + \overline{x}_{w2})$$

$$= 40 \text{ mm} - (3 \text{ mm} + 27 \text{ mm} + 4 \text{ mm})$$

$$= 6 \text{ mm}$$

Again, the preceding law holds only for component processes in statistical control.

Law of the Addition of Component Dimension Standard Deviations

If component parts are assembled at random (for example, so that each component part is drawn randomly from its own bin with no selection criteria), the standard deviation of the assembly will be the square root of the sum of the component variances, regardless of whether the components are added or subtracted from each other. This law applies to assemblies in which the component parts combine linearly and are statistically independent.

For example, consider again the bolt projection in Figure 10.3. Recall that $\overline{x}_s = 12$ mm and $\overline{x}_p = 8$ mm; consequently, we found from Equation 10.2 that $\overline{x}_{\text{bolt projection}} = 4$ mm. Further, assume that the standard deviation of the bolt shank length, σ_s, is 0.010 mm and the standard deviation of the steel plate width, σ_p, is 0.008 mm. The standard deviation of the bolt projection thus is

$$\sigma_{\text{bolt projection}} = [\sigma_s^2 + \sigma_p^2]^{1/2} \qquad (10.4)$$
$$= [(0.010)^2 + (0.008)^2]^{1/2}$$
$$= 0.0128 \text{ mm}$$

We must realize that the standard deviation of the bolt projection is not 0.018 mm, the sum of the individual component standard deviations. The square root of the sum of the individual component variances will always be less than the sum of the individual component standard deviations. This means that the assembly-to-assembly variation among random assemblies will be less than would be indicated by summing the individual components' unit-to-unit variations. Again, the preceding law holds only for component processes in statistical control.

Law of the Average for Created Areas and Volumes

If areas and volumes are created by the assembly of component parts, then the average area or volume of the assembly will equal the product of the individual component average dimensions if the component processes are stable and independent. Figure 10.5 illustrates this concept.[6] If a boxlike container is constructed with two short sides, two long sides, and two top/bottom sides, then the average internal volume of the boxlike container equals the product of the average length of the short side, the average length of the long side, and the average width of the sides:

$$\bar{x}_v = (\bar{x}_s)(\bar{x}_l)(\bar{x}_w) \qquad (10.5)$$

where

$\bar{x}_v$ = Average internal volume of the constructed container,

$\bar{x}_s$ = Average length of short side,

$\bar{x}_l$ = Average length of long side, and

$\bar{x}_w$ = Average width of the sides.

The average length of the short side is 3.0 mm, the average length of the long side is 8.0 mm, and the average width of the sides is 2.0 mm. Consequently, the

FIGURE 10.5   Created Volume of a Container

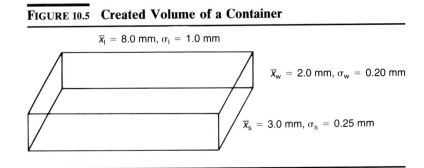

$\bar{x}_l$ = 8.0 mm, σ_l = 1.0 mm

$\bar{x}_w$ = 2.0 mm, σ_w = 0.20 mm

$\bar{x}_s$ = 3.0 mm, σ_s = 0.25 mm

average internal volume of the constructed container is 48 mm³ (3 mm × 8 mm × 2 mm).

$$\overline{x}_v = (\overline{x}_s)(\overline{x}_1)(\overline{x}_w)$$

$$= (3 \text{ mm})(8 \text{ mm})(2 \text{ mm})$$

$$= 48 \text{ mm}^3$$

Again, the preceding law holds only for component processes in statistical control.

Law of the Standard Deviation for Created Areas and Volumes

If areas and volumes are created by the assembly of component parts, we can calculate the standard deviation of the created areas or volumes:

$$\sigma_{area} = [\overline{x}_s^2\sigma_1^2 + \overline{x}_1^2\sigma_s^2 + \sigma_1^2\sigma_s^2]^{1/2} \tag{10.6}$$

where

$\overline{x}_s$ = Average length of the short side,

$\overline{x}_1$ = Average length of the long side,

σ_s = Standard deviation of length of the short side,

σ_1 = Standard deviation of length of the long side, and

σ_{area} = Standard deviation of the created internal area.

Hence

$$\sigma_{volume} = [\overline{x}_s^2\overline{x}_w^2\sigma_1^2 + \overline{x}_1^2\overline{x}_w^2\sigma_s^2 + \overline{x}_s^2\overline{x}_1^2\sigma_w^2$$
$$+ \overline{x}_s^2\sigma_1^2\sigma_s^2 + \overline{x}_1^2\sigma_s^2\sigma_w^2 + \overline{x}_w^2\sigma_s^2\sigma_1^2 \tag{10.7}$$
$$+ \sigma_1^2\sigma_s^2\sigma_w^2]^{1/2}$$

where

$\overline{x}_w$ = Average width,

σ_w = Standard deviation of the width,

and $\overline{x}_s$, $\overline{x}_1$, σ_s, and σ_1 are defined as in Equation 10.6.

Equations 10.6 and 10.7 assume that the component processes are stable and independent. Figure 10.5 illustrates this concept.[7] If the means and standard deviations for the boxlike container's dimensions are as shown in Figure 10.5, then the standard deviation of the internal volume for the assembled container is

$$\sigma_{volume} = [\overline{x}_s^2\overline{x}_w^2\sigma_1^2 + \overline{x}_1^2\overline{x}_w^2\sigma_s^2 + \overline{x}_s^2\overline{x}_1^2\sigma_w^2$$
$$+ \overline{x}_s^2\sigma_1^2\sigma_w^2 + \overline{x}_1^2\sigma_s^2\sigma_w^2 + \overline{x}_w^2\sigma_s^2\sigma_1^2$$
$$+ \sigma_1^2\sigma_s^2\sigma_w^2]^{1/2}$$

$$= [3^2 2^2 1^2 + 8^2 2^2 (.25)^2 + 3^2 8^2 (.20)^2$$
$$\quad + 3^2 1^2 (.20)^2 + 8^2 (.25)^2 (.20)^2 + 2^2 (.25)^2 (1^2)$$
$$\quad + (1^2)(.25)^2 (.20)^2]^{1/2}$$
$$= [36 + 16 + 23.04 + .36 + .16 + .25 + .0025]^{1/2}$$
$$= (75.8125)^{1/2}$$
$$= 8.71 \text{ mm}^3$$

Again, the preceding law holds only for component processes in statistical control.

Summary

An organization's quality consciousness can be better understood by examining the types of specifications it uses in production and service. If a firm uses individual unit and/or AQL specifications as guidelines to separate good product/service from bad, the firm is operating in a "defect detection" mode. If a firm uses individual unit specification as guidelines to determine the percentage of its output that's out-of-specification so that the difference between customer needs and process performance can be decreased, the firm has advanced to a "defect prevention" mode. Finally, if a firm uses distribution or performance specifications in its never-ending pursuit of total process improvement, it's operating in a "never-ending improvement" mode. The goals of never-ending improvement are the constant reduction of unit-to-unit variation and the movement of the process average toward nominal. Conformance to technical specifications ("Zero Defects") isn't an acceptable form of quality consciousness.

This chapter introduced the statistical laws that govern product dimensions created in the act of assembly: the law of the addition of component dimension averages, the law of the differences of component dimension averages, the law of the sums and differences of component dimension averages, the law of the addition of component dimension standard deviations, the law of the average for created areas and volumes, and the law of the standard deviation for created areas and volumes. These laws offer important insights into how to continually reduce the difference between customer needs (specifications for created dimensions) and process performance (for created dimensions).

We must understand the information in Chapter 10 to be able to determine a process's capability or identity. The next chapter focuses on process capability studies.

Exercises

10.1 a. Discuss the significance of the phrase "defect detection" to the state of an organization's quality consciousness.

b. Discuss the significance of the phrase "defect prevention" to the state of an organization's quality consciousness. Relate your discussion to the type of data an organization would use in its improvement efforts.

c. Discuss the significance of the phrase "never-ending improvement" to the state of an organization's quality consciousness. Relate your discussion to the type of data an organization would use in its quality improvement efforts.

10.2 What's the basic function of a performance specification?

10.3 Explain the purpose and describe the construction of the three types of technical specifications: individual unit specifications, acceptable quality level specifications, and distribution specifications.

10.4 In an assembly operation steel sheet A is glued onto steel sheet B to create a double-sheet thickness. Assume that the glue has no discernable thickness and that the unit-to-unit variation in thickness for both types of steel sheets is stable over time. The resulting thickness of the combined steel sheets is the quality characteristic of interest. The following process statistics have been collected concerning both types of steel sheets:

Steel Sheet A	Steel Sheet B
Mean = 2.50 inches	Mean = 4.75 inches
Std. dev. = 0.25 inches	Std. dev. = 0.50 inches

a. Compute the mean of the double-sheet thickness.
b. Compute the standard deviation of the double-sheet thickness.

10.5 Rectangular sheets of material are produced in an assembly operation. Their dimensions are 9.0 inches in width by 14.0 inches in length. Assume that unit-to-unit variations among the widths and lengths of the rectangular sheets are stable over time. The area of the sheets is the quality characteristic of interest. The following process statistics have been collected for the widths and lengths of the rectangular sheets:

Width	Length
Mean = 9.0 inches	Mean = 14.0 inches
Std. dev. = 0.10 inches	Std. dev. = 0.40 inches

a. Compute the mean area of the rectangular sheets.
b. Compute the standard deviation of the area of the rectangular sheets.

Endnotes

1. A Camp, University of Miami School of Business Administration, *Methods for the Improvement of Quality and Productivity,* Class Project, 1986.
2. AT&T, *Statistical Quality Control Handbook,* 10th printing, May 1984 (Indianapolis: AT&T, 1956), pp. 119–27.
3. Ibid.
4. Ibid.
5. Ibid.
6. Acheson Duncan, *Quality Control and Industrial Statistics,* 5th ed. (Homewood, IL.: Richard D. Irwin, 1986), pp. 107–12.
7. Ibid.

Process Capability and Improvement Studies

Introduction

Process capability studies determine whether a process is unstable, investigate any sources of instability, determine their causes, and take action to eliminate such sources of instability. After all sources of instability have been eliminated from a process, the natural behavior of the process is called its *process capability*. A process must have an established process capability before it can be improved. Consequently, a process capability study must be successfully completed before a process improvement study can have any chance for success.

Process improvement studies follow the Deming cycle of *Plan, Do, Study, Act*. First, managers construct a plan for process improvement, or a plan to decrease the difference between customer needs and process performance (Plan). Second, they test the plan's validity using a planned experiment (Do). Third, they collect data and study the results of the planned experiment to determine if the plan will decrease the difference between customer needs and process performance (Study). Fourth, if the data collected about the plan show that the plan will achieve its objective(s), it's standardized through "best practices" and training (Act); and the managers responsible for the plan return to the Plan phase of the Deming cycle to find other variables that will further reduce the difference between customer needs and process performance. If the data collected about the plan show that the plan won't achieve its objective(s), the managers responsible for the plan return to the Plan phase of the Deming cycle to find other variables that will reduce the difference between customer needs and process performance. The Deming cycle follows a never-ending path of process and quality improvement.

This chapter is divided into three sections: process capability studies, process improvement studies, and quality improvement stories. The quality improvement story is an effective format for quality control practitioners to present process capability and process improvement studies to management.

Process Capability Studies

There are two types of process capability studies: attribute process capability studies and variables process capability studies. This chapter details both types.

Attribute Process Capability Studies

Attribute process capability studies determine a process's capability in terms of fraction of defective output or some other measure of process performance. The major tools used in attribute process capability studies are attribute control charts and the tools discussed in Chapter 9. The process capability for a p chart is $\bar{p}$ (the average fraction of defective units generated by the process). The process capability for the np chart is $n\bar{p}$ (the average number of defective units generated by the process for a given subgroup size, n). The process capability for a c chart is $\bar{c}$ (the average number of defects per unit generated by the process for a given area of opportunity). Finally, the process capability for a u chart is $\bar{u}$ (the average number of defects per unit generated by the process where the area of opportunity varies from subgroup to subgroup).

A shortcoming of this type of study is that it begins with a specification, but it's not specific about the reason for failure to meet that specification. The p chart doesn't indicate if defective units result from the process being off nominal and too close to the specification limit, or because the process has too much unit-to-unit variation, or because the process isn't stable with respect to its mean and/or variance. Further, as p charts are relatively insensitive to shifts or trends in the process, problems can go undetected for so long that they cause defectives before they're checked. However, p charts are frequently based on readily available data, which explains their popularity and use.

Variables Process Capability Studies

Variables process capability studies determine a process's capability in terms of the distribution of process output or some other process quality characteristic. The major tools used in variables process capability studies are variables control charts and the tools discussed in Chapter 9. Variables control charts are used to stabilize a process so we can determine meaningful upper and lower natural limits. Natural limits are computed for stable processes by adding and subtracting three times the process's standard deviation to the process centerline. In general, for any variables control chart, the upper and lower natural limits are

$$\text{UNL} = \bar{\bar{x}} + 3\sigma \tag{11.1}$$

$$\text{LNL} = \bar{\bar{x}} - 3\sigma \tag{11.2}$$

Specifically, for x-bar and R charts, the upper and lower natural limits are

$$\text{UNL} = \bar{\bar{x}} + 3(\bar{R}/d_2) \tag{11.3}$$

$$\text{LNL} = \bar{\bar{x}} - 3(\bar{R}/d_2) \tag{11.4}$$

For x-bar and s charts, the upper and lower natural limits are

$$\text{UNL} = \bar{\bar{x}} + 3(\bar{s}/c_4) \tag{11.5}$$

$$\text{LNL} = \bar{\bar{x}} - 3(\bar{s}/c_4) \tag{11.6}$$

For individuals charts, the upper and lower natural limits are

$$\text{UNL} = \bar{x} + 3(\bar{R}/d_2) \tag{11.7}$$

$$\text{LNL} = \bar{x} - 3(\bar{R}/d_2) \tag{11.8}$$

Natural limits shouldn't be confused with control limits, and as a rule, natural limits shouldn't be shown on variables control charts because natural limits apply to individual units of output and control limits apply to subgroup statistics—so the comparison would be misleading. One notable exception to this rule is the individuals control chart for variables. In that case, the subgroups consist of individual units, and natural limits and control limits are the same.

Interpretation of the natural limits requires stability of the process under study and the Empirical Rule discussed in Chapter 4. If the output distribution of a process is stable, then for Equations 11.1 through 11.8 we can say that virtually all process output will be between the natural limits. For example, if samples of four steel ingots are drawn from an ingot-producing process every hour, and the process is stable with a process average subgroup weight of 42.0 pounds ($\bar{\bar{x}} = 42.0$ pounds) and an average range of 0.6856 pounds, we can say the following about the process using Equations 11.3 and 11.4:

1. The process's upper natural limit is

$$\text{UNL} = \bar{\bar{x}} + 3(\bar{R}/d_2) = 42.0 + 3(0.6856/2.059)$$

$$= 42.0 + 3(0.333) = 42.999$$

$$\cong 43.0 \text{ pounds}$$

2. The process's lower natural limit is

$$\text{LNL} = \bar{\bar{x}} - 3(\bar{R}/d_2) = 42.0 - 3(0.6856/2.059)$$

$$= 42.0 - 3(0.333) = 41.001$$

$$\cong 41.0 \text{ pounds}$$

3. Virtually all of the steel ingots produced will weigh between 41.0 and 43.0 pounds. This is what the steel ingot process is capable of producing; it's the process's identity.

The disadvantage of variables process capability studies is that they frequently require that special data be collected. The advantages of variables process capability studies are that they provide information such as whether the process is centered on nominal, exhibiting too much unit-to-unit variation, or unstable with respect to its mean and/or variation—and furthermore, these studies are sensitive to shifts in the process and are helpful in detecting trends or shifts in the process before they cause trouble. Finally, variables process capability studies allow for the examination of specification limits to determine whether the specification limits were reasonable in the first place.

Data Requirements for Process Capability Studies

Attribute Studies.   Attribute process capability studies require a great deal of data. As a rule of thumb, the study should cover at least three distinct time periods, where each time period should contain 20 to 25 samples and each sample should have between 50 and 100 units. This rule of thumb is based on experience as well as statistical theory.

Variables Studies.   Variables process capability studies require far less data than attribute studies. However, a variables study may be required for each quality characteristic that can cause a unit to be defective. As a rule of thumb, a variables study should cover at least three distinct time periods. The first period should contain about 50 samples of between three and five units each, and the second and third time periods should contain 25 samples of between three and five units each.

Addition of New Data onto a Process Capability Chart.   After initial control limits have been calculated, the question arises as to what to do with additional data: should revised control limits be computed, or should the old control limits be extended across the control chart and new points plotted against the old limits? Recall from our discussion in Chapter 7 on revising control limits that if the process is stable and hasn't changed significantly, new limits shouldn't be calculated because they can cause action based on common variation.[1] In this case, the best procedure is to plot the new data against the old limits and search for a change in the data pattern. If the process has changed significantly, new limits should then be calculated from the additional data. These new limits allow for analysis of the process's new capability.

Process Capability Studies on Unstable Processes

Process statistics, such as the measures of location, dispersion, and shape discussed in Chapter 4, can't be estimated from a process capability study performed on an unstable or chaotic process; nevertheless, useful information is still available. In such cases, the study often reveals information about the sources of special variation that affect the process, and it provides an opportunity to better understand the process.[2]

Process Capability Studies on Stable Processes

A process capability study on a stable process sets the stage for the estimation of the process's central tendency, $\bar{\bar{x}}$, and standard deviation, σ. These statistics allow (1) comparisons between the process's performance and specifications, and (2) use of centerlines, or process averages, on which to establish budgets and forecasts,[3] all required to fuel the Deming cycle. Note that predicting a stable process's behavior in the near future assumes that the process will remain stable. Unfortunately, it's impossible to know if this will be the case, so caution is advised.

An Example of an Attribute Process Capability Study

The centerline on a stable attribute control chart should be used as an estimate of the "overall" process capability. But there's one important proviso: an estimate of overall process capability isn't specific as to the potential cause or causes of defective output. To identify these, we must separate out all possible sources of defects (such as operators, machines, and vendors) and perform individual process capability studies for each source. In such a case, x-bar and R charts are often more cost-effective, in terms of sample size and information, than attribute charts, if they can be used.

To illustrate an attribute capability study, let's consider the case of a manager of a data entry department who has taken a survey indicating customer dissatisfaction. The manager wants to determine the capability of the data entry operation in her department in terms of the proportion of defective cards produced.[4] She decides to take samples of the first 200 lines of code from each day's output, inspect them for defects, and construct an initial p chart. Figure 11.1 shows the

FIGURE 11.1   Attribute Process Capability Study on Data Entry Operation

	Raw Data for Construction of Control Chart		
Day	*Number of Lines Inspected*	*Number of Defective Lines*	*Fraction of Defective Lines*
1	200	6	.030
2	200	6	.030
3	200	6	.030
4	200	5	.025
5	200	0	.000
6	200	0	.000
7	200	6	.030
8	200	14	.070
9	200	4	.020
10	200	0	.000
11	200	1	.005
12	200	8	.040
13	200	2	.010
14	200	4	.020
15	200	7	.035
16	200	1	.005
17	200	3	.015
18	200	1	.005
19	200	4	.020
20	200	0	.000
21	200	4	.020
22	200	15	.075
23	200	4	.020
24	200	1	.005
Totals	4,800	102	

FIGURE 11.2   p Chart from Raw Data

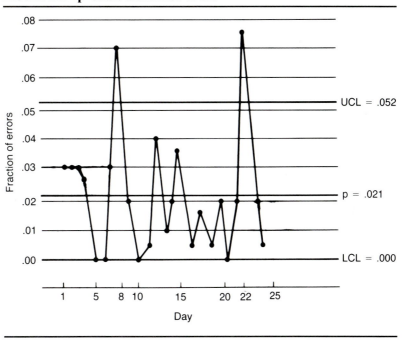

raw data, and Figure 11.2 shows the initial process control chart. The latter reveals that on days 8 (14 defective lines out of 200 inspected) and 22 (15 defective lines out of 200 inspected) something special happened, not attributable to the system, to cause defective lines to be entered.

The manager calls a meeting of the 10 operators to brainstorm for possible special causes of variation on days 8 and 22. Results of the brainstorming session are put onto the cause-and-effect diagram in Figure 11.3. The 10 group members vote that their best guess for the problem on day 8 was a new untrained operator (see cause in Figure 11.3—circled in a cloud) who had been added to the work force, and that the one day it took the worker to acclimate to the new environment probably caused the unusually high number of errors. To ensure that this special cause won't be repeated, the manager institutes a one-day training program for all new employees. The 10 group members also vote that their best guess for the problem on day 22 was that on the previous evening the department had run out of paper from the regular vendor, didn't expect a new shipment until the morning of day 23, and consequently purchased a one-day supply of paper from a new vendor. The operators found this paper was of inferior quality, which caused the large number of defective entries. To correct this special cause of variation, the manager revises the firm's relationship with its regular paper vendor and operationally defines acceptable quality for paper.

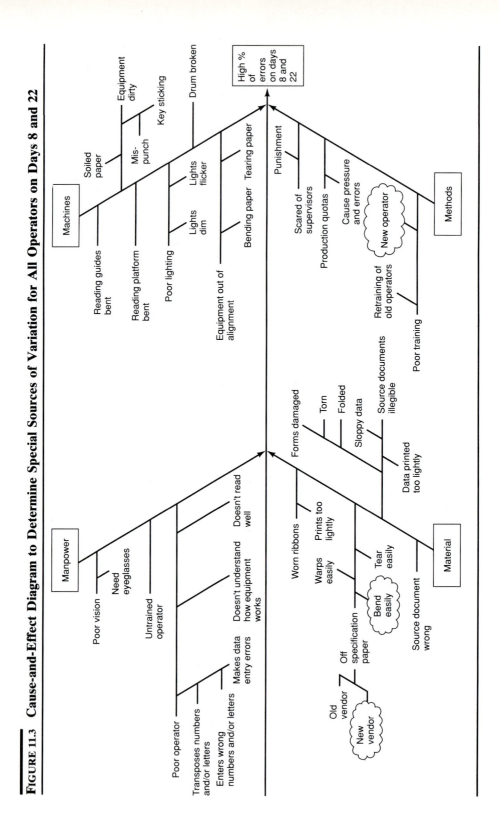

After eliminating the days for which special causes of variation are found, the manager recomputes the control chart statistics using Equations 5.1 through 5.3:

$$\text{Centerline(p)} = \text{Average fraction of defective lines} = \bar{p}$$

$$\bar{p} = 73/4,400 = 0.01659 \cong 0.017$$

$$\text{UCL(p)} = 0.044$$

$$\text{LCL(p)} = 0.000$$

Figure 11.4 shows the revised control chart. The process appears stable. The centerline and control limits were extended out into the future for 25 days. Data from daily samples of 200 lines of code were collected for these 25 days and plotted with respect to the forecasted centerline and control limits. The process was found to be stable. The capability of the process is such that it will produce an average of 1.7 percent defective lines per day. Further, the percentage defective will rarely surpass 4.4 percent. Although the process's capability is now known, the manager isn't satisfied with its capability and shouldn't stop attempting further improvement. We'll see how this is done in this chapter's section on process improvement studies.

An Example of a Variables Process Capability Study

Many authors attempt to use process capability studies to predict the fraction of process output, as measured by variables data, that will be out of specification. Such predictions

come in two flavors: ordinary fantasies and outright hallucinations. . . . A specific fraction defective will always require the use of some assumed distribution. . . . Such

FIGURE 11.4   Revised p Chart Following Removal of Special Causes

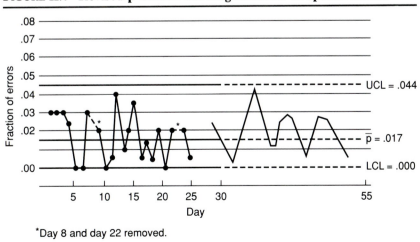

*Day 8 and day 22 removed.

assumptions are essentially unverifiable. . . . Assumptions certainly can not support computations in the parts per million range. . . . Rough guidelines such as the Empirical Rule may be used to characterize the approximate percentages which will fall outside certain intervals.[5]

A stable process may not be capable because of three conditions. The first condition occurs when the process exhibits too much unit-to-unit variation, which causes output to exceed specification limits. Condition 2 occurs when the process mean isn't centered on nominal, which causes output to exceed specification limits. Condition 3 occurs when some combination of the previous two conditions exists.

We can determine the existence of condition 1 by calculating the C_p index. The C_p index is used to summarize a process's ability to meet acceptable tolerances (USL − LSL). It assumes that the process under study is stable, is centered on nominal, and is being measured using variables data. C_p is computed as follows:

$$C_p = \frac{\text{USL} - \text{LSL}}{6\sigma} \tag{11.9}$$

C_p measures the ratio of the range of the process output desired by customers as defined by specifications (called the voice of the customer), to the range of the process output, given by the Empirical Rule as three standard deviations above and below the mean, or a range of 6σ (called the voice of the process). If C_p is greater than or equal to 1, the process will produce virtually 100 percent conforming output according to the Empirical Rule; that is, the range of the process output permitted by the specifications is larger than the actual range of the process output. If C_p is less than 1, the process will produce some nonconforming output; in other words, the process generates output over a larger range than the specification limits permit. As C_p increases, the proportion of nonconforming output decreases. Later in this chapter we discuss in detail C_p.

We can determine condition 2's existence by calculating Z_{USL} and/or Z_{LSL}. The Z_{USL} and Z_{LSL} values are used to summarize a process's ability to meet upper and lower specifications, respectively. Z_{USL} and Z_{LSL} assume that the process under study is stable and is being measured using variables data. Z_{USL} and Z_{LSL} are computed using

$$Z_{\text{USL}} = \frac{\text{USL} - \bar{\bar{x}}}{\sigma} \tag{11.10}$$

and

$$Z_{\text{LSL}} = \frac{\bar{\bar{x}} - \text{LSL}}{\sigma} \tag{11.11}$$

Z_{USL} measures the distance from the upper specification limit to the process average in process standard deviation units. Z_{LSL} measures the distance from the lower specification limit to the process average in process standard deviation units. If either Z_{USL} or Z_{LSL} is less than or equal to 3, then according to the Empirical Rule, the process will produce some nonconforming output. If Z_{USL} or Z_{LSL} is greater than 3, then according to the Empirical Rule, the process will produce virtually 100 percent conforming output. As Z_{USL} or Z_{LSL} increases over 3, the proportion of nonconforming output decreases. Z_{USL} and Z_{LSL} are discussed in detail later in this chapter.

To illustrate a variables process capability study, consider an auto manufacturer who wishes to purchase camshafts from a vendor.[6] The buyer is concerned with the finish grind, diameters, and case hardness as well as other quality characteristics of the camshaft. For illustrative purposes, this discussion will focus only on the case hardness depth of the camshafts.

The contract between the auto manufacturer and camshaft vendor calls for camshafts that have an average case hardness depth of 7.0 mm and are distributed around the average with a dispersion not to exceed 3.5 mm; this is a distribution specification. Further, the contract requires that the vendor produce a process capability study demonstrating statistical control of his process. Consequently, management's objective is to reduce camshaft-to-camshaft variation for case hardness depth and to move the process's average case hardness depth to the desired nominal of 7.0 mm.

A *camshaft* is a rod with elliptical lobes along its length. As the rod rotates, so do the elliptical lobes, and this ultimately causes intake and exhaust valves to open and close. The intake valves permit a mixture of fuel and air to enter the cylinders, where combustion takes place. The exhaust valves permit the waste gases to exit the cylinders after combustion. The surfaces of the elliptical lobes must be hardened (made brittle) to reduce wear, as in Figure 11.5. This hardening is called *case hardening* and is accomplished by immersing the camshaft in oil, placing electric bearing coils around the lobes, and passing electric current through the coils. This process heat treats the lobes and makes them brittle. The depth to which the brittleness extends is called *case hardness depth*. The case hardness depth must be tightly controlled since if the case hardness depth is too deep, the lobes will be too brittle and will tend to crack, while if the case hardness depth is too shallow, the lobes will be too soft and will wear quickly.

Pursuant to the terms of the contract calling for a process capability study, a sample of five camshafts is drawn from the vendor's process every day. Each shaft is measured with respect to each of the relevant quality characteristics.

Figure 11.6 shows the initial data collected in the process capability study. These data reveal that the vendor's process isn't in statistical control; this is indicated by points A through E in Figure 11.6 (June 9, 22, 23, and 28 and July 16). Consequently, corrective action on the process is required by vendor management.

An engineer from the vendor's plant forms a brainstorming group comprised of workers in the Induction Hardening and Quench Department—the department that performs case hardening on the camshafts. The brainstorming group's aim is to determine causes for the out-of-control points in Figure 11.6. The brainstorming session's results appear in the cause-and-effect diagram in Figure 11.7. The group decides (votes) that probable causes for out-of-control points were

1. Point A. Low power in the coil resulted in increased variability and less stable depth in the case hardness. (See cloud 1 in Figure 11.7.)

FIGURE 11.5   Camshaft in an Engine

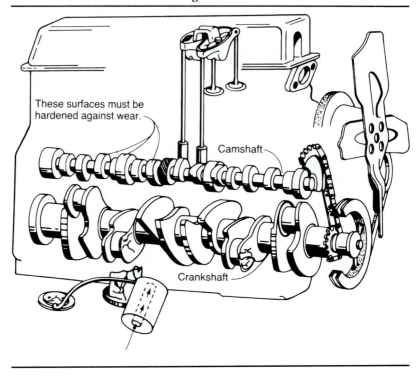

These surfaces must be
hardened against wear.

Camshaft

Crankshaft

2. Point B. A temporary operator was used because the regular operator
   was sick. (See cloud 2 in Figure 11.7.)
3. Point C. The case hardness setting on the machine was incorrect. (See
   cloud 3 in Figure 11.7.)
4. Point D. Low power in the coil resulted in increased variability and less
   depth in case hardness. (See cloud 1 in Figure 11.7.)
5. Point E. Low power in the coil resulted in an out-of-control situation.
   (See cloud 1 in Figure 11.7.)

This analysis leads vendor management to take action on the process by
repairing the voltage meter on the induction hardening machine (See points A, D,
and E.) and training all personnel in the proper operation of the machine. (See
points B and C.)

After the preceding policies are instituted, the engineer collects 30 additional
days of data, as shown in Figure 11.8. We see that the vendor's process is in
statistical control, with values for $\bar{\bar{x}}$ and $\bar{R}$ of 4.43 and 1.6, respectively. Recall
from Chapter 5 that $\bar{R} = d_2\sigma$, so we can calculate σ as $\bar{R}/d_2$, where d_2 is found in
Table 8 for a subgroup of size 5. However, from Figure 11.9's calculations, we see
that $Z_{LSL} = 1.35$; that is, the process average is only 1.35 process standard devia-

FIGURE 11.6   Process Capability Study of the Camshaft

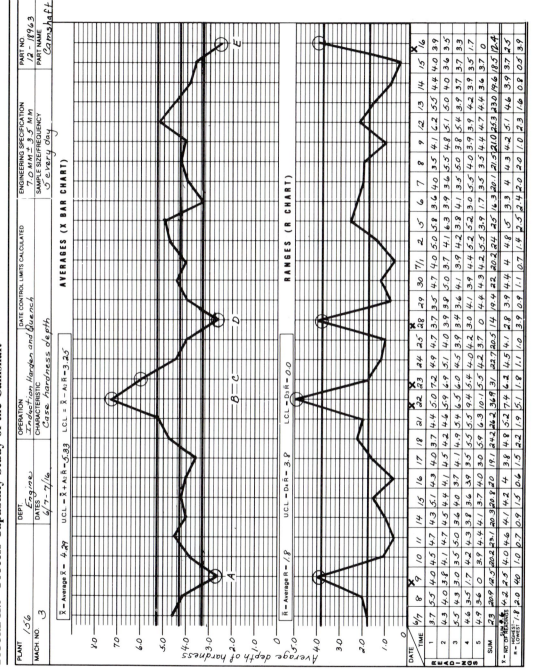

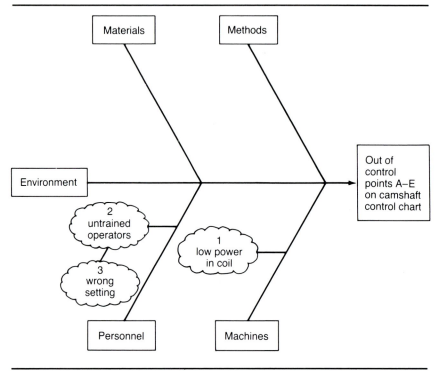

FIGURE 11.7   Cause-and-Effect Diagram to Diagnose Reasons for Out-of-Control Points

tions above the lower specification limit. Since Z_{LSL} is less than 3, this indicates that the process may produce some nonconforming product. We determine the proportion of nonconforming product by examining the histogram of actual process output in Figure 11.9. As we see, one camshaft had a case hardness depth less than the lower specification limit. In other words, the process generated 0.67 percent nonconforming camshafts in the study period.

It's interesting to note from the histogram that the case hardness depths aren't normally distributed—a disproportionate number of values lie near the lower specification limit. Asymmetrical situations like this could imply that there's some form of sorting prior to inspection or possibly that inspectors are accepting unsatisfactory product because they're fearful of failing to meet some production quota.

Process Capability Studies on Created Dimensions

In Chapter 10 we discussed created dimensions. In this section, we discuss how to determine the process capability for created dimensions. For example, suppose that the process capability studies for the thickness of the sheet, pin, and both

FIGURE 11.8 Additional Data for Camshaft Process Capability Study

PLANT 156	DEPT Engine	OPERATION Induction harden and Quench	DATE CONTROL LIMITS CALCULATED	ENGINEERING SPECIFICATION 7.0 MM ± 3.5 MM	PART NO. 12-18963
MACH NO. 3	DATES 7/19 – 8/27	CHARACTERISTIC Case hardness depth		SAMPLE SIZE/FREQUENCY 5 everyday	PART NAME Camshaft

$\bar{\bar{X}}$ – Average $\bar{X}$ = 4.43   UCL = $\bar{\bar{X}} + A_2\bar{R}$ = 5.35   LCL = $\bar{\bar{X}} - A_2\bar{R}$ = 3.51

AVERAGES (X BAR CHART)

6.0 5.0 4.0 3.0 2.0 1.0 0.0

$\bar{R}$ – Average R = 1.6   UCL = $D_4\bar{R}$ = 3.38   LCL = $D_3\bar{R}$ = 0.0

RANGES (R CHART)

4.0 3.0 2.0 1.0 0.0

DATE / TIME	7/19	20	21	22	23	26	27	28	29	30	8/2	3	4	5	6	9	10	11	12	13	16	17	18	19	20	23	24	25	26	27
READING 1	5.0	5.5	3.6	5.5	4.3	3.8	4.6	4.0	3.9	5.5	5.3	3.9	5.3	4.4	5.5	3.9	3.5	5.3	3.9	5.1	3.5	3.7	4.0	5.5	3.7	4.9	3.5	4.4	4.0	5.5
READING 2	4.1	4.0	4.3	4.0	3.5	4.4	4.4	3.7	5.0	3.5	5.1	4.4	5.1	4.9	5.0	4.7	5.0	5.1	4.9	4.4	5.2	5.1	4.4	4.0	5.2	3.6	5.1	4.4	4.0	5.5
READING 3	4.0	3.9	4.0	3.7	5.0	4.4	4.4	5.2	5.0	3.9	5.0	3.5	4.7	5.1	4.7	5.1	3.9	4.7	3.5	5.3	3.6	4.9	3.6	5.2	5.1	5.1	4.7	4.6	3.7	4.4
READING 4	3.7	3.6	3.9	3.7	3.7	5.0	4.5	3.5	3.7	3.7	5.3	5.3	3.9	3.9	3.9	5.1	4.2	5.0	4.2	3.9	3.7	3.6	4.9	4.3	4.0	4.7	4.0	4.3	3.7	4.9
READING 5	3.9	4.4	5.0	4.1	5.1	5.5	4.4	3.7	5.0	3.8	3.5	4.2	5.0	3.5	3.7	3.9	5.1	4.4	4.2	5.1	5.4	4.9	3.7	4.0	3.5	5.5	3.5	4.0	3.7	1.9
SUM	20.7	21.4	20.9	21.2	21.7	22.7	22.5	20.1	24.6	20.5	22.3	21.3	21.7	24.6	23.5	23.2	21.7	25.4	21.3	22.2	22.3	22.3	21.0	20.5	20.7	23.6	20.7	24.3	19.5	22.4
$\bar{X}$ – NO. OF READINGS	4.1	4.3	4.2	4.2	4.3	4.5	4.5	4.0	4.9	4.1	4.5	4.3	4.3	4.9	4.7	4.6	4.3	5.1	4.3	4.4	4.5	4.5	4.2	4.1	4.1	4.7	4.1	4.9	3.9	4.5
R – HIGHEST – LOWEST	1.3	1.9	1.4	1.8	1.6	1.7	0.2	1.7	1.6	2.0	0.6	1.6	1.8	0.9	1.6	1.4	1.6	1.6	1.6	1.8	1.8	2.0	1.2	1.9	1.6	1.8	1.9	1.6	1.2	1.9

FIGURE 11.9   Fraction of Camshafts out of Specification

$$Z_{LSL} = \frac{\bar{\bar{x}} - LSL}{\sigma}$$

$$Z_{LSL} = \frac{4.43 - 3.50}{\sigma}$$

$$\sigma = \frac{\bar{R}}{d_2} = \frac{1.60}{2.326} = 0.688 \, mm.$$

$$Z_{LSL} = \frac{(4.43 - 3.50)}{0.688} = 1.35$$

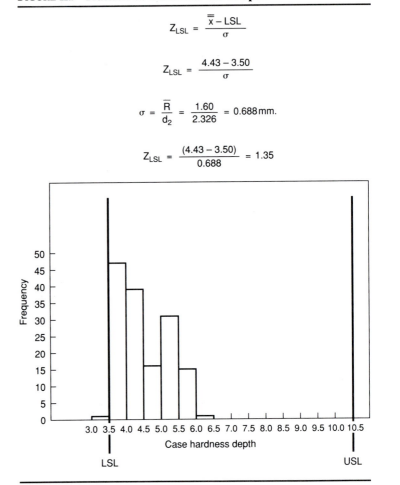

washer subcomponents of the assembly in Figure 10.4 were based on subgroups of five subcomponents and yielded

$$\bar{x}_s = 40 \text{ mm} \quad \text{and} \quad \sigma_s = 0.0050$$

$$\bar{x}_{w1} = 3 \text{ mm} \quad \text{and} \quad \sigma_{w1} = 0.0007$$

$$\bar{x}_{w2} = 4 \text{ mm} \quad \text{and} \quad \sigma_{w2} = 0.0008$$

$$\bar{x}_p = 27 \text{ mm} \quad \text{and} \quad \sigma_p = 0.0030$$

Further, assume that all component processes are independent of each other and stable.

The average bolt projection, from Equation 10.3, is

$$\bar{x}_{\text{bolt projection}} = 40 \text{ mm} - 3 \text{ mm} - 27 \text{ mm} - 4 \text{ mm} = 6 \text{ mm}$$

The standard deviation of the bolt projection, from Equation 10.4, is

$$\sigma_{\text{bolt projection}} = [(0.0050)^2 + (0.0007)^2 + (0.0008)^2 + (0.0030)^2]^{1/2}$$

$$= 0.00593 \cong .006$$

The process capability of the created bolt projection dimension is computed using Equations 11.1 and 11.2:

$$\text{UNL} = 6 \text{ mm} + 3(0.006) = 6 \text{ mm} + 0.018 \text{ mm} = 6.018 \text{ mm}$$

$$\text{LNL} = 6 \text{ mm} - 3(0.006) = 6 \text{ mm} - 0.018 \text{ mm} = 5.982 \text{ mm}$$

From the Empirical Rule, virtually all of the bolt projections will be between 5.982 and 6.018 mm.

Bolt projection specifications are set at 5.99 ± 0.02 mm. The following calculations summarize the process's capability:

$$C_p = \frac{(\text{USL} - \text{LSL})}{6\sigma}$$

$$C_p = \frac{(6.01 - 5.97)}{6(0.006)} = \frac{0.04}{0.036} = 1.11$$

The process generates output that's within the desired tolerance limits. But we also wish to check that the output doesn't exceed specification limits (condition 2 discussed earlier) by calculating Z_{USL} and Z_{LSL}:

$$Z_{\text{USL}} = \frac{(\text{USL} - \bar{x})}{\sigma}$$

$$Z_{\text{USL}} = \frac{(6.01 - 6.00)}{0.006} = \frac{0.010}{0.006} = 1.67$$

The process average is only 1.67 standard deviations below the upper specification limit, indicating that some output will be nonconforming. The fraction of output that was nonconforming in the period under study can be determined by examining the histogram of the output, as in our camshaft example.

$$Z_{LSL} = \frac{(\overline{x} - LSL)}{\sigma}$$

$$Z_{LSL} = \frac{(6.00 - 5.97)}{0.006} = \frac{0.030}{0.006} = 5.00$$

The process average is 5.00 standard deviations above the lower specification limit, indicating the likely production of only conforming output.

Finally, created dimensions should be control charted because a created dimension can be out of control while component dimensions are in control. Never-ending improvement can't progress without reducing unit-to-unit variations and moving the process toward nominal for created dimensions.

The Relationships between Control Limits, Natural Limits, and Specification Limits for Variables Control Charts

Natural Limits and Control Limits.   *Natural limits* are used with respect to individual observations—and consequently on run charts. *Control limits* are used with respect to subgroup statistics—and consequently on control charts. Figure 11.10 shows the relationships between natural limits and control limits. We see that for x-bar charts, if A_2 (x-bar and R chart) or A_3 (x-bar and s chart) is multiplied by the square root of the subgroup size ($\sqrt{n}$), and these new quantities ($A_2\sqrt{n}$) and ($A_3\sqrt{n}$) are added and subtracted from the process average ($\overline{\overline{x}}$), the control

---

FIGURE 11.10   Relationship between Control Limits and Natural Limits for Variables Location Charts

Chart Type	Control Limits	Natural Limits	Comments
$\overline{x}$ (with R chart)	$\overline{\overline{x}} \pm A_2\overline{R} =$ $\overline{\overline{x}} \pm 3\left(\dfrac{\overline{R}/d_2}{\sqrt{n}}\right)$ $\left(A_2 = \dfrac{3}{d_2\sqrt{n}}\right)$	$\overline{\overline{x}} \pm 3\,(\overline{R}/d_2)$	If A_2 and/or A_3 are multiplied by the square root of the subgroup size ($\sqrt{n}$), and this new quantity is added or subtracted from $\overline{\overline{x}}$, the new limits are the natural limits.
$\overline{x}$ (with s chart)	$\overline{\overline{x}} \pm A_3\overline{s} =$ $\overline{\overline{x}} \pm 3\left(\dfrac{\overline{s}/c_4}{\sqrt{n}}\right)$ $\left(A_3 = \dfrac{3}{c_4\sqrt{n}}\right)$	$\overline{\overline{x}} \pm 3\,(\overline{s}/c_4)$	
x (individuals chart with moving range chart)	$\overline{x} \pm E_2\overline{R} =$ $\overline{x} \pm 3\,(\overline{R}/d_2)$ $(E_2 = 3/d_2)$	$\overline{x} \pm 3\,(\overline{R}/d_2)$	Control limits and natural limits are equivalent.

limits are transformed into natural limits. In the case of control limits for individuals charts, the control limits and the natural limits are identical because the subgroup size is one.

Natural Limits and Specification Limits. Natural limits and specification limits are comparable quantities for stable processes because they're both measured with respect to the individual units of output generated by the process under study. There are four basic relationships between natural limits and specification limits for stable processes. Each relationship is portrayed using a normal distribution. However, from the Empirical Rule, the assumption of normality is not necessary.

Relationship 1. The process's natural limits are inside the specification limits and the process is centered on nominal. This is illustrated in the run chart in Figure 11.11(a).

Relationship 2. The process's natural limits are inside the specification limits and the process isn't centered on nominal. This is illustrated in the run chart in Figure 11.11(b).

Relationship 3. The process's natural limits are outside the specification limits and the process is centered on nominal. This is illustrated in the run chart in Figure 11.11(c).

Relationship 4. The process's natural limits are outside the specification limits and the process isn't centered on nominal. This is illustrated in the run chart in Figure 11.11(d).

In Chapter 5 we discussed the four states of a process.[7] Relationships 1 and 2 represent a process in its ideal state; given the same variation between relationships 1 and 2, relationship 1 is preferable. Relationships 3 and 4 represent a process in the threshold state; given the same variation between relationships 3 and 4, relationship 3 represents the more desirable situation.

Control Limits and Specification Limits. In no case should specification limits be shown on x-bar charts. This is because while control limits apply to process statistics ($\bar{x}$), specification limits apply to individual units of process output (or some other quality characteristic). Nevertheless, specification limits are sometimes shown on control charts for individuals because in this special case the control limits are based on subgroups of size one—hence, on individual values.

Process Capability Indices for Variables Data

A common desire of many control chart users is to be able to state a process's ability to meet specifications in one summary statistic.[8] Such statistics are available and are called *process capability indices*. We use these indices to summarize internal processes as well as vendor processes.

Assumptions.   All process capability indices we'll discuss here require variables data and stability of the process characteristic under study. Additionally, it's common practice to assume that this process characteristic is normally distributed. Unfortunately, the assumption of normality isn't realistic even when dealing with processes that are both stable and capable. This is because capability calculations in this situation are based on the extreme tails of the normal distribution, or the portion of the normal distribution that's beyond the specification limits. Unfortunately, the extreme tails of the normal distribution are mathematical quantities that rarely, if ever, characterize the real world. In the outer 5 percent of each tail, considerable discrepancies will occur between the theoretical fraction nonconforming and the actual fraction nonconforming.[9] Given this caveat, let's now discuss capability indices.

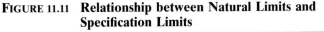

FIGURE 11.11   Relationship between Natural Limits and Specification Limits

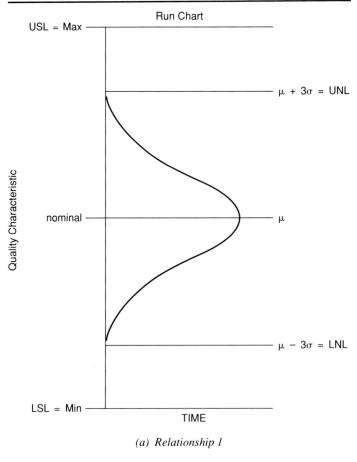

(a) Relationship 1

FIGURE 11.11   (*continued*)

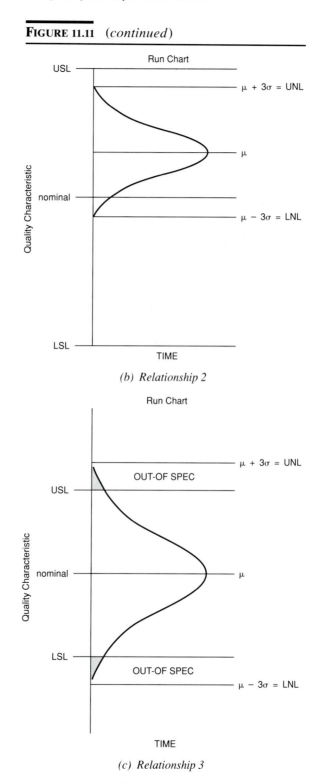

(b) *Relationship 2*

(c) *Relationship 3*

FIGURE 11.11 *(concluded)*

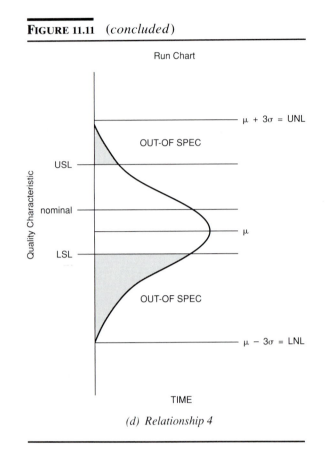

Run Chart

(d) Relationship 4

Indices. Four process capability indices are commonly used: C_p, CPU, CPL, and C_{pk}. Figure 11.12 summarizes them.

C_p. The C_p index, introduced earlier in this chapter, is used to summarize a process's ability to meet two-sided specification limits. As we saw in Equation 11.9, C_p is computed as

$$C_p = \frac{USL - LSL}{6\sigma}$$

In addition to the general assumptions stated in the prior section, the C_p index also assumes that the process average ($\bar{\bar{x}}$) is centered on the nominal value, m.

Recall that a process's capability is defined to be the range in which virtually all of the output will fall; usually, this is described as plus or minus three standard deviations from the process's mean, or within an interval of six standard deviations (6σ). Consequently, if a process's USL = UNL = $\bar{\bar{x}} + 3\sigma$ and its LSL = LNL = $\bar{\bar{x}} - 3\sigma$, the process's capability is 1.0:

FIGURE 11.12   Process Capability Indices

Index	Estimation Equation	Equation No.	Purpose	Assumptions about Process
C_p	$\dfrac{USL - LSL}{6\sigma}$	11.9	Summarize process potential to meet acceptable tolerances (USL − LSL).	1. Stable process. 2. Variables data. 3. Centered process. (Process average equals nominal.)
CPU	$\dfrac{USL - \bar{\bar{x}}}{3\sigma}$	11.12	Summarize process potential to meet only a one-sided upper specification limit.	1. Stable process. 2. Variables data.
CPL	$\dfrac{\bar{\bar{x}} - LSL}{3\sigma}$	11.13	Summarize process potential to meet only a one-sided lower specification limit.	(same as CPU)
C_{pk}	$C_p - \dfrac{\|m - \bar{\bar{x}}\|}{3\sigma}$ where m = Nominal value of the specification	11.14	1. Summarize process potential to meet two-sided specification limits. 2. $\|m - \bar{\bar{x}}\|/3\sigma$ is a penalty factor for the process's being off nominal. It's stated in terms of the number of natural limit units the process is off nominal.	(same as CPU)

$$C_p = \frac{USL - LSL}{6\sigma} = \frac{(\bar{\bar{x}} + 3\sigma) - (\bar{\bar{x}} - 3\sigma)}{6\sigma} = \frac{6\sigma}{6\sigma} = 1.0$$

A process capability index of 1.0 indicates that a process will generate virtually all of its output within specification limits, according to the Empirical Rule.

For centered processes, given the preceding assumptions, Figure 11.13(a) shows a process with $C_p = 1.0$, indicating that the UNL = USL, and the LNL = LSL.

Figure 11.13(b) shows a process with $C_p = 2.0$, indicating that the UNL is midway between nominal and the USL, and the LNL is midway between the LSL and nominal. Clearly all process output will be well within specification limits.

Figure 11.13(c) shows a process with $C_p = 0.5$, where the USL is midway between nominal and the UNL, and the LSL is midway between the LNL and nominal. Here, process output will be outside the specification limits.

CPU.   The CPU index is used to summarize a process's ability to meet a one-sided upper specification limit. In many situations process owners are concerned only that a process not exceed an upper specification limit. For example, for products that can warp in only one direction, there's no LSL for warpage; the

---

FIGURE 11.13   Process Capability Indices

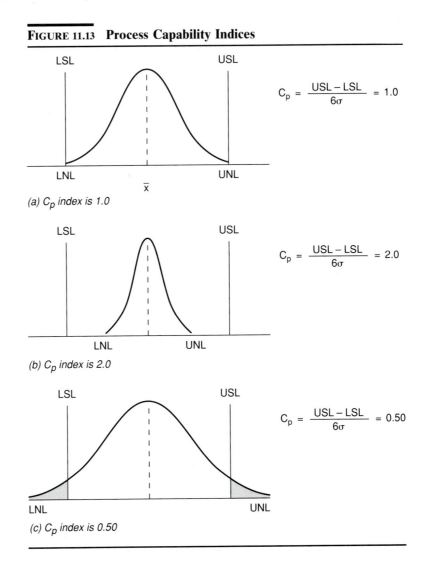

(a) C_p index is 1.0

(b) C_p index is 2.0

(c) C_p index is 0.50

---

lower the warpage the better. However, there's a USL for warpage, the value for warpage that will critically impair the product's ability to meet customer needs. It can also be used in situations where we want to examine one side of a two-sided specification limit.

CPU is computed as

$$CPU = \frac{USL - \bar{\bar{x}}}{3\sigma} \qquad (11.12)$$

The CPU index measures how far the process average ($\bar{\bar{x}}$) is from the upper specification limit in terms of one-sided natural tolerance limits (3σ). Natural tolerances, when added and subtracted from the process mean ($\bar{x}$), yield the range in which a process is capable of operating—the process's capability, $\bar{x} \pm 3\sigma = \bar{x} \pm$ natural tolerance.

If a process's USL = UNL (= $\bar{\bar{x}} + 3\sigma$), the CPU is 1.0:

$$CPU = \frac{USL - \bar{\bar{x}}}{3\sigma} = \frac{USL - \bar{\bar{x}}}{UNL - \bar{\bar{x}}} = 1.0$$

A CPU of 1.0 or more indicates that a process will generate virtually all of its output within the upper specification limit.

If a process's UNL is greater than the USL, the CPU is less than 1. As UNL − USL increases, the fraction of process output that's out of specification will increase geometrically. Conversely, if a process's UNL is less than the USL, then the CPU is greater than 1. As USL − UNL increases, the fraction of process output that's out of specification will decrease geometrically. To determine the fraction of process output that will be out of specification, examine the histogram of process output with respect to the upper specification limit, as illustrated earlier.

CPL. The CPL index is used to summarize a process's ability to meet a one-sided lower specification limit. The CPL operates exactly like the CPU. CPL is computed as

$$CPL = \frac{\bar{\bar{x}} - LSL}{3\sigma} \tag{11.13}$$

If a process's LNL is greater than the LSL, then CPL is less than 1. As LNL − LSL increases, the fraction of process output that's out of specification will increase geometrically. Conversely, if a process's LNL is less than the LSL, then CPL is greater than 1. As LSL − LNL increases, the fraction of process output that's out of specification will decrease geometrically. To determine the fraction of process output that will be out of specification, we examine the histogram of process output with respect to the lower specification limit.

C_{pk}. The C_{pk} index is used to summarize a process's ability to meet two-sided specification limits when the process isn't centered on nominal. The C_{pk} index uses the C_p index as a starting point for stating a process's capability, but it penalizes C_p if the process isn't centered on nominal, m:

$$C_{pk} = C_p - \left\{ \frac{|m - \bar{\bar{x}}|}{3\sigma} \right\} \tag{11.14}$$

The term in brackets in Equation 11.14 is always positive, and hence lowers the value of C_p, which indicates that the process is less able to produce within specifications. The bracketed term is a measure of how many natural tolerance units (3σ) the process mean ($\bar{\bar{x}}$) is from nominal (m). The further off-center the process, the more C_p is penalized by the bracket factor. Hence, C_{pk} is a two-sided capability index that accounts for process centering.

A firm that exists in a defect detection mode won't know the process capability indices for its various processes. On the other hand, a firm operating in a defect prevention mode will know the values for its various processes and will be striving for C_p or C_{pk} approximately equal to 1.0. Finally, if a firm is pursuing never-ending improvement, it will be striving to move C_p and C_{pk} higher and higher. As C_p and C_{pk} become increasingly greater than 1, the specification limits from which they were computed become increasingly irrelevant.

Limitations of Capability Indices. Several potential problems exist when using the C_p and C_{pk} indices. First, if a process isn't stable, C_p and C_{pk} are meaningless statistics. Second, not all processes meet the assumption of normality. Hence, the naive user of capability indices may incorrectly assess the actual fraction of process output that will be out of specification. Last, experience shows that naive users of capability indices frequently confuse C_p and C_{pk}; they think they yield the same information about a process. Of course, this can result in a great deal of confusion.

An Example. Each process capability index discussed earlier in this chapter is calculated using the camshaft example in Figures 11.6, 11.8, and 11.9. Figure 11.6 shows the camshaft operation is out of control. After special sources of variation are removed from the process, it becomes stable, as shown in Figure 11.8. From Figure 11.8 we see that the average case hardness depth is 4.43 mm, the average range for case hardness depth is 1.60 mm, the upper specification limit is 10.5 mm, and the lower specification limit is 3.5 mm.

$$\bar{\bar{x}} = 4.43 \text{ mm}$$

$$\bar{R} = 1.60 \text{ mm; hence } \sigma = \bar{R}/d_2 = 1.60/2.326 = 0.688 \text{ mm}$$

$$\text{USL} = 10.5 \text{ mm}$$

$$\text{LSL} = 3.5 \text{ mm}$$

Given these figures, the C_p, CPU, CPL, and C_{pk} can be computed and interpreted.

C_p. We compute C_p as

$$C_p = \frac{\text{USL} - \text{LSL}}{6\sigma} = \frac{10.5 - 3.5}{6(0.688)} = 1.70$$

This C_p indicates an extremely capable process that will almost never produce out-of-specification product. However, from Figure 11.9 we see that while virtually all the camshafts have case hardness depths within acceptable tolerances, the C_p index, which assumes the process is centered, has failed to detect that the process, with an average of 4.43, is 2.57 mm off nominal ($|\bar{\bar{x}} - m| = |4.43 - 7.00| = 2.57$ mm).

CPU. We compute CPU as

$$\text{CPU} = \frac{\text{USL} - \bar{\bar{x}}}{3\sigma} = \frac{10.5 - 4.43}{3(0.688)} = 2.94$$

The CPU accurately indicates that the process is operating well within the USL of 10.5 mm.

CPL.   We compute CPL as

$$CPL = \frac{\bar{\bar{x}} - LSL}{3\sigma} = \frac{4.43 - 3.5}{3(0.688)} = 0.45$$

The CPL accurately indicates that the process isn't operating within the LSL of 3.5 mm.

***C*$_{pk}$.**   We compute C$_{pk}$ as

$$C_{pk} = C_p - \frac{|m - \bar{\bar{x}}|}{3\sigma}$$

$$= \frac{USL - LSL}{6\sigma} - \frac{|m - \bar{\bar{x}}|}{3\sigma}$$

$$= \frac{10.5 - 3.5}{6(0.688)} - \frac{|7.0 - 4.43|}{3(0.688)}$$

$$= 1.70 - 1.25$$

$$= 0.45$$

This C$_{pk}$ correctly indicates a process that will produce out-of-specification product, because, unlike C$_p$, it has taken into account that the process isn't centered on nominal.

In the end, capability indices can sometimes potentially cause more problems than they can provide benefits. Consequently, some practitioners recommend that they not be used.

Process Improvement Studies

In Chapter 5 we discussed the Deming cycle as a vehicle for process improvements. The Deming cycle requires that managers understand and utilize both enumerative and analytic studies of design/redesign, conformance, and performance to pursue continuous and never-ending improvement of an organization's interdependent system of stakeholders. There are many tools and techniques available so that managers can conduct the previously mentioned studies; these tools and techniques are the subject of this book. In addition, successful process improvement endeavors require a managerial view that embraces the tools and techniques discussed in this book. The Deming theory of management philosophy is such a managerial view.

As with process capability studies, there are two types of process improvement studies: attribute improvement studies and variables improvement studies. In the pursuit of continuous and never-ending improvement, it's natural that attribute improvement studies lead to variables improvement studies.

Attribute Improvement Studies

Returning to this chapter's data entry example, recall that the percentage of defective entries was stabilized with an average of 1.7 percent defective and would rarely go above 4.4 percent defective.

At this point the manager decides that to improve the process further she must study each operator individually. However, she must determine which operators to work with first. She makes a check sheet to record the number and fraction of defective lines of code by operator for December 1994 (Figure 11.14). Next, she constructs a c chart for the number of defective lines of code per operator for December 1994, as in Figure 11.15(a). From the c chart, she notes that operators 004 and 009 are out of control. She revises the c chart to determine if any other operators are out of control after having removed the impact of operators 004 and 009; she finds none, as we see in Figure 11.15(b). Next, she constructs a Pareto diagram for the number of defects per operator, as shown in Figures 11.16(a) and 11.16(b). From the Pareto diagram, she determines that 72 percent of all defective lines of code are produced by operators 004 and 009.

The manager decides to perform separate analyses for operators 004 and 009. She begins with operator 004 by setting up a check sheet (Figure 11.17) to determine the sources for operator 004's defects. The corresponding Pareto diagram, in Figures 11.18(a) and 11.18(b), shows 82 percent of operator 004's defects resulted from "transposed numbers" and "wrong character." Subsequently, the manager forms a brainstorming group composed of three select employees to do a cause-and-effect analysis of these two problems (Figure 11.19). The group members vote to attack both problems simultaneously. The cause-and-effect diagram leads the manager to send operator 004 to have her eyes checked. The optometrist finds that operator 004 is legally blind in her right eye. Eyeglasses correct her vision.

FIGURE 11.14 Check Sheet of Defective Lines of Code by Operator (12/1/94 – 12/31/94)*

Operator	Tally	Frequency
001	II	2
002	III	3
003	I	1
004	₩₩ ₩₩ ₩₩ IIII	19
005		0
006	II	2
007	I	1
008	III	3
009	₩₩ ₩₩ ₩₩ II	17
010	II	2
Total		50

* All operators produced approximately the same number of lines of code during the period under study.

FIGURE 11.15 c Chart for Number of Defective Lines of Code by Operator

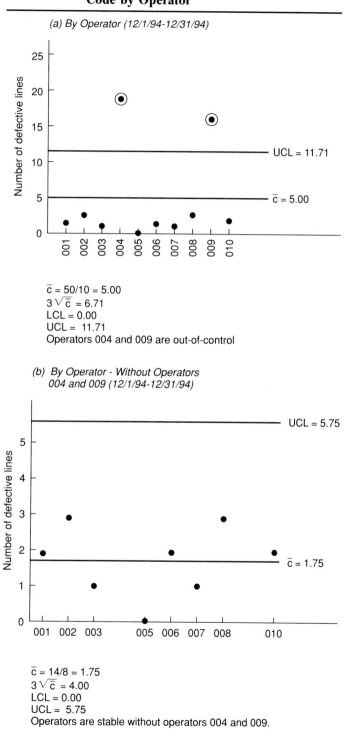

(a) By Operator (12/1/94-12/31/94)

$\bar{c}$ = 50/10 = 5.00
$3\sqrt{\bar{c}}$ = 6.71
LCL = 0.00
UCL = 11.71
Operators 004 and 009 are out-of-control

(b) By Operator - Without Operators
004 and 009 (12/1/94-12/31/94)

$\bar{c}$ = 14/8 = 1.75
$3\sqrt{\bar{c}}$ = 4.00
LCL = 0.00
UCL = 5.75
Operators are stable without operators 004 and 009.

FIGURE 11.16   Pareto Analysis of Defective Lines of Code

(a) By Operator

Operator	Frequency	Fraction	Cumulative Fraction
004	19	.38	.38
009	17	.34	.72
002	3	.06	.78
008	3	.06	.84
001	2	.04	.88
006	2	.04	.92
010	2	.04	.96
003	1	.02	.98
007	1	.02	1.00
005	0	.00	1.00
Totals	50	1.00	

(b) Pareto Diagram

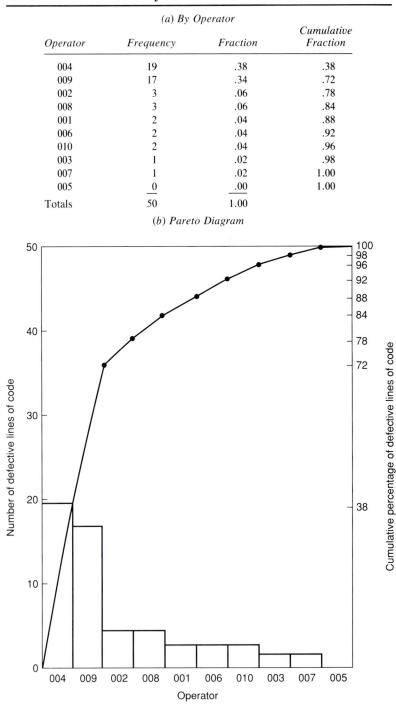

72% of all defective lines of code are produced
by operators 004 and 009.

FIGURE 11.17   Check Sheet to Determine Defects for Operator 004 (1/95–4/95)

Major Causes of Defective Entries	Month				
	1/95	2/95	3/95	4/95	Totals
Transposed numbers	7	10	6	5	28
Out of field	1		2		3
Wrong character	6	8	5	9	28
Data printed too lightly		1	1		2
Torn document	1	1		2	4
Creased document			1	1	2
Illegible source document	—	—	1	—	1
Totals	15	20	16	17	68

FIGURE 11.18   Pareto Analysis to Determine Sources of Defects for Operator 004

(a) Causes (1/95–4/95)

Major Causes of Defective Entries	Frequency	Fraction	Cumulative Fraction
Transposed numbers	28	.412	.412
Wrong character	28	.412	.824
Torn document	4	.059	.883
Out of field	3	.044	.927
Data printed too lightly	2	.029	.956
Creased document	2	.029	.985
Illegible source document	1	.015	1.000
Totals	68	100.0	

(b) Pareto Diagram

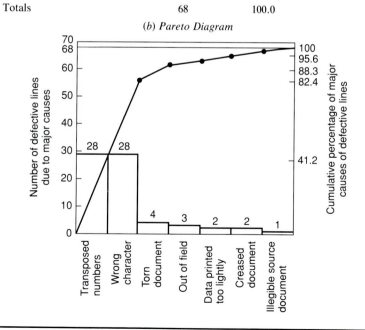

FIGURE 11.19 Cause-and-Effect Diagrams for Operator 004

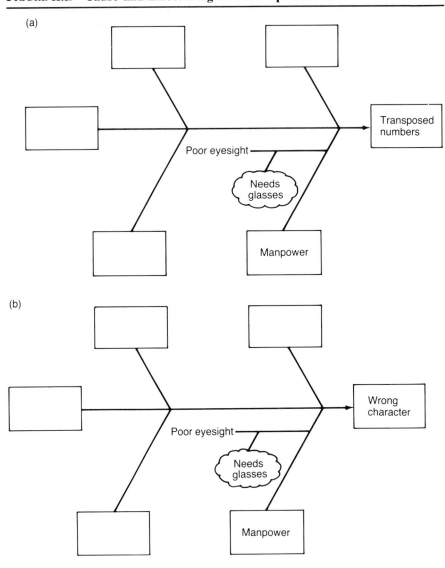

Next, the manager collects 24 more daily samples of 200 lines of code and constructs a p chart for the fraction of defectives (Figure 11.20). From the p chart, the manager finds that operator 004's work is now stable, has an average defective rate of 0.8 percent (8 in 1,000 lines), and rarely goes above 2.6 percent defective.

The manager realizes that if she wants to improve operator 004's performance further, she must switch from an attribute process improvement study to a variables process improvement study. Her next step is to plan her future courses of action: (1) study operator 009 and (2) review the entire department.

FIGURE 11.20   p Chart for Operator 004 Following Fitting with Eye Glasses

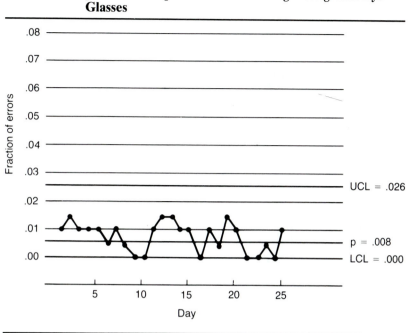

Variables Improvement Studies

Returning to our camshaft example, recall that the camshafts' case hardness depth was stable, with an average of 4.43 mm and standard deviation of 0.688 mm.

At this point, the engineer assigned to study the induction quench-and-harden operation decides that to improve the process further, the induction coil must be changed. The old coil is pitted and consequently emits an erratic electrical output, causing increased variability in case hardness depth between camshafts. The induction coil is changed on the evening of August 29, and then 30 more days of data are collected (August 30–October 8) and control charted (Figure 11.21). The process is in statistical control, with an average case hardness depth of 5.45 mm and a standard deviation of 0.434 mm ($\overline{R}/d_2 = 1.01/2.326 = 0.434$). Note that the process has been shifted toward nominal (7.0 mm) and its unit-to-unit variation has been reduced.

Next, the process is studied to determine the number of standard deviations the specification limits are from the process average. This is done by computing Z_{LSL} and Z_{USL}. Remember, if the process average is more than three process standard deviations from both specification limits, then virtually all of the process's output will be within the specification limits. The necessary calculations are

$$Z_{LSL} = \frac{(\overline{\overline{x}} - LSL)}{\sigma}$$

$$= \frac{(5.45 - 3.50)}{0.434} = 4.49$$

FIGURE 11.21 Case Hardness Depth after Coil Change

PLANT	DEPT.	OPERATION	DATE CONTROL LIMITS CALCULATED	ENGINEERING SPECIFICATION	PART NO.
156	Engine	Induction harden and Quench		7.0 MM ± 3.5 MM	12-18963
MACH NO.	DATES	CHARACTERISTIC		SAMPLE SIZE/FREQUENCY	PART NAME
3	8/30 - 10/8/	Case hardness depth		5 everyday	Cam shaft

X̄ = Average x̄ = 5.45   UCL = X̄ + A₂R̄ = 6.03   LCL = X̄ - A₂R̄ = 4.87

AVERAGES (X BAR CHART)

R̄ = Average R = 1.01   UCL = D₄R̄ = 2.14   LCL = D₃R̄ = 0.0

RANGES (R CHART)

DATE	8/30	31	9/1	2	3	4	5	6	7	8	9	10	13	14	15	16	17	20	21	22	23	24	27	28	29	30	10/1	4	5	6	7	8
TIME																																
R 1	4.9	5.8	5.9	5.9	6.2	5.9	5.5	6.0	6.2	6.1	5.8	5.2	5.2	6.2	4.9	6.1	5.3	6.3	5.4	5.6	4.7	6.0	5.1	5.6	5.5	4.9	4.7	6.0	5.5	5.2	5.0	5.5
E 2	5.5	5.5	6.2	6.2	5.9	5.9	5.0	5.7	4.6	5.8	6.1	5.4	4.6	5.8	4.9	6.2	5.4	5.6	5.3	5.0	4.7	5.3	6.2	6.1	6.0	4.8	4.9	5.1	5.0	4.9	5.2	5.1
A 3	5.3	5.6	5.9	6.4	5.7	5.6	5.2	5.7	6.4	6.2	5.7	5.2	6.1	5.1	4.9	5.9	5.4	5.9	5.4	5.2	5.6	5.6	5.0	5.8	5.7	6.1	5.2	5.3	5.3	5.2	6.0	5.3
D 4	5.6	6.3	5.8	5.3	4.9	5.3	5.2	6.3	6.1	5.9	6.5	5.8	6.1	5.2	5.2	4.5	5.4	4.7	4.5	4.9	5.0	5.0	5.7	5.0	5.3	5.3	6.0	4.9	5.7	5.0	4.9	4.1
S 5	5.1	5.7	5.4	5.9	5.2	5.8	5.2	6.0	6.2	5.7	4.6	5.8	4.6	5.4	4.8	5.2	5.2	5.6	4.7	5.6	5.8	5.8	5.7	5.8	4.9	5.7	5.7	5.2	5.0	6.1	5.0	5.0
SUM	26.4	28.9	29.2	28.7	27.9	28.5	26.1	29.7	26.5	29.7	28.6	26.2	26.5	27.7	24.4	29.3	26.7	28.7	26.3	25.4	25.9	27.1	27.2	29.2	27.1	26.3	26.5	27.4	27.2	27.2	25.6	25.0
x̄ -	5.3	5.8	5.8	5.7	5.7	5.7	5.2	5.9	5.3	5.7	5.9	5.2	5.3	5.5	4.9	5.7	5.3	5.7	5.3	5.1	5.2	5.4	5.4	5.8	5.4	5.3	5.3	5.5	5.4	5.4	5.1	5.0
R - HIGHEST-LOWEST	0.7	0.8	1.2	1.3	1.3	0.6	0.5	0.6	1.5	1.1	1.3	1.2	1.5	1.1	0.1	1.7	0.2	1.6	0.9	0.9	0.9	1.0	1.2	0.5	1.1	1.2	1.3	1.2	1.0	0.4	1.2	1.4

and

$$Z_{USL} = \frac{(USL - \bar{\bar{x}})}{\sigma}$$

$$= \frac{(10.5 - 5.45)}{0.434} = 11.6$$

The process is operating well within specification limits. But in the spirit of never-ending improvement, the engineer assigned to study the induction quench-and-harden operation should continually work to reduce unit-to-unit variation and move the process average toward nominal.

This example highlights the benefits of process improvement. The case hardness process was moved from chaos to the threshold state to the ideal state. The move from the threshold state to the ideal state resulted in

1. Increased quality.
2. Increased productivity.
3. Lower unit cost. (It costs less to make good items than defective items because good items don't require rework.)
4. Increased price flexibility resulting from lower unit costs.
5. Increased market share resulting from increased quality and price flexibility.
6. Increased profit resulting from lower unit costs and greater market share.
7. More secure jobs for all employees.

Quality Improvement Stories

Employees trying to improve quality have found that their ideas and recommendations are more persuasive when based on facts rather than opinions and guesses. The Quality Improvement (QI) story is an efficient format for employees to persuasively present process improvement studies to management. QI stories standardize quality control reports, avoid logical errors in analysis, and make reports easy for all to understand. Further, QI stories can be used to facilitate the transition from the old to the new style of management.

Relationship between QI Stories and the PDSA Cycle

A seven-step procedure should be utilized to construct a QI story; these seven steps follow the Deming cycle of *Plan, Do, Study, and Act*. The Plan phase of the Deming cycle involves three steps: (1) selecting a theme for the QI story (obtaining all the background information necessary to understand the selected theme, including a process flow chart; explaining the reason for selecting the theme; and determining the organization and department objective(s) that should be influenced by the theme), (2) getting a full grasp of the present situation surrounding the theme, and (3) analyzing the present situation to identify appropriate action(s) (called *countermeasures*) to the process—that is, constructing a plan of action.

The Do phase of the Deming cycle involves a further step: (4) testing the proposed countermeasures on a small scale using a planned experiment.

The Study phase of the Deming cycle involves yet another step: (5) studying, creatively thinking about, collecting, and analyzing data from the planned experiment concerning the effectiveness of the countermeasure(s). Do the countermeasures reduce the difference between customer needs and process performance? Before-and-after comparisons of the experimental countermeasures' effects on the targeted department and organization objectives must be presented.

The Act phase of the Deming cycle requires two final steps: (6) determining if the countermeasure(s) were effective in pursuing department and organization objectives. If not, we go back to the Plan stage to find other countermeasures that will be effective in pursuing department and organization objectives. If the countermeasures *were* effective in pursuing department and organization objectives, we either go to the Plan stage to seek the optimal settings of the countermeasure(s) or formally establish revised standard operating procedures based on the data about the experimental countermeasures and train all relevant personnel in the new "best practice" method. Further actions must be taken to prevent backsliding for the countermeasures set into motion. This phase also includes (7) identifying remaining process problems, establishing a plan for further actions, and reflecting on the positive and negative aspects of past countermeasures.

To summarize, the seven steps of a QI story are

1. *Select a theme* for the QI story.
2. Get a full *grasp of the present situation* surrounding the theme.
3. *Analyze the present situation* to identify appropriate countermeasures.
4. *Set the countermeasures into action* in a planned experiment.
5. Study data concerning the *effectiveness of the countermeasures.*
6. Establish revised *standard operating procedures.*
7. Establish a *plan for future actions.*

These steps form the basis of a QI story; all organization personnel should adhere to the QI story procedure, and each section of a QI story should be clearly numbered and labeled so that it can be related to one of the seven steps, as Figure 11.22 shows.

Potential Difficulties

Two areas of potential difficulty when applying QI stories are qualitative (nonnumerical) themes and exogenous problems. Themes that are difficult to describe with numerical values should be analyzed by focusing on the magnitude of the gap between actual performance and the desired performance.

If a problem's cause (e.g., cold weather or no rain) is beyond the control of anyone in the organization, we don't conclude that it's impossible to take countermeasures to remedy the exogenous problem. Instead, we attempt to determine why there are so many occurrences of the exogenous problem in area A versus

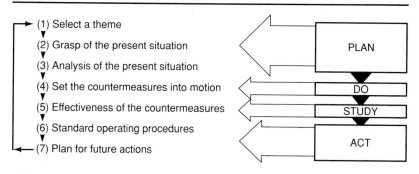

FIGURE 11.22   Relationship between the QI Story and the PDSA Cycle

(1) Select a theme

(2) Grasp of the present situation

(3) Analysis of the present situation

(4) Set the countermeasures into motion

(5) Effectiveness of the countermeasures

(6) Standard operating procedures

(7) Plan for future actions

PLAN

DO

STUDY

ACT

SOURCE: Adapted from Masaaki Imai, *Kaizen—The Keys to Japan's Competitive Success* (New York: Random House Business Division, 1986), p. 76.

area B, given that the areas have equal opportunities for the exogenous problem's occurrence.

Pursuit of Objectives

Unfortunately, QI stories will be initially selected because they're nearly complete resolutions to departmental problems and may not relate to organizational and departmental objectives. As employees gain experience with QI stories, they'll want to select themes related to organization and department objectives. Consequently, quantitative measures must be established that clearly recognize the relationship between organization objectives and department objectives. For example, a purchasing department is aiming at "reducing processing time for purchase orders." It claims this aim relates to organization objectives concerning "creating a great place to work free from system impediments." However, there's no quantitative measure that shows the relationship between "reducing processing time for purchase orders" and "creating a great place to work free from system impediments." These quantitative relationships must be shown in the QI story. If QI story activities aren't consistent with department objectives, and department objectives aren't consistent with organization objectives, then quality improvement efforts may not be in line with an organization's goals.

Quality Improvement Story Case Study

A QI story drawn from a data processing department demonstrates the role of QI stories in an organization's improvement efforts. Figure 11.23 shows the QI story in QI story boards 1 through 14.

The QI story goes through two iterations of the PDSA cycle; nevertheless, a never-ending set of PDSA iterations will follow as the data processing department pursues continuous improvement in its daily work. The first iteration of the PDSA cycle focuses on all data entry operators in the data processing department. In this iteration of the PDSA cycle, *Select a Theme* is presented in QI story board 1; this includes showing the background of theme selection and the reason for selecting the theme in relation to the organization's and department's objectives. *A Grasp of the Present Situation* is presented in QI story board 2. An *Analysis of the Present Situation* (QI story board 3) is performed to determine appropriate countermeasures that pursue the theme and the organization and department objectives. *Set the Countermeasures into Motion* on a trial basis is presented in QI story board 4. The *Effectiveness of the Countermeasures* on the theme and the organization and department objectives are measured as shown in QI story board 5. *Standard Operating Procedure* is set that formalizes the countermeasures and prevents backsliding (QI story board 6). *A Plan for Future Actions* is presented in QI story board 7.

The second iteration of the PDSA cycle focuses attention on an individual data entry operator. In this iteration, *Select a Theme* is accomplished when the data processing manager realizes that future process improvements will require her to identify and train operators whose performance is out of control on the high side (QI story board 8). In this iteration of the PDSA cycle, a *Grasp of the Present Situation* determines that data entry operators 004 and 009 are out of control on the high side and why operator 004 is out of control on the high side; this is presented in QI story board 9. An *Analysis of the Present Situation* (QI story board 10) determines the countermeasures necessary to improve operator 004's work. The manager *Sets the Countermeasures into Motion* as shown in QI story board 11. The positive *Effectiveness of the Countermeasures* on operator 004 and on the organization and department objectives is confirmed; see QI story board

---

FIGURE 11.23   Quality Improvement Story

QI STORY BOARD 1

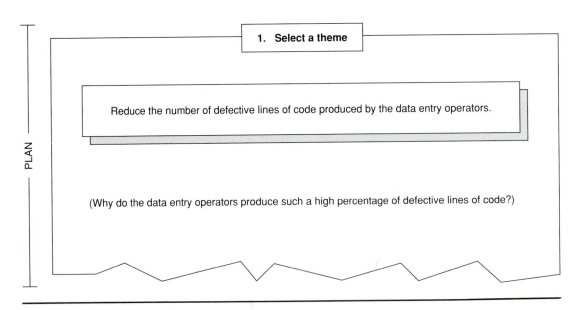

FIGURE 11.23 (*continued*)

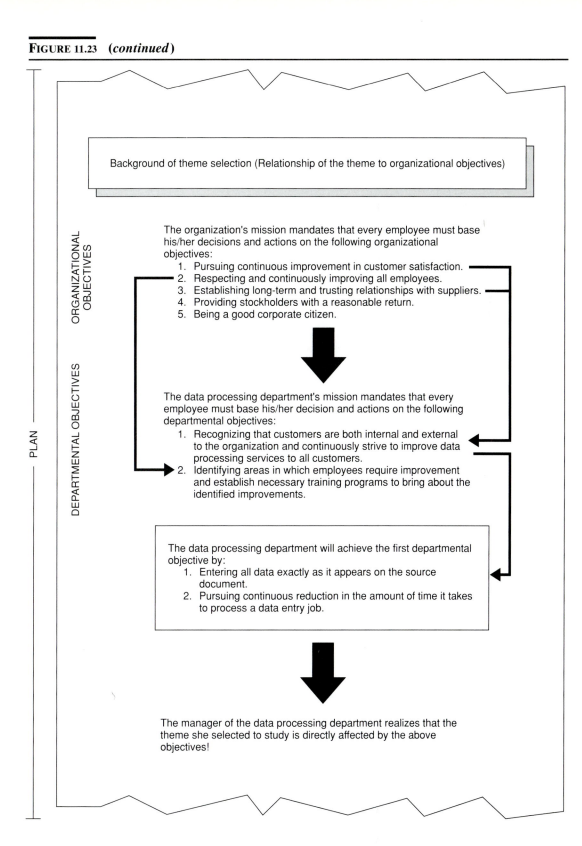

PLAN

ORGANIZATIONAL OBJECTIVES

DEPARTMENTAL OBJECTIVES

Background of theme selection (Relationship of the theme to organizational objectives)

The organization's mission mandates that every employee must base his/her decisions and actions on the following organizational objectives:
1. Pursuing continuous improvement in customer satisfaction.
2. Respecting and continuously improving all employees.
3. Establishing long-term and trusting relationships with suppliers.
4. Providing stockholders with a reasonable return.
5. Being a good corporate citizen.

The data processing department's mission mandates that every employee must base his/her decision and actions on the following departmental objectives:
1. Recognizing that customers are both internal and external to the organization and continuously strive to improve data processing services to all customers.
2. Identifying areas in which employees require improvement and establish necessary training programs to bring about the identified improvements.

The data processing department will achieve the first departmental objective by:
1. Entering all data exactly as it appears on the source document.
2. Pursuing continuous reduction in the amount of time it takes to process a data entry job.

The manager of the data processing department realizes that the theme she selected to study is directly affected by the above objectives!

FIGURE 11.23 (*continued*)

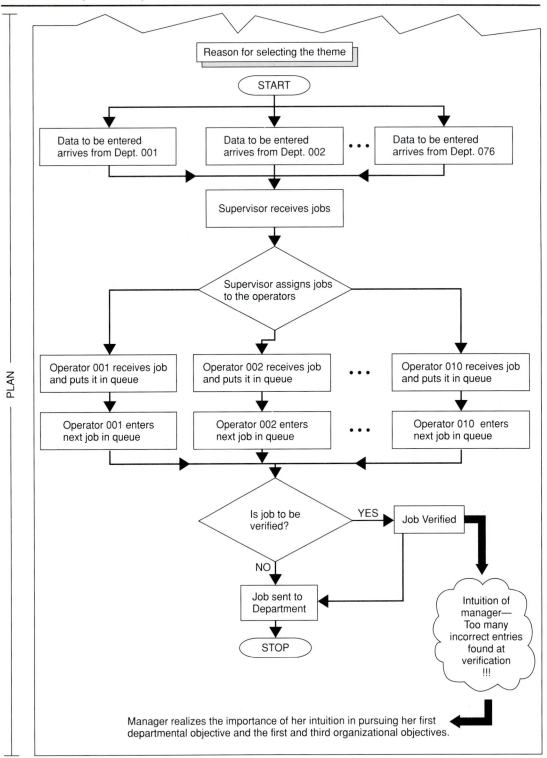

FIGURE 11.23 *(continued)*

QI STORY BOARD 2

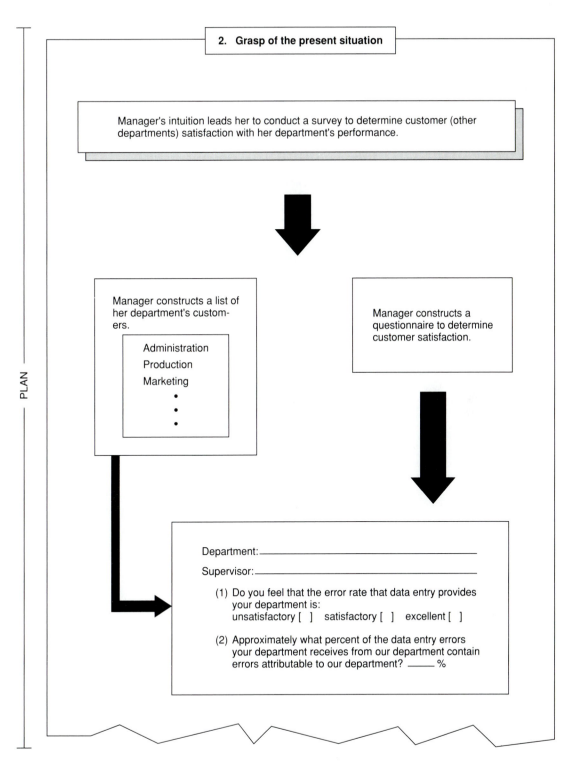

FIGURE 11.23 *(continued)*

PLAN

Questionnaires were sent to all of the departments and all of the departments responded. Analysis of the questionnaires yielded the following results:

Findings:

• Do you feel that the error rate that data entry provides your department is:

unsatisfactory?	(72%)
satisfactory?	(20%)
excellent?	(8%)

• Appproximately 2% of the data entry errors received by the various departments contain errors attributable to the data processing department.

FIGURE 11.23 (*continued*)

PLAN

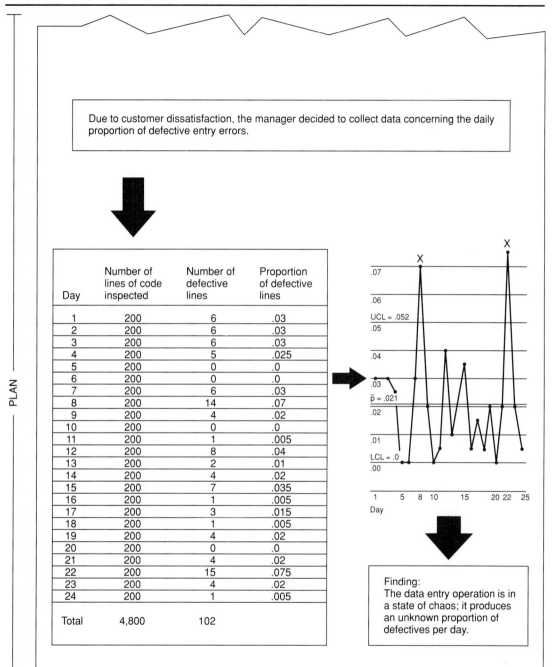

Due to customer dissatisfaction, the manager decided to collect data concerning the daily proportion of defective entry errors.

Day	Number of lines of code inspected	Number of defective lines	Proportion of defective lines
1	200	6	.03
2	200	6	.03
3	200	6	.03
4	200	5	.025
5	200	0	.0
6	200	0	.0
7	200	6	.03
8	200	14	.07
9	200	4	.02
10	200	0	.0
11	200	1	.005
12	200	8	.04
13	200	2	.01
14	200	4	.02
15	200	7	.035
16	200	1	.005
17	200	3	.015
18	200	1	.005
19	200	4	.02
20	200	0	.0
21	200	4	.02
22	200	15	.075
23	200	4	.02
24	200	1	.005
Total	4,800	102	

Finding:
The data entry operation is in a state of chaos; it produces an unknown proportion of defectives per day.

FIGURE 11.23 (*continued*)

QI STORY BOARD 3

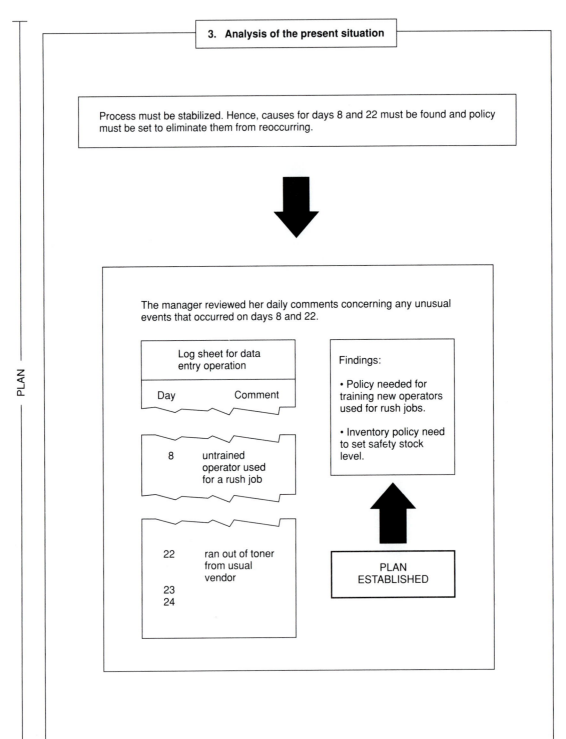

FIGURE 11.23 *(continued)*

QI STORY BOARD 4

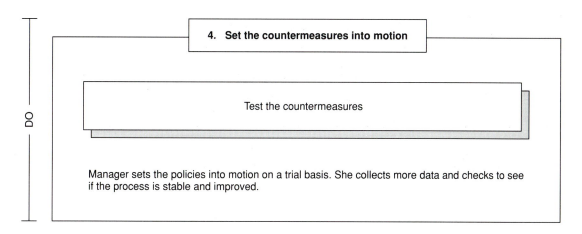

DO

| 4. Set the countermeasures into motion |

Test the countermeasures

Manager sets the policies into motion on a trial basis. She collects more data and checks to see if the process is stable and improved.

QI STORY BOARD 5

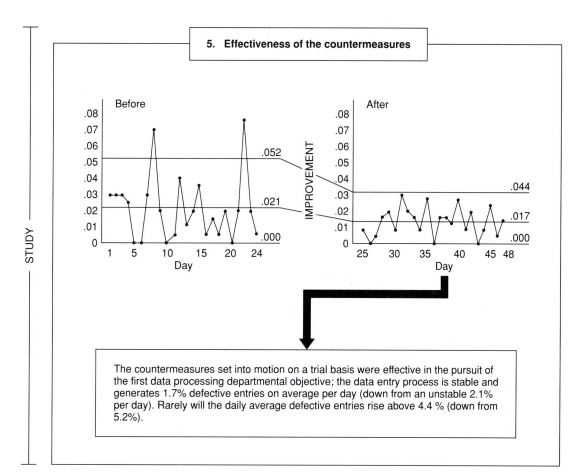

STUDY

| 5. Effectiveness of the countermeasures |

The countermeasures set into motion on a trial basis were effective in the pursuit of the first data processing departmental objective; the data entry process is stable and generates 1.7% defective entries on average per day (down from an unstable 2.1% per day). Rarely will the daily average defective entries rise above 4.4 % (down from 5.2%).

FIGURE 11.23 *(continued)*

QI STORY BOARD 6

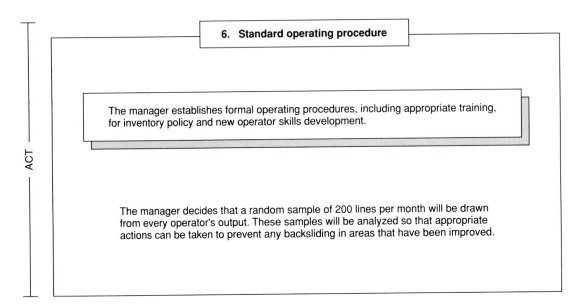

6. **Standard operating procedure**

ACT

The manager establishes formal operating procedures, including appropriate training, for inventory policy and new operator skills development.

The manager decides that a random sample of 200 lines per month will be drawn from every operator's output. These samples will be analyzed so that appropriate actions can be taken to prevent any backsliding in areas that have been improved.

QI STORY BOARD 7

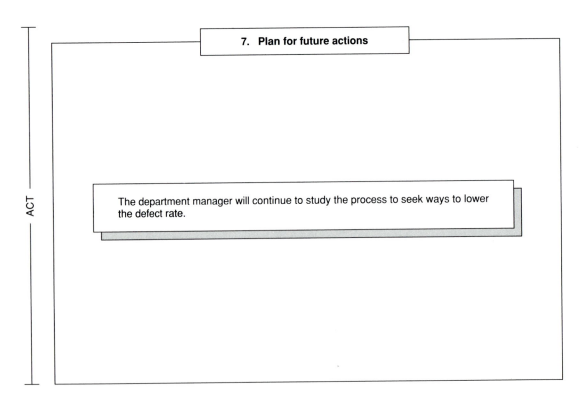

7. **Plan for future actions**

ACT

The department manager will continue to study the process to seek ways to lower the defect rate.

FIGURE 11.23 (*continued*)

QI STORY BOARD 8

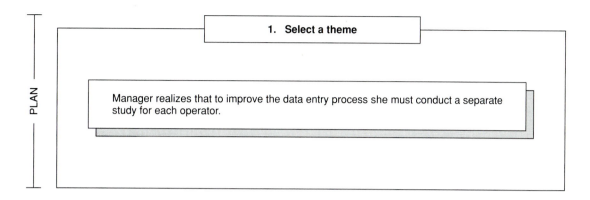

QI STORY BOARD 9

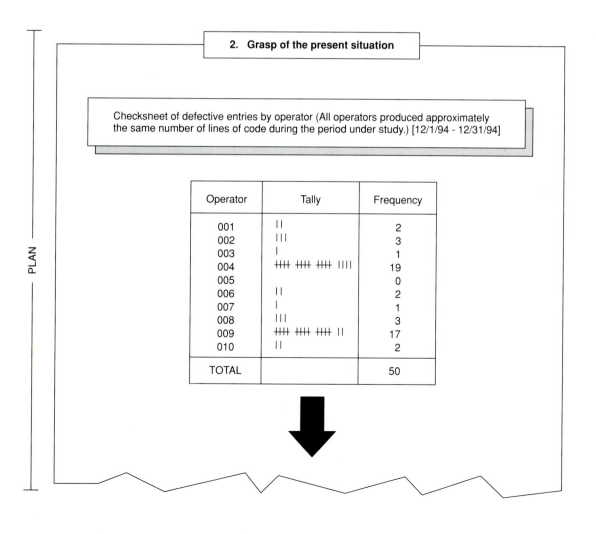

FIGURE 11.23 (*continued*)

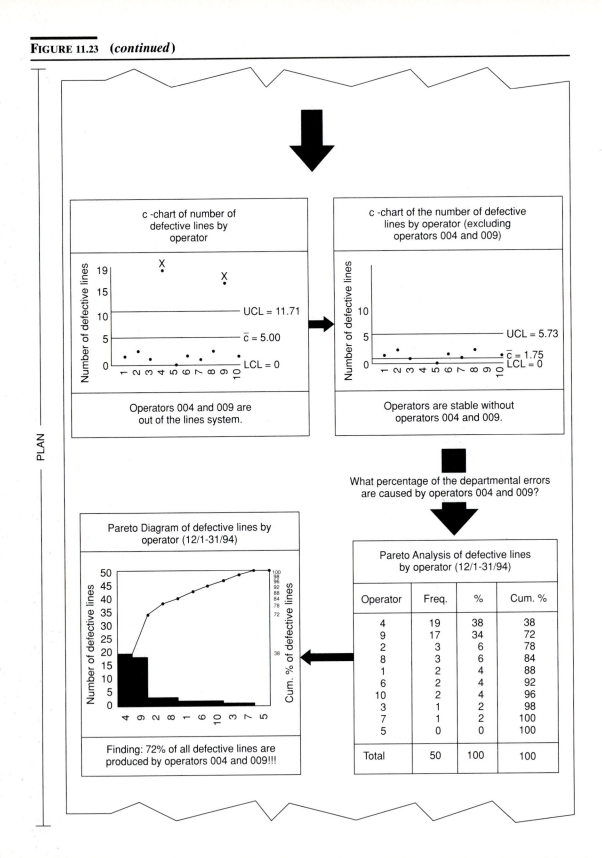

PLAN

c -chart of number of defective lines by operator

Number of defective lines

19
15
10
5
0

X
X

UCL = 11.71
c̄ = 5.00
LCL = 0

1 2 3 4 5 6 7 8 9 10

Operators 004 and 009 are out of the lines system.

c -chart of the number of defective lines by operator (excluding operators 004 and 009)

Number of defective lines

10
5
0

UCL = 5.73
c̄ = 1.75
LCL = 0

1 2 3 4 5 6 7 8 9 10

Operators are stable without operators 004 and 009.

What percentage of the departmental errors are caused by operators 004 and 009?

Pareto Diagram of defective lines by operator (12/1-31/94)

Number of defective lines

50
45
40
35
30
25
20
15
10
5
0

Cum. % of defective lines

100
98
96
92
88
84
78
72

38

4 9 2 8 1 6 10 3 7 5

Finding: 72% of all defective lines are produced by operators 004 and 009!!!

Pareto Analysis of defective lines by operator (12/1-31/94)

Operator	Freq.	%	Cum. %
4	19	38	38
9	17	34	72
2	3	6	78
8	3	6	84
1	2	4	88
6	2	4	92
10	2	4	96
3	1	2	98
7	1	2	100
5	0	0	100
Total	50	100	100

FIGURE 11.23 (*continued*)

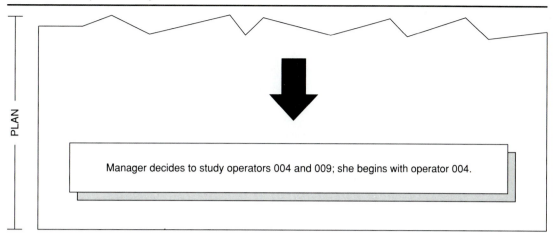

PLAN

Manager decides to study operators 004 and 009; she begins with operator 004.

QI STORY BOARD 10

PLAN

3. Analysis of the present situation

Checklist to determine the sources of operator 004's defective lines (1/95-4/95)

Major causes of defective lines	Month				
	1/95	2/95	3/95	4/95	Total
Transposed numbers	7	10	6	5	28
Out of field	1		2		3
Wrong charac.	6	8	5	9	28
Data printed too lightly		1	1		2
Torn document	1	1		2	4
Creased document			1	1	2
Illegible source doc.			1		1
TOTAL	15	20	16	17	68

FIGURE 11.23 (*continued*)

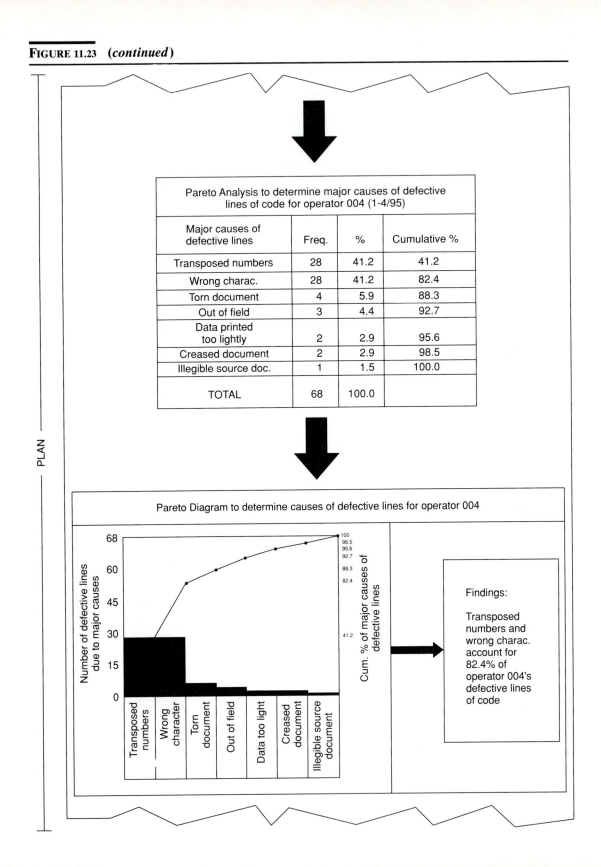

FIGURE 11.23 (*continued*)

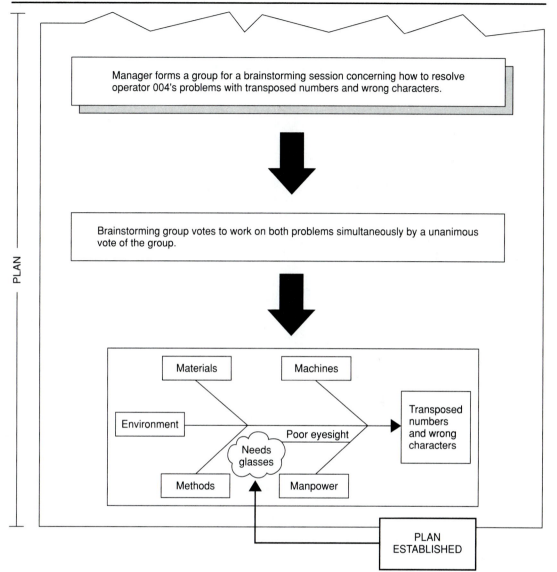

Manager forms a group for a brainstorming session concerning how to resolve operator 004's problems with transposed numbers and wrong characters.

Brainstorming group votes to work on both problems simultaneously by a unanimous vote of the group.

Materials

Machines

Environment

Poor eyesight

Needs glasses

Transposed numbers and wrong characters

Methods

Manpower

PLAN ESTABLISHED

PLAN

QI STORY BOARD 11

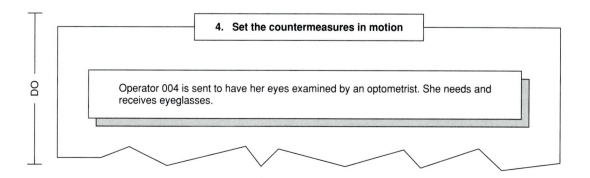

4. Set the countermeasures in motion

DO

Operator 004 is sent to have her eyes examined by an optometrist. She needs and receives eyeglasses.

FIGURE 11.23 *(continued)*

QI STORY BOARD 12

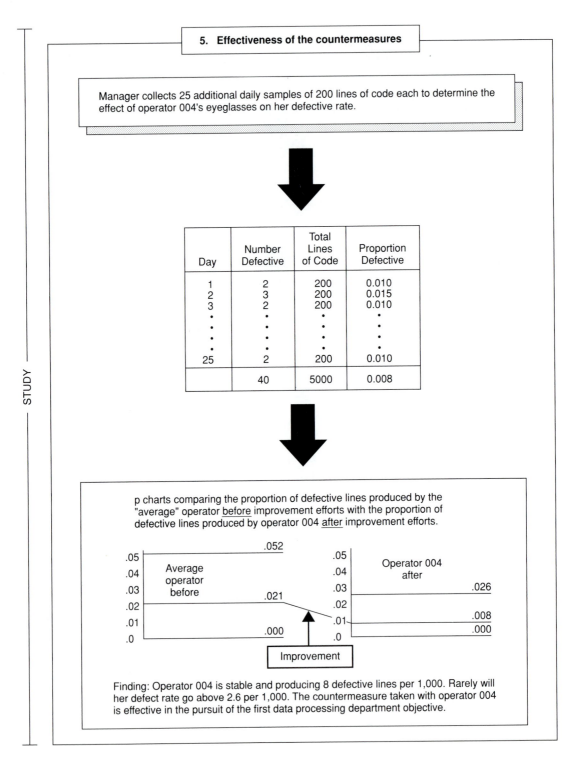

STUDY

5. Effectiveness of the countermeasures

Manager collects 25 additional daily samples of 200 lines of code each to determine the effect of operator 004's eyeglasses on her defective rate.

Day	Number Defective	Total Lines of Code	Proportion Defective
1	2	200	0.010
2	3	200	0.015
3	2	200	0.010
.	.	.	.
.	.	.	.
.	.	.	.
25	2	200	0.010
	40	5000	0.008

p charts comparing the proportion of defective lines produced by the "average" operator before improvement efforts with the proportion of defective lines produced by operator 004 after improvement efforts.

Finding: Operator 004 is stable and producing 8 defective lines per 1,000. Rarely will her defect rate go above 2.6 per 1,000. The countermeasure taken with operator 004 is effective in the pursuit of the first data processing department objective.

FIGURE 11.23   *(continued)*

QI STORY BOARD 13

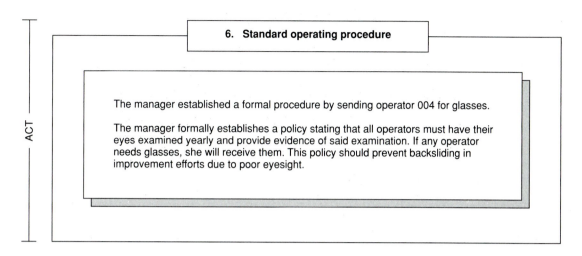

6.   Standard operating procedure

The manager established a formal procedure by sending operator 004 for glasses.

The manager formally establishes a policy stating that all operators must have their eyes examined yearly and provide evidence of said examination. If any operator needs glasses, she will receive them. This policy should prevent backsliding in improvement efforts due to poor eyesight.

ACT

QI STORY BOARD 14

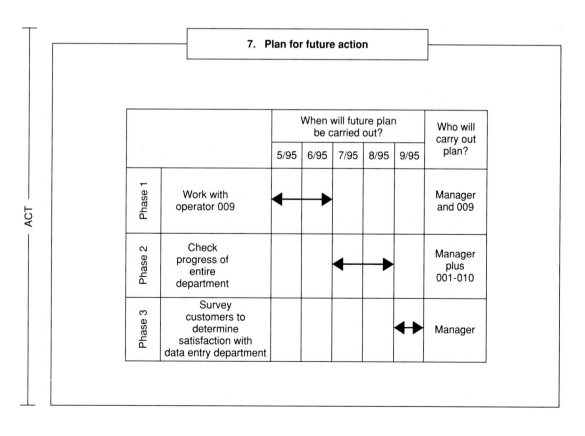

7.   Plan for future action

ACT

		When will future plan be carried out?					Who will carry out plan?
		5/95	6/95	7/95	8/95	9/95	
Phase 1	Work with operator 009	◄──────►					Manager and 009
Phase 2	Check progress of entire department			◄──────►			Manager plus 001-010
Phase 3	Survey customers to determine satisfaction with data entry department					◄►	Manager

12. *Standard Operating Procedure* is set, which formalizes the countermeasure to all operators and prevents backsliding (QI story board 13). Finally, *A Plan for Future Action* is specified in QI story board 14.

Summary

This chapter discussed process capability studies, process improvement studies, and quality improvement stories. Process capability studies determine if a process is unstable, investigate any sources of instability and determine their causes, and take action to eliminate these sources of instability. After all sources of instability have been eliminated, the natural behavior of the process is called its process capability.

We illustrated the two types of process capability studies: attribute studies and variables studies. For each, we considered data requirements and possible actions that can be taken on the process as a result of the process capability study. For variables process capability studies, we discussed determining process capability for a created dimension; the relationship between control limits, natural limits, and specification limits; and process capability indices.

This chapter presented case studies of attribute and variables process improvement studies. It also discussed quality improvement stories and showed an attribute capability study in the context of a QI story.

Exercises

The ABC Company produces steel tubes. The steel tube process is a stable cut-to-length operation that generates tubes that have a mean of 12.00 inches and standard deviation of 0.10 inches.

11.1 The XYZ Company wishes to buy tubes from the ABC Company. The XYZ Company requires steel tubes between 11.77 inches and 12.23 inches in length.
   a. Compute C_p.
   b. Compute CPU.
   c. Compute CPL.
   d. Compute C_{pk}.
   e. Compute Z_{LSL}.
   f. Compute Z_{USL}.
   g. Compare and contrast the preceding capability indices with respect to their ability to explain the capability of the ABC Company's steel tube process.
   h. Discuss the managerial implications of the capability indices you computed in parts a through f.

11.2 The LMN Company wishes to buy tubes from the ABC Company. The LMN Company requires steel tubes 11.95 inches long with a tolerance of 0.30 inches.

    a. Compute C_p.

    b. Compute CPU.

    c. Compute CPL.

    d. Compute C_{pk}.

    e. Compute Z_{LSL}.

    f. Compute Z_{USL}.

    g. Compare and contrast the preceding capability indices with respect to their ability to explain the capability of the ABC Company's steel tube process.

    h. Discuss the managerial implications of the capability indices you computed in parts a through f.

The Arco Company produces plastic containers. The plastic container process is a stable operation that generates containers with a mean volume of 12,500.00 cubic inches and standard deviation of 10.00 cubic inches.

11.3 The Beta Company wishes to buy plastic containers from the Arco Company. The Beta Company requires plastic containers with a volume between 12,495.00 cubic inches and 12,545.00 cubic inches.

    a. Compute C_p.

    b. Compute CPU.

    c. Compute CPL.

    d. Compute C_{pk}.

    e. Compute Z_{LSL}.

    f. Compute Z_{USL}.

    g. Compare and contrast the preceding capability indices with respect to their ability to explain the capability of the Beta Company's plastic container process.

    h. Discuss the managerial implications of the capability indices you computed in parts a through f.

11.4 The Largo Corporation wishes to buy plastic containers from the Arco Company. The Largo Corporation requires plastic containers with a volume of 12,495.00 cubic inches and a tolerance of 20.00 cubic inches.

    a. Compute C_p.

    b. Compute CPU.

    c. Compute CPL.

    d. Compute C_{pk}.

    e. Compute Z_{LSL}.

    f. Compute Z_{USL}.

    g. Compare and contrast the preceding capability indices with respect to their ability to explain the capability of the Beta Company's plastic container process.

    h. Discuss the managerial implications of the capability indices you computed in parts a through f.

11.5 How do you determine a process's capability, given that the only information available comes from an attribute process capability study?

11.6 a. Discuss the purpose of a quality improvement (QI) story.

b. List the seven steps in a QI story.

c. Explain the relationship between the seven steps in a QI story and the four stages of the PDSA cycle.

11.7 Define *capability of a process* in statistical terms. Consider the Empirical Rule in your definition.

11.8 a. Discuss the data requirements to conduct an attribute process capability study. Consider the number of time periods and the number of subgroups per time period.

b. Discuss the data requirements to conduct a variables process capability study. Consider the number of time periods and the number of subgroups per time period.

Endnotes

1. AT&T, *Statistical Quality Control Handbook*, 10th printing, May 1984 (Indianapolis: AT&T, 1956), pp. 34–37, 45–73.

2. Ibid.

3. H Gitlow and S Gitlow, *The Deming Guide to Quality and Competitive Position* (Englewood Cliffs, NJ: Prentice-Hall, 1987), p. 161.

4. H Gitlow and P Hertz, "Product Defects and Productivity," *Harvard Business Review*, September-October 1983, pp. 131–41.

5. Donald Wheeler and David Chambers, *Understanding Statistical Process Control*, 2d ed. (Knoxville, TN: SPC Press, 1992), p. 129.

6. This example was modified from Operations Support Staffs, "Statistical Process Control Case Study," *Introduction to Ford's Operating Philosophy and Principles and Statistical Management Methods—Participant Notebook* (Ford Motor Company, September 1983), pp. 7.E.9.–7.E.18.

7. Donald Wheeler and David Chambers, *Understanding Statistical Process Control* (Knoxville, TN: Statistical Process Controls, 1986), pp. 12–21.

8. V Kane, "Process Capability Indices," *Journal of Quality Technology* 18 (January 1986), pp. 41–52.

9. Adapted from Wheeler and Chambers, *Understanding Statistical Process Control*, 2d ed., p. 130.

PART V Process/Product Design

Chapter 12 presents contributions of Japanese statistician Genichi Taguchi to quality management and statistics. In particular, the chapter considers the relationship between the quality of a manufactured product and the total loss created by that product to society; the necessity of continuous quality improvement and cost reduction for an organization's health in a competitive economy; the need for never-ending reduction of variation in product and/or process performance around nominal, or target, values; the relationship between society's loss due to performance variation and the deviation of the performance characteristic from its nominal value; the impact of product and process design on a product's quality and cost; the nonlinear effects between a product's and/or process's parameters and the product's desired performance characteristics; the identification of product and/or process parameter settings that reduce performance variation; and quality function deployment.

12 Taguchi Methods— Quality Improvement in Product and Process Design

Introduction

Traditionally, quality control activities have centered on control charts and process control; this is called *on-line* quality control. Dr. Genichi Taguchi, a Japanese statistician and Deming Prize winner, has extended quality improvement activities to include product and process design; this is called *off-line* quality control. Taguchi's methods provide a system to develop specifications, design those specifications into a product and/or process, and produce products that continuously surpass said specifications.[1]

There are seven aspects[2] to off-line quality control:

1. The quality of a manufactured product is measured by the total loss created by that product to society.
2. Continuous quality improvement and cost reduction are necessary for an organization's health in a competitive economy.
3. Quality improvement requires the never-ending reduction of variation in product and/or process performance around nominal values.
4. Society's loss due to performance variation is frequently proportional to the square of the deviation of the performance characteristic from its nominal value.
5. Product and process design can have a significant impact on a product's quality and cost.
6. Performance variation can be reduced by exploiting the nonlinear effects between a product's and/or process's parameters and the product's desired performance characteristics.
7. Product and/or process parameter settings that reduce performance variation can be identified with statistically designed experiments.

This chapter covers these seven points. For a more detailed discussion, see Taguchi and Wu[3] or Kackar.[4]

Point 1

The quality of a manufactured product is measured by the total loss created by that product to society.

Taguchi defines *quality* in terms of the loss imparted to society from the time a product is shipped. Many people find his definition unusual because it presents quality in a negative fashion; the basis of Taguchi's words is that the smaller the loss caused to society by a product, the better the product's quality.

Viewing quality from a societal perspective is profound because it includes customers, manufacturers, and the community in the definition of quality. According to this perspective on quality, quality improvement saves society more resources than it costs, and it benefits everyone: customers, manufacturers, and the community. Hence, investment in quality improvement is worthwhile so long as it reduces the loss to society from the time a product is shipped.

The total loss to society to produce a product with given parameter values (nominal settings) consists of two component parts: (1) the production cost to the

FIGURE 12.1   Cost of Producing a Vinyl Sheet

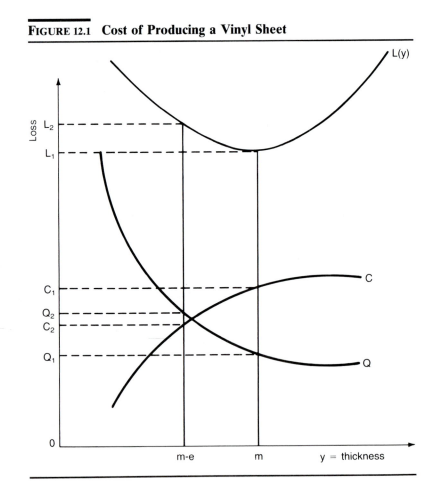

manufacturer of producing a product with given parameters and (2) the inferior quality cost to the customer and community of producing a product with given parameters.

For example, one important characteristic of the vinyl sheets used to build houses for agricultural production is the thickness of the sheets.[5] In Figure 12.1, y is the thickness setting on the machine producing the vinyl sheet. From the manufacturer's perspective, if the thickness of the vinyl sheet is increased, the production cost (including raw material cost, processing costs, and inventory costs) increases, as shown by curve C in Figure 12.1. From the customer's and community's perspectives, if the thickness of the vinyl sheets is increased, the cost of inferior quality decreases because the sheets become sturdier and farmers must replace or repair them less frequently, as shown by curve Q. The total loss (cost) to society, L(y), is computed by summing the cost to manufacturers (C curve) and the cost to customers and society (Q curve) at every setting of vinyl sheet thickness, y, as shown by curve L(y).

The optimal thickness setting is selected by picking the thickness setting, y, for which the loss to society, L(y), is a minimum; this is thickness setting y = m in Figure 12.1. If the thickness setting is changed from y = m to y = m−e, then a disproportionately large loss is created for customers; the cost to the customer and society increases from Q_1 to Q_2, or $(Q_2 − Q_1)$, while the cost to the manufacturer decreases from C_1 to C_2, or $(C_1 − C_2)$. As $(Q_2 − Q_1)$ is greater than $(C_1 − C_2)$, there's a net loss to society. This net loss is seen in that the total cost of producing the vinyl sheets has increased from L_1 to L_2. The thickness setting that minimizes the loss to society from the time the product is shipped is y = m.

Point 2

Continuous quality improvement and cost reduction are necessary for an organization's health in a competitive economy.

High quality and low cost are strategic factors in any plan for corporate health. Companies that realize these strategic factors' significance know that quality can always be improved and costs can always be reduced. A major contention of Deming and Taguchi is that products and processes must be improved in a relentless and never-ending manner. This reasoning is demonstrated in Figure 12.2. Distribution A depicts unit-to-unit performance variation before a product and/or process is improved. The costs (losses to society) incurred using the system that produced distribution A are represented by the area under distribution A beneath the total cost curve, L(y). Distribution B depicts unit-to-unit performance variation after the product and/or process is improved; the costs (losses to society) incurred using the improved system that produced distribution B are represented by the area under distribution B beneath the total cost curve, L(y). The losses to society incurred under the system with lower unit-to-unit performance variation, system B, are clearly lower than the losses to society incurred with system A.

FIGURE 12.2   Loss and Variation Reduction

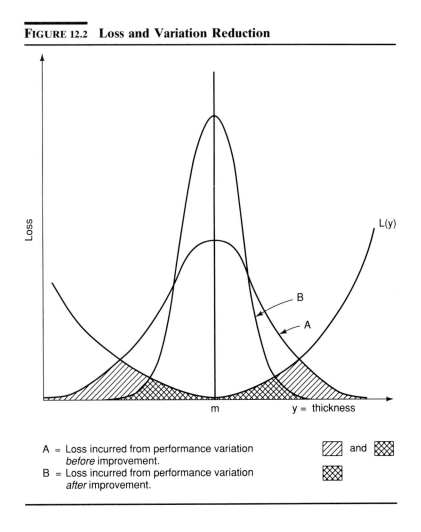

A = Loss incurred from performance variation
    before improvement.
B = Loss incurred from performance variation
    after improvement.

Point 3

Quality improvement requires the never-ending reduction of variation in product and/or process performance around nominal values.
Following the logic stated in Point 2, it's always economical to reduce unit-to-unit performance variation around nominal, even when we're within specification limits.

As an example illustrating this view, a car battery is charged by an alternator that has a voltage regulator that controls the charge to the battery. The alternator-voltage regulator assembly must put out a charge of 13.2 volts to keep the battery's charge at 12 volts. If the alternator produces a charge of fewer than 13.2 volts, eventually the electrolyte (acid) will turn into water and lose its charge, and the battery will die. If the alternator produces a charge of more than 13.2 volts, the battery plates will warp from excessive heat, the electrolyte will evaporate, and the battery will die.

This example demonstrates that *any* deviation from nominal causes a loss. Just being within specification limits is not the minimum loss position.

Point 4

Society's loss due to performance variation is frequently proportional to the square of the deviation of the performance characteristic from its nominal value. Any variation in a product's performance characteristic about its nominal value, at any randomly selected position in the product's life cycle, causes a loss to society, as discussed in the prior section. Again, let L(y) equal the total cost to society as a result of a product's having a value of y for a specified performance characteristic, given that the nominal value for the performance characteristic is m.

There are many possible forms for L(y). The two most common forms are

$$L(y) = \begin{cases} A & \text{if} \quad y < LSL \text{ or } y > USL \\ 0 & \text{if} \quad LSL \leq y \leq USL \end{cases} \tag{12.1}$$

and

$$L(y) = \begin{cases} k(y - m)^2 & \text{if} \quad |y - m| > 0 \\ 0 & \text{if} \quad y - m = 0 \end{cases} \tag{12.2}$$

Equation 12.1 simply states that the loss caused by a product's characteristic value, y, deviating from nominal, m, is zero when y is within specification limits; when y is either lower than the lower specification limit or greater than the upper specification limit, the loss is a constant value, A, as shown in Figure 12.3's shaded areas.

Equation 12.2 states that the loss caused by a product's characteristic value, y, deviating from nominal, m, is proportional to the squared distance between y and m, or $k(y - m)^2$, where k is a proportionality constant, or loss coefficient. If a product's characteristic value, y, is the same as the nominal value, m, then $y - m = 0$ and $L(y) = k(y - m)^2 = 0$. But if a product's characteristic value isn't the same as the nominal value, then $L(y) = k(y - m)^2$. In other words, the farther from nominal a product characteristic's value lies, the larger the loss. This quadratic loss function is illustrated in Figure 12.1.

The value of k is set by establishing the value of L(y) at a specification limit as A, so that $L(y) = A$ if $y = USL$ or $y = LSL$. Thus

$$L(y) = k(y - m)^2$$

$$A = k_{USL}(USL - m)^2 \text{ and } A = k_{LSL}(LSL - m)^2$$

Then

$$k_{USL} = \frac{A}{(USL - m)^2} \quad \text{and} \quad k_{LSL} = \frac{A}{(LSL - m)^2} \tag{12.3}$$

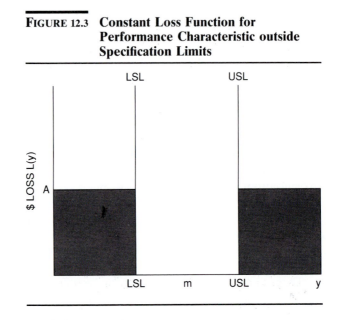

FIGURE 12.3   Constant Loss Function for Performance Characteristic outside Specification Limits

Using the loss function in Equation 12.2 is the same as stating that the minimum loss to society, L(y), occurs when y = m (when a product's characteristic value is at nominal). This minimum-loss situation occurs with increasing frequency if a process's average is centered on nominal and its unit-to-unit variation is continuously reduced. In other words, when the average value of the product characteristic is m and its standard deviation about m approaches zero, the loss to society is minimized.

Point 5

Product and process design can have a significant impact on a product's quality and cost.

The number of manufacturing imperfections in a product—hence the manufacturing cost of a product—is significantly affected by the product's design and the design of the process used to produce the product. Figure 12.4 shows the relationship between the number of manufacturing imperfections in a product and the degree of process control used to regulate that number.

Curve A shows the relationship between the number of manufacturing imperfections and degree of process control for a given product/process design configuration. Curve A indicates that greater process control results in fewer manufacturing imperfections, and vice versa.

Curve B shows the relationship between the number of manufacturing imperfections and degree of process control for an improved product/process design configuration. The product/process design configuration in curve B is superior to the product/process design configuration in curve A because the number of manu-

FIGURE 12.4   Manufacturing Imperfections for Various Degrees of Process Control

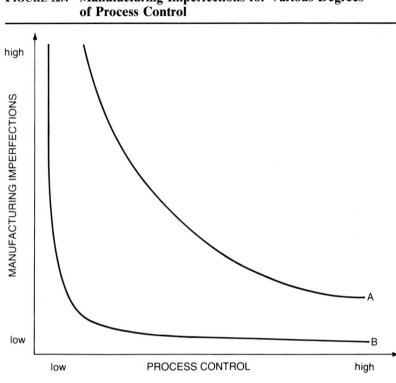

facturing imperfections is lower in curve B than in curve A for every level of process control. Curve B indicates that the number of manufacturing imperfections decreases as process control increases; however, the relationship isn't linear. Hence, given an improved product/process design, the number of manufacturing imperfections can be dramatically reduced with a low level of process control. To state this another way, medium to high degrees of process control have little effect on reducing the number of manufacturing imperfections in the product in curve B because the number of manufacturing imperfections is already low as a result of the improved product/process design.

Point 6

Performance variation can be reduced by exploiting the nonlinear effects between a product's and/or process's parameters and the product's desired performance characteristic.

Quality control activities must begin with *quality-of-design/redesign studies*. These studies lead to the development of product and process parameters, which create products that surpass customers' needs. Product and process parameters must be stated in terms of specifications with nominal values and tolerances around these nominal values.

Taguchi has developed a three-part procedure for constructing nominal values and tolerances for product and process parameters that will create products that surpass customers' needs. The three-part procedure includes systems design and quality function deployment, parameter design, and allowance design.

Systems Design and Quality Function Deployment

Systems design requires engineering and scientific knowledge to create an initial product prototype. Parameter settings of the initial product prototype's specifications are defined by a set of nominal values and their respective tolerances based on the results of a Quality Function Deployment study.

Quality Function Deployment (QFD)[6] is a frequently employed tool to do systems design; that is, to set specifications that consider the needs and wants of different market segments and satisfy requirements for manufacturability. Also, it provides a clear line of traceability from a specific customer need or want to a specific manufacturing process.

QFD is used to (1) define the ever-changing market segments for customers, (2) determine and prioritize the customer requirements of each market segment, (3) identify the processes (methods) used to respond to the customer requirements of each market segment, (4) construct a matrix explaining the relationships between the customer requirements of each market segment and the processes (methods) used to respond to the customer requirements, (5) prioritize the processes (methods) used to respond to customer requirements, and (6) establish specifications for the processes used to respond to customer requirements.

Use 1: Define the Ever-Changing Market Segments for Customers.   The term *market segment* describes the dynamic and changing homogeneous groupings of customers with respect to the demographic, psychographic, and purchasing behavior variables that affect their decision to purchase and/or use a good or service. Focus groups as well as other methods should be used to identify customer requirements for each market segment. Special care should be taken to identify and define the customer requirements of noncustomers and future market segments.

Use 2: Determine and Prioritize the Customer Requirements of Each Market Segment.   Management must collect and analyze observational, survey, and experimental data to understand the "voice of the customer" by market segment. The question asked of a sample of customers from each market segment is: From your perspective, what requirements must the organization surpass in order to pursue the mission statement?

Figure 12.5 shows a prioritized list of customer (student) requirements for MBA students collected at a local university.[7] The procedure for prioritizing the customer (student) requirements important to a market segment requires further discussion. For each market segment, each customer requirement is scored on three scales measuring (1) importance to the customer, (2) current level of perfor-

FIGURE 12.5   Basic Statistics Course for MBAs

Processes used to respond to customer requirements

Customer (Student) Requirements	Lecture	Exams/Term Papers	Scheduling Time	Maintenance	Professor	Importance to Customer	Current Level of Importance in the Eyes of the Customer	Desired Level of Performance by Management	Total Weight
Learns requisite knowledge	◎ 15	◎ 15	△ 5		◎ 15	5	3	3	5.0
Easy to understand	◎ 37.5		○ 25		◎ 37.5	5	2	5	12.5
Convenient time slot			◎ 3		○ 2	4	4	1	1.0
Pleasant classroom			○ 3	◎ 4.5		3	4	2	1.5
Unnormalized weights	52.5	15.0	36.0	4.5	54.5	162.5			
Normalized weights	.32	.09	.22	.03	.34				
Nominal	75 min	Exams = 2 Paper = 1	10:00 AM to 4:00 PM	30 min./ daily	Systems Design				

LEGEND:
◎ strong relationship = 3
○ moderate relationship = 2
△ weak relationship = 1
⊘ no relationship = 0

mance in the eyes of the customer, and (3) desired level of performance by management to optimize the interdependent system of stakeholders. The total weight is computed for each customer requirement in each market segment. Total weight is a measure of the need to take action on a customer requirement,[8] as shown in Figure 12.5.

The *importance to customer* scale quantifies the importance of customer requirements for each market segment; it does not quantify how well the organization is currently handling customer requirements or how much improvement is required with respect to customer requirements. In this 1-to-5 scale, 1 = very unimportant and 5 = very important. Importance to the customer scores are obtained by computing the average ratings for each customer requirement for each market segment from survey and/or focus group data.

The *current level of performance* scale quantifies the gap between customer requirements and organization performance. In this 1-to-5 scale, 1 = very large gap and 5 = very small gap. Current level of performance scores are obtained through *gap analysis*. Gap analysis is a procedure for studying the root cause(s) of the difference between customer requirements and organization performance. It's based on the analysis of relevant data. For each customer requirement and market segment, gap analysis requires a measure of customer requirements (see the importance to the customer scale) and a measure of current organization performance (see current level of importance scale) to highlight the customer requirements that should be studied further with gap analysis. The measure of current performance scale is obtained by computing the average ratings from survey or focus group data for each customer requirement for each market segment from the question, How is the organization doing with respect to exceeding customer requirement x? Customer requirements that show a "special cause" gap are targeted for further study via gap analysis. For example, a group of staff personnel might study the root causes of the gap for a particular customer requirement for a particular market segment. The group might study the gap over time and determine that it's stable and contains only common variation. Next, the group could construct a Pareto diagram of the common causes of the gap, isolate the most significant common cause, and develop a cause-and-effect diagram of its causes. Next, the staff would study the correlation between the suspected root cause and the most significant cause of the gap. If they found the correlation to be significant, they would recommend a plan of action for improving the current level of performance for said customer requirement.

The *desired level of performance* scale quantifies the desired level of performance for each customer requirement for each market segment. The scale is again a 1-to-5 scale, where 1 = small improvement in the organization's ability to exceed a customer requirement, and 5 = large improvement in the organization's ability to exceed a customer requirement. Desired level of performance scores are developed by staff personnel. They analyze the levels of performance required for each customer requirement for each market segment to exceed customer needs and wants (benchmarking) and future customer requirements.

The *total weight* score measures the gap's importance for each customer requirement in each market segment. Total weight scores are computed using the formula

$$\text{Total weight} = \frac{\text{Importance to the customer} \times \text{Desired level of performance}}{\text{Current level of performance}}$$

Use 3: Identify the Processes (Methods) Used to Respond to the Needs and Wants of Each Market Segment and Indirect Customer.   It's critical that the needs and wants of each appropriate market segment or indirect customer be serviced by identifiable methods. Customer needs and wants must be translated into improved and innovated methods. This translation is accomplished by asking what methods are necessary to respond to each customer need or want, as Figure 12.5 shows.

Use 4: Construct a Matrix that Explains the Relationships (Represented as Cells of the Matrix) between Customer Requirements (Represented by the Rows of the Matrix) and the Processes, or Methods, Used to Respond to Customer Requirements (Represented by the Columns of the Matrix) for Each Market Segment and Each Indirect Customer.   All matrices should have the same columns; that is, the matrices for all market segments and indirect customers should have the same "processes used to respond to customers' needs and wants" as their columns.

The relationships shown in each cell of each matrix are determined by a group of staff personnel. The staff use their own expertise as well as that of others, combined with their knowledge of the organization and customers to determine the relationships. Relationships are measured on the following scale: 3 = strong relationship, 2 = moderate relationship, 1 = weak relationship, and blank = no relationship. Sometimes a doughnut symbol is used for a 3, a circle for a 2, and a triangle for a 1. These numbers or symbols are used whether the relationships are positive or negative. A matrix showing the needs and wants of a particular market segment and the methods needed to respond to the needs and wants of the customers in the market segment appears in Figure 12.5.

Every customer requirement must be adequately serviced by one or more methods. If a customer requirement isn't being met by any method (or isn't adequately serviced), then one or more methods must be developed that will satisfy the customer requirement. If a method isn't meeting (directly or indirectly) at least one customer requirement, then the method should be dropped or receive decreased attention by stakeholders.

Use 5: Prioritize the Processes (Methods) Used to Respond to Customer Requirements for Attention as Input into Setting the Strategic Objectives of the Organization.   Prioritization of methods is accomplished by the following procedure:

1. Compute the unnormalized weights (as in Figure 12.5) for each process' for a given market segment or indirect customer. For a given process (column in Figure 12.5), multiply the total weight score for each customer requirement by the relationship score between said process and all customer requirements. For example, we compute the unnormalized weights for the processes "lecture" and "exams/term papers" as follows:

Lecture: $52.5 = [5.0(3) + 12.5(3)]$, where

   5.0 = Total weight for "learns requisite knowledge,"
   12.5 = Total weight for "easy to understand,"
   3 = The relationship between "lecture" and "learns requisite knowledge," and
   3 = The relationship between "lecture" and "easy to understand."

Exams/term papers: $15.0 = (5.0 \times 3)$, as shown in Figure 12.5.

2. Normalize the weighted values by dividing the individual weighted values by the sum of all weighted values, as in Figure 12.5.

3. Prioritize the normalized weighted values over all methods to provide input into the selection of strategic objectives for the organization. For example, the preceding analysis indicates that "lecture" would receive a higher priority for attention than "exams/term papers."

Use 6: Establish Nominal Values and Specifications for the Processes Used to Respond to Customer Requirements.   This phase of QFD completes the systems design.

Parameter Design

Parameter design involves determining the specification settings for product and process parameters in terms of nominal values so that the final product will be less sensitive to sources of variation caused by environmental factors, product deterioration, and manufacturing variations.

> **Environmental Factors.**   *Environmental factors* are conditions in the environment in which the product will be used by the customer. Such factors include human variations in operating the product.
>
> **Product Deterioration.**   *Product deterioration* consists of changes in product parameters over time from wear and tear on the product during its life cycle.
>
> **Manufacturing Variations.**   *Manufacturing variations* are manufacturing conditions that cause the production of product that deviates from its nominal values.

These three sources of product variation are usually common sources of variation because they're chronically present and affect the product's performance.

Let's consider the design of an electrical circuit to illustrate and appreciate the purpose of parameter design. Kackar offers the following illustration.[9] Suppose the performance characteristic of interest is the output voltage of an electric circuit, y, and its nominal value is y_0. Assume that the circuit's output voltage is largely determined by the gain of a transistor X in the circuit, and the circuit designer is at liberty to choose the nominal value of this gain. Suppose also that the transistor gain's effect on the output voltage is nonlinear. This relationship is shown in Figure 12.6.

To obtain an output voltage of y_0, the circuit designer can select the nominal value of transistor gain to be x_0. If the actual transistor gain deviates from the nominal value x_0, the output voltage will deviate from y_0. The transistor gain can deviate from x_0 because of manufacturing imperfections in the transistor, deterioration during the circuit's life span, and environmental variables. If the distribution of transistor gain is as shown in Figure 12.6, the output voltage will have a large variation. One way of reducing the output variation is to use an expensive transistor whose gain has a very narrow distribution around x_0. Another way of

FIGURE 12.6   Effect of Transistor Gain on Output Voltage

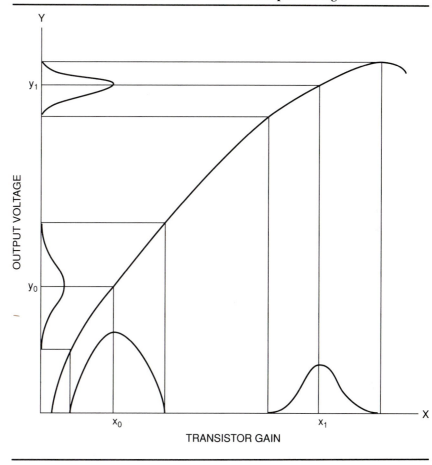

reducing output variation is to select a different value of transistor gain. For instance, if the nominal transistor gain is x_1, the output voltage will have a much smaller variance. But the mean value y_1 associated with the transistor gain x_1 is far from the target value y_0. Now suppose there's another component in the circuit, such as a resistor, that has a linear effect on the output voltage, and the circuit designer is at liberty to choose the nominal value of this component. The circuit designer can then adjust this component to move the mean value of voltage from y_1 to y_0. Adjusting the mean value of a performance characteristic to its target value is usually a much easier engineering problem than reducing performance variation. When the circuit is designed so that the nominal gain of transistor X is x_1, an inexpensive transistor having a wide distribution around x_1 can be used. Of course, this change wouldn't necessarily improve circuit design if it were accompanied by an increase in the variance of another performance characteristic of the circuit.

This example demonstrates that exploiting the nonlinear effects of product or process parameters on product performance characteristics can be an effective method for reducing the sensitivity of product performance to environmental factors, product deterioration, and manufacturing variations.

Allowance Design

The amount of allowable tolerance around a nominal value is determined by balancing the customer's loss from increased performance variation resulting from wide tolerances and the manufacturer's loss from increased costs resulting from narrow tolerances. Allowance design involves establishing the allowable size of a tolerance around the nominal setting of product or process parameters determined in the parameter design stage.

Design engineers must resort to allowance design when the influences of environmental factors and product deterioration can't be successfully reduced through parameter design. Allowance design should be performed after parameter design as it's less expensive to reduce performance variation through parameter design than it is to control performance variation by establishing tolerances.

The establishment of tolerances isn't contradictory to the philosophy of never-ending improvement because engineers will continuously endeavor to produce a better design (by establishing nominal values that reduce performance variation), and production personnel will continuously strive to reduce variation around the nominal values established in the parameter design stage.

Point 7

Product and/or process parameter settings that reduce performance variation can be identified with statistically designed experiments

Parameter design can be accomplished through a statistically designed experiment. The experiment requires that the variables affecting the performance variation of the product under study be classified into two categories: design parameter variables and noise variables.

Design Parameter Variables

Design parameter variables are variables that can be set at one of two or more possible parameter settings. For example, water level in a cooling system can be set at either a high (= 1) or a low (= 0) parameter setting. Design parameter variables represent the nominal values of product or process parameter settings that can be determined by a design engineer who has conducted and utilized the information from quality-of-design and quality-of-performance studies.

Noise Variables

Noise variables include all the factors that cause the product's performance characteristics to deviate from their nominal value or cause actual performance to differ from desired performance. The noise variables that most significantly affect the product's performance characteristics should be determined and included in the parameter design experiment. The experimenter should systematically vary the levels of the most significant noise variables, or their surrogates, to determine their effects on the product's performance characteristics.

In reality, it may not be possible to consider every noise variable in a parameter design experiment. It may be impossible to conduct an experiment with all known noise variables because of the large data requirements of the experiment; or the importance of a particular noise variable may be unknown to the person conducting the parameter design experiment. Consequently, the experimenter must be wary of potential problems from unknown noise variables.

The purpose of a parameter design experiment is to determine the nominal values for the design parameter variables that yield the lowest impact on the product's performance characteristics by the noise variables. Nominal values of the parameter design variables are established by systematically varying their settings in conjunction with a selected combination of the settings of the noise variables, and then comparing the resultant performance characteristics, as shown in Figure 12.7.

Selecting the parameter design variable settings and noise variable settings for a parameter design experiment isn't a trivial task. Taguchi has developed a recommended procedure for performing a parameter design experiment, utilizing a design parameter matrix and a noise factor matrix.[10] The design parameter matrix lists the design variables in its columns and the appropriate combinations of parameter design variable settings in its rows. The noise factor matrix lists the noise variables in its columns and the appropriate combinations of noise variable settings in its rows. The design parameter experiment consists of running tests for every row in the design parameter matrix (a given set of product design parameter nominal values) under the conditions specified in every row of the noise factor

FIGURE 12.7   An Example of a Taguchi-Type Parameter Design Matrix

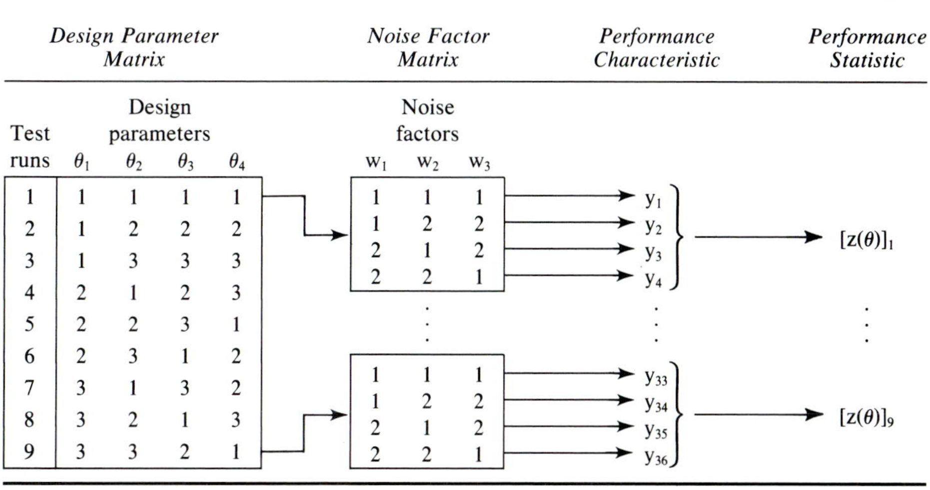

SOURCE: Kackar, "Taguchi's Quality Philosophy: Analysis and Commentary," *Quality Progress,* December 1986, p. 27.

matrix (a given set of noise variable conditions); this is shown in Figure 12.7. The resulting performance characteristics are recorded for each specified combination of design parameter variables and noise variables.

For each configuration of the design parameter variables (a row in the design parameter matrix, representing a particular set of nominal values for the design parameter variables), all the test run settings (rows) of the noise variables shown in the noise variable matrix are used to compute a performance statistic. The performance statistic estimates the noise variables' effects on the performance characteristics for a given design (a particular set of nominal values for the design parameter variables). The setting of the parameter design variables that yield the best performance statistic is deemed the best product design.

Taguchi recommends using a performance statistic called a *signal-to-noise ratio*. Three types of signal-to-noise ratios exist for variables type performance statistics[11]:

1. The smaller the ratio, the better the product design. For example, if friction is the performance characteristic under study, low friction is desirable.

2. The larger the ratio, the better the product design. For example, if adhesion is the performance characteristic under study, high adhesion is desirable.

3. A nominal value is best. For example, if the gap size of a created dimension is the performance characteristic under study, and nominal is a gap size of 3 mm, a gap size of 3 mm is most desirable.

Signal-to-noise ratios can similarly be defined for attribute-type performance characteristics. Consequently, statistically designed experiments using a signal-to-noise ratio as a measure of product or process performance can be used to reduce variation by careful selection of parameter settings.

CASE STUDY 12.1
SUMITOMO ELECTRIC INDUSTRIES, LTD.

Fine lines (lines under a micron in width) fabricated on IC (integrated circuit) wafers are essential to the proper functioning of ICs. They offer microminiaturization, high integration, and high performance of the transistors that compose ICs. Variability of fine line width causes malfunctions in ICs so it's important to reduce that variability.

The line fabrication system for the circuits is composed of six processes: resist-spin-coating, exposure, development, descumming, metal evaporating, and resist-pattern removal. Taguchi methods are used to optimize the line fabrication system.

Source: This case study is taken from T Fukuzawa, Y Goda, M Nishiguchi, N Nishizawa, and K Sakurai, "Quality Engineering Approach of Line Width Stability for IC's Fabrication," Sumitomo Electric Industries, Ltd, *American Supplier Institute,* pp. 381–89. The authors have rewritten the case study to make it congruent with the writing style of this chapter. They thank Shin Taguchi and the American Supplier Institute for permission to publish this paper in this text.

FIGURE 12.8   IC's Line Fabrication System

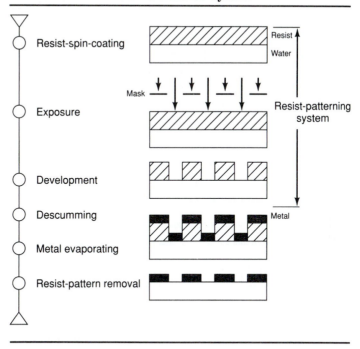

Resist-Patterning System and Experimental Specimen

Resist-Patterning System.   The line fabrication system is shown in Figure 12.8. The "resist" (fine line) is spin-coated on a wafer through a mask. This establishes the "fine line" (resist) after exposure, development, descumming, metal evaporation, and resist-pattern removal. The variability of the line fabrication system depends strongly on the resist-patterning system.

FIGURE 12.9   Control Factors and Levels in First Experiment

				Level	
Process	*Factors*		*1*	*2*	*3*
Resist-spin-coating	A: Resist type		A	B	—
	B: Resist thickness	(μm)	0.7	1.0	1.3
Exposure	C: Exposure time	(msec)	350	400	450
	D: Focus depth	(μm)	−(thickness)	0	+(thickness)
Development	E: Method		dipping	paddle	—
	F: Temperature	(C)	20	24	28
	G: Time	(sec) E1	60	80	100
		E2	40	60	80
Descumming	H: Time	(min.)	0	1	5

FIGURE 12.10   **Measured Data and Error Factors**

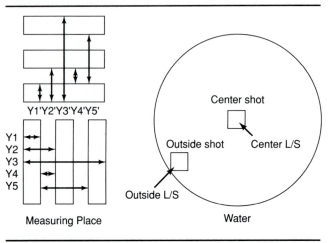

First Experiment

Design Parameters.   The design parameters that influence the variance of the resist patterns are resist-spin-coating, exposure, development, and descumming (Figure 12.9). Levels of the design parameters (A through H) are also shown in Figure 12.9.[12]

FIGURE 12.11   **Experiment Design Layout for Parameter Design (Orthogonal Array L18)**

				Factor					S/N Ratio
Number	*E*	*H*	*A*	*B*	*C*	*D*	*F*	*G*	η(dB)
1	1	1	1	1	1	1	1	1	12.479
2	1	1	2	2	2	2	2	2	10.365
3	1	1	1	3	3	3	3	3	16.165
4	1	2	1	1	2	2	3	3	11.174
5	1	2	2	2	3	3	1	1	12.499
6	1	2	1	3	1	1	2	2	15.019
7	1	3	1	2	1	3	2	3	10.308
8	1	3	2	3	2	1	3	1	5.457
9	1	3	1	1	3	2	1	2	1.982
10	2	1	1	3	3	2	2	1	11.426
11	2	1	2	1	1	3	3	2	18.817
12	2	1	1	2	2	1	1	3	2.512
13	2	2	1	2	3	1	3	2	5.530
14	2	2	2	3	1	2	1	3	10.729
15	2	2	1	1	2	3	2	1	11.810
16	2	3	1	3	2	3	1	2	4.782
17	2	3	2	1	3	1	2	3	5.163
18	2	3	1	2	1	2	3	1	12.027

FIGURE 12.12   Measured Data (First Column of L18)

Error Factors	Signal Factors				
	M1 (1.2 μm)	*M2* (2.4 μm)	*M3* (6.0 μm)	*M4* (1.2 μm)	*M5* (3.8 μm)
Center					
Vertical	1.00	2.19	5.44	1.21	3.40
Horizontal	1.06	2.33	5.63	1.29	4.03
Outside					
Vertical	0.84	2.18	5.32	1.35	3.60
Horizontal	1.00	2.35	5.53	1.37	3.60

Noise Factors.   The noise factors in this study are locations of shots on a wafer (center or outside), locations of patterns in a shot (center or outside), and directions of patterns (vertical or horizontal), as shown in Figure 12.10.

Performance Characteristics.   The resist-patterning system is measured at five levels—performance characteristics Y1 through Y5 in Figure 12.10.

Experimental Method.   Eighteen combinations of the levels of the design parameters A through H are assigned by the orthogonal arrays in an L18 matrix, as shown in Figure 12.11. The noise factors are shown in the rows of Figure 12.12. A scanning electron microscope is used to measure the dimensions of Y1 through Y5, called M1 through M5 (the columns of Figure 12.12). Figure 12.12 shows the experimental data for the first row of the L18 matrix.

The S/N ratio for each row of the L18 matrix is shown in the last column of Figure 12.11. The S/N ratio for the first column of the L18 matrix is 12.479 dB.[13] The means for each level of the design parameters are shown in Figure 12.13. Figure 12.14 plots graphically the S/N ratio for each level of each design parameter.

FIGURE 12.13   S/N Ratio for Each Level of Control Factors (First Experiment)

Factor	S/N Ratio, η (dB)		
	Level 1	*Level 2*	*Level 3*
A	9.60	10.51	—
B	10.24	8.87	10.59
C	13.23	7.68	8.79
D	7.69	9.62	12.40
E	10.60	9.20	—
F	7.50	10.68	11.53
G	10.95	9.42	9.34
H	11.96	11.13	6.62

FIGURE 12.14   Effect Responses of First Experiment

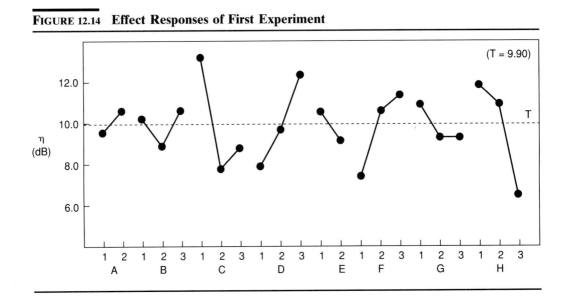

Summary of First Experiment.   We can draw the following conclusions from Figure 12.14:

1. Factor A (resist type): A2 is selected. (original parameter setting A1)
2. Factor B (resist thickness): B3 is selected. (originally B3)
3. Factor C (exposure time): Re-experiments should be done near the C1 level. (originally C2)
4. Factor D (focus depth): Re-experiments should be done near the D3 level. (originally D2)
5. Factor E (development method): E1 is selected. (originally E1)
6. Factor F (developer temperature): F2 is selected. (originally F2)
7. Factor G (development time): G1 is better, but the effect is very small so re-experiments should be done with wide levels. (originally G2)
8. Factor H (descumming): H1 is better, but scums have still existed. Re-experiments should be done to confirm the effect. (originally H2)

A second experiment is done to select the best levels for C and D and to confirm the effects for G and H. Factor A yields the smallest difference in S/N and is eliminated from the second experiment to reduce the size of the experiment to an L9 matrix. It's set at A2. Factors B, E, and F are set at B3, E1, and F2, respectively, and also are eliminated as variables in the second experiment.

Second Experiment

Experimental Method and Analysis of Experimental Data.   From the results of the first experiment, engineers determine the levels of the design parameters C, D, G, and H (Figure 12.15). The original settings of the design parameters (A1, B3, C2, E1, F2, G2, and H2) are used for comparative purposes.

FIGURE 12.15 **Control Factors and Levels in Second Experiment**

			Level	
Factors		*1*	*2*	*3*
C: Exposure time	(msec)	275	300	325
D: Focus depth	(μm)	0	+(thickness)	+2×(thickness)
G: Development time	(sec)	50	80	110
H: Descumming time	(min.)	0	1	—

Seven dimensions, Y1 through Y7, are measured in the second experiment (Figure 12.16). Levels of the design parameters are assigned by an L9 matrix (Figure 12.17). The noise factors and performance characteristics are assigned outside in the same way as in the first experiment, as we see in Figure 12.18.

Figure 12.18 presents the measured data from the first column of L9. The S/N ratios for each row in the L9 matrix are shown in the last columns of Figure 12.17. The means of each level of the design parameters are shown in Figures 12.19 and 12.20, respectively.

Prediction of Optimum Conditions.   The optimum conditions of the resist-patterning system are selected from Figure 12.20. Design parameters C and G are found to have the largest impact on the S/N ratio. The predicted improvement in the S/N ratios by using C1 and G1 over the original conditions is obtained by using the equation

$$\mu = C1 + G1 - T$$

where

> T = The average S/N ratio over all design parameters
> and levels in the second experiment
>
> = 19.00 + 20.94 + 16.86
>
> = 23.08 dB

FIGURE 12.16 Measured Data (Second Experiment)

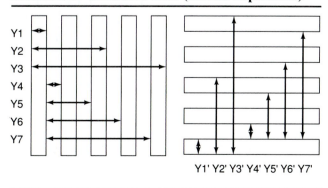

Y1' Y2' Y3' Y4' Y5' Y6' Y7'

FIGURE 12.17   Experiment Design Layout for Parameter Design (Orthogonal Array L9)

		Factor			S/N Ratio
Number	*D*	*C*	*G*	*H*	$\eta(dB)$
1	1	1	1	1	23.342
2	1	2	2	2	13.356
3	1	3	3	1'	11.750
4	2	1	2	1'	19.246
5	2	2	3	1	15.584
6	2	3	1	2	17.024
7	3	1	3	2	14.405
8	3	2	1	1'	22.445
9	3	3	2	1	14.575
Present conditions					14.784

FIGURE 12.18   Measured Data (First Column of L9)

	Signal Factors						
Error Factors	*0.7 μm*	*3.5 μm*	*8.3 μm*	*0.7 μm*	*2.1 μm*	*3.5 μm*	*4.9 μm*
Center							
Vertical	0.61	3.26	5.97	0.75	2.07	3.41	4.74
Horizontal	0.67	3.40	6.17	0.77	2.11	3.47	4.87
Outside							
Vertical	0.65	3.33	6.00	0.73	2.06	3.38	4.72
Horizontal	0.67	3.37	6.18	0.71	2.01	3.35	4.79

FIGURE 12.19   S/N Ratio for Each Level of Control Factors (Second Experiment)

	S/N Ratio, η (dB)		
Factor	*Level 1*	*Level 2*	*Level 3*
C	19.00	17.13	14.45
D	16.15	17.28	17.14
G	20.94	15.73	13.91
H	17.82	14.93	—

FIGURE 12.20 Effect Responses of Second Experiment

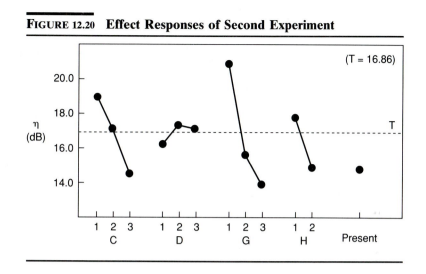

The S/N ratio for the original conditions in the second experiment is 14.784 dB. The gain in the S/N due to using C1 and G1 is obtained from the equation

$$\text{Gain} = 23.08 - 14.78$$

where

$$14.78 = \text{The S/N ratio for the original levels of}$$
$$\text{the design parameters}$$

$$= 8.30 \text{ dB}$$

Confirmation Experiment

Confirmation experiments are done at both the original conditions and the optimum conditions. The optimum conditions show an estimated gain of 8.30 dB and a confirmed gain of 8.05 dB over the original conditions shown in Figure 12.21. *Consequently, the standard deviation of line width using the optimum condition is 38 percent of the standard deviation using the original conditions.[14] Every time the S/N ratio gains 6 dB, the standard deviation is cut in half.*

FIGURE 12.21 Results of Estimation and Confirmation Experiment

	Optimum Condition	Present Condition	Gain
Estimation	23.08 dB	14.78 dB	8.30 dB
Confirmation Experiment	20.42	12.37	8.05

Summary

Genichi Taguchi has made a large contribution to quality management and statistics. His views on the definition of quality, quality loss functions, never-ending improvement, and parameter design experiments, to name a few of his contributions, have advanced the worldwide movement for quality.

In this chapter we discussed several aspects of Taguchi's views of quality:

1. The quality of a manufactured product is measured by the total loss created by that product to society.
2. Continuous quality improvement and cost reduction are necessary for an organization's health in a competitive economy.
3. Quality improvement requires the never-ending reduction of variation in product and/or process performance around nominal values.
4. Society's loss due to performance variation is frequently proportional to the square of the deviation of the performance characteristic from its nominal value.
5. Product and process design can have a significant impact on a product's quality and cost.
6. Performance variation can be reduced by exploiting the nonlinear effects between a product's and/or process's parameters and the product's desired performance characteristics.
7. Product and/or process parameter settings that reduce performance variation can be identified with statistically designed experiments.

Exercises

12.1 Explain Taguchi's statement that the quality of a manufactured product is measured by the total loss created by that product to society.

12.2 a. Why is continuous quality improvement critical to any organization's health and competitive position?
   b. Explain the relationship between quality improvement and cost reduction for any organization.

12.3 Explain why being within specification limits isn't enough to be competitive in today's world economy.

12.4 Explain why the concept of "zero defects" is flawed and will harm an organization in the long run.

12.5 Discuss two possible models for quantifying society's loss from poor quality: the linear loss function and the quadratic loss function. Explain the rationale behind each.

12.6 A firm produces steel rods with a length specification of 6.0 inches, plus or minus 0.10 inches. If a steel rod exceeds either specification limit, it's melted down for stock to produce new rods. The cost of a rod's being out of specification is $0.75 per rod. Given this information, construct a quadratic loss function.

12.7 Explain why a product's design and the design of the process used to produce the product significantly impact the product's quality and cost. Discuss the cost of process control versus the number of manufacturing imperfections in the product under question.

12.8 a. Explain the purpose of system design.
   b. Discuss the relevance of quality-of-design/redesign studies and quality-of-performance studies to system design.

12.9 a. Explain the purpose of parameter design.
   b. Define the term *design parameter variable*.
   c. Define the term *noise variable*.

12.10 a. Explain the purpose of allowance design.
   b. Explain why allowance design does not contradict the philosophy of continuous and never-ending improvement.

12.11 a. Explain the purpose of a signal-to-noise ratio.
   b. Describe the three types of signal-to-noise ratios for non-negative variables type data.

Endnotes

1. Taguchi's methods have been improved and extended in recent years by Box, Hunter, and others. This chapter focuses on an introduction to Taguchi's work.

2. This list of seven items has been paraphrased from R Kackar, "Taguchi's Quality Philosophy: Analysis and Commentary," *Quality Progress,* December 1986, pp. 21–29.

3. G Taguchi and Y Wu, *Introduction to Off-Line Quality Control* (Nagoya, Japan: Central Japan Quality Control Association, 1980).

4. Kackar, "Taguchi's Quality Philosophy," pp. 21–29.

5. Abstracted from Taguchi and Wu, *Introduction to Off-Line Quality Control,* pp. 7–9.

6. See J Hauser and D Clausing, "House of Quality," *Harvard Business Review,* May/June 1988, pp. 63–73; B King, *Better Design in Half the Time* (Methuen, MA: GOAL/QPC, 1987); Y Akao, *Quality Function Deployment: Integrating Customer Requirements into Product Design* (Cambridge, MA: Productivity Press, 1990); and W Eureka and N Ryan, *The Customer Driven Company: Managerial Perspectives on QFD* (Dearborn, MI: SAI Press, 1988).

7. This figure is taken from *Florida Power and Light's Total Quality Management* (Miami, FL), Unit 12, page 13.

8. This procedure for prioritizing customer requirements is adapted from the quality function deployment methods of Dr Akao.

9. Kackar, "Taguchi's Quality Philosophy," p. 26.

10. Genichi Taguchi has developed sets of matrices for performing parameter design experiments. The matrices are called *orthogonal design matrices*.

11. The measurement variable must be non-negative to use the performance statistics described.

12. A sliding level technique is applied to setting levels because developments are performed under some combination of conditions E and G.

13. The calculation of the S/N ratio for the first column of the L18 matrix [12.479 (dB)] is:

Total variation:

$$ST = 1.00^2 + 1.06^2 + 0.84^2 + \ldots + 3.60^2 = 205.0514 \text{ (df = 20)}$$

Effective divisor:

$$r = 4x(1.2^2 + 2.4^2 + 6.0^2 + 1.2^3 + 3.6^2) = 230.4$$

Linear term to signal:

$$S_\beta = [1.2x(1.00 + 1.06 + 0.84 + 1.00) + \ldots + 3.6x(3.40 + 4.03 + 3.60 + 3.00)]^2/r$$
$$= 204.1007 \text{ (df = 1)}$$

Error variation:

$$S_e = ST - S_\beta = 0.9507 \text{ (df = 19)}$$

Error variance:

$$V_e = S_e/19 = 0.05004$$

S/N ratio:

$$\eta = (S_\beta - V_e)/r/V_e = 17.6986 = 12.479 \text{ (dB)}$$

14. We estimate the reduction in the standard deviation as 8.30 dB/6 dB = 1.383 and $(0.5)^{1.383} = .38$.

PART VI Inspection Policy

Chapter 13 discusses policies and procedures for inspection of incoming, intermediate, and final goods and services. Three options exist for inspection of goods and services: (1) no inspection, (2) 100 percent inspection, and (3) sampling inspection.

The first part of the chapter focuses on sampling inspection, commonly called acceptance sampling, as a method to determine whether to accept, reject, or screen goods and services. Three types of acceptance sampling plans are discussed: lot-by-lot plans, continuous plans, and special plans.

The second part of the chapter presents a theoretical argument against using acceptance sampling plans, followed by a discussion of the kp rule, an alternative inspection procedure that minimizes the total cost of inspection for incoming, intermediate, and final goods and services. The chapter ends with two mathematical proofs of the arguments made.

CHAPTER 13 Inspection Policy

Inspecting Goods and Services

Goods or services enter an organization from a vendor or are passed on internally from one section of the organization to another, such as department to department, or operation to operation within a department. These goods or services move inter- or intra-organizationally either in discrete lots or in continuous flows and have certain customer-specified quality characteristics.

Organizations or their subcomponents must have some method for minimizing the total cost of inspecting incoming and intermediate goods or services plus the cost to repair and test these goods and services in process; or final goods or services that fail to meet specifications because of a defective good or service used in production. Three alternatives exist for inspecting of goods or services: (1) no inspection (send items straight into use with no screening), (2) 100 percent inspection (screen all goods or services to weed out defectives), or (3) sampling inspection, also known as acceptance sampling (screen a sample of goods or services to determine if the remainder should be accepted, rejected, or screened). Historically, acceptance sampling has been considered useful if the inspection test is destructive (100 percent inspection will destroy all goods or services), the cost of 100 percent inspection is high, or too many units have to be inspected.

Acceptance Sampling

The purpose of acceptance sampling is to determine the disposition of goods or services (accept, reject, or screen). We do this by selecting the disposition that minimizes the cost of inspection to achieve a desired level of quality (called the Acceptable Quality Level or AQL, as discussed in Chapter 10) or to financially penalize the vendor (external or internal) if his quality is poor. Figure 13.1 shows several types of acceptance sampling plans.[1]

FIGURE 13.1   Acceptance Sampling Plans

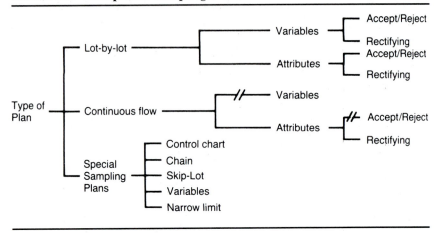

Lot-by-Lot Acceptance Sampling

Lot-by-lot acceptance sampling plans are used to inspect goods or services whenever the goods or services can conveniently be grouped into lots. All lot-by-lot acceptance plans are based on accepting, rejecting, or screening the remainder of the lot based on the number of defects found in the sample. Lot-by-lot plans exist for attribute data and variables data. Such plans can also be broken down into *acceptance/rejection plans* (plans in which a sample of items is drawn from the lot and the remainder of the lot is accepted or rejected based on an analysis of the sample) or *rectifying plans* (plans that call for either total or partial screening of remainders). The most common acceptance/rejection lot-by-lot acceptance sampling plan for variables used in American industry today is Military Standard 414.[2] This plan is used to control the fraction of incoming material that doesn't conform to specifications for variables data. The most common acceptance/rejection lot-by-lot acceptance sampling plan for attributes is Military Standard 105D.[3] This plan is used to constrain suppliers so that they'll deliver at least an Acceptable Quality Level of goods or services for attribute data.

Continuous Flow Acceptance Sampling

Continuous flow acceptance sampling plans are used to inspect goods or services whenever the goods or services can't be grouped into lots—for example, goods on conveyor belts or goods on a continuous moving line. Only rectifying attribute sampling plans exist for continuous flow processes. The most common acceptance sampling plan in this category used today is Military Standard 1235B.[4] All continuous flow acceptance plans are based on a *clear sampling,* the number of conforming units observed between the occurrence of two defective units. If the number of units between two defective units is greater than the number of units specified in the clearing sample, the units will be accepted and shipped; otherwise, they'll be 100 percent inspected.

Special Sampling Plans

Other types of acceptance sampling plans have been developed that are of the lot-by-lot type but are applied to a series of lots considered as a group. They're called *special sampling plans*. These plans include control chart plans,[5] chain sampling plans,[6] skip-lot plans,[7] variables plans,[8] and narrow limit plans.[9]

Much attention is given to acceptance sampling plans in textbooks and courses. However, these plans don't minimize the total cost of inspection of (1) incoming and intermediary goods or services plus the cost to repair and test these goods and services in process or (2) final goods or services that fail to meet specifications because of a defective good or service that was used in production. They also emphasize inspection, not process improvement to remove the need for inspection. Furthermore, there's a strong theoretical basis for not using acceptance sampling plans, as we shall now relate.

A Theoretical Invalidation of Acceptance Sampling

A discussion of the invalidity of acceptance sampling must consider the stability, or lack of stability, of the process's output undergoing inspection. Let's first consider the case of the stable process.

Stable Process

Suppose a lot of N independent items is drawn from a stable process that generates 100p percent defective output, and that x of the items are defective and $N - x$ are conforming. Here the number of defectives is binomially distributed with fraction defective p. This is extremely common in stable processes. Then

$$N = \text{Total number of items in the lot,}$$

$$x = \text{Number of defective items in the lot,}$$

$$N - x = \text{Number of conforming items in the lot, and}$$

$$E(x/N) = p = \text{Fraction defective items in the process.}$$

Suppose a sample of n items is drawn from the lot of N items (without replacement, due to the finite nature of the lot), such that r of the items are defective and $n - r$ of the items are conforming. Then,

$$n = \text{Total number of items in the sample,}$$

$$r = \text{Number of defective items in the sample,}$$

$$n - r = \text{Number of conforming items in the sample, and}$$

$$E(r/n) = p = \text{Estimated fraction defective items in the process.}$$

The selection of the sample from the lot creates a new entity we'll call the *remainder* or the rest of the lot. The remainder is composed of N − n items, such that x − r of the items are defective and (N − n) − (x − r) are conforming. Then

$$N - n = \text{Total number of items in the remainder,}$$

$$x - r = \text{Number of defective items in the remainder,}$$

$$(N - n) - (x - r) = \text{Number of conforming items in the remainder, and}$$

$$E[(x - r)/(N - n)] = p = \text{Estimated fraction defective items in the process.}$$

Figure 13.2 illustrates this sequence of item groupings.

If it can be shown that the number of defectives in the sample is independent of (not correlated with) the number of defectives in the remainder, then acceptance sampling plans that determine the disposition of a remainder (accept, reject, or screen) based on the number of defectives in a sample are invalid; this proof appears in Appendix 13.1 at the end of the chapter.

In other words, the number of defectives in the sample and in the remainder are both binomially distributed with the same mean fraction, p, and are independent. For example, if a lot of 1,000 fair coins was tossed repeatedly and a sample of 50 of the 1,000 was drawn each time for inspection, the fraction of heads in the sample and in the remainder would both be distributed around p = 0.5, but the number of heads in the samples would be independent of the number of heads in the remainders. This means the distribution of heads in the remainders associated with samples yielding 0 defectives would be the same as the distribution of heads in the remainders associated with samples yielding 50 defectives. As a direct result, acceptance sampling plans that determine the disposition of remainders based on samples are invalid for a stable process[10]—a shocking result to many.[11]

FIGURE 13.2   Selection of Samples from Lots Drawn from a Process

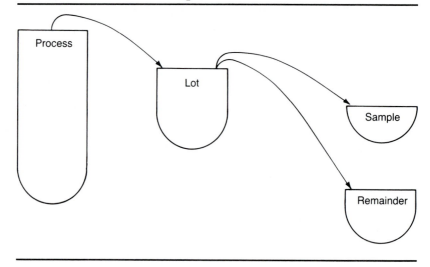

An alternative to acceptance sampling from stable processes must be found. W Edwards Deming offered, as an alternative, the *kp rule,* to be discussed later in this chapter.

Chaotic Process

Suppose a lot of N items is drawn from a chaotic or unknown process. In the lot, x of the items are defective and N − x are conforming, and the process fraction defective p wanders from lot to lot (or day to day) and isn't predictable. Then

$$N = \text{Total number of items in the lot,}$$

$$x = \text{Number of defective items in the lot, and}$$

$$N - x = \text{Number of conforming items in the lot.}$$

Suppose a sample of n items is drawn from the lot of N items (without replacement, as a result of the finite nature of the lot) such that r of the items are defective and n − r of the items are conforming. Then

$$n = \text{Total number of items in the sample,}$$

$$r = \text{Number of defective items in the sample, and}$$

$$n - r = \text{Number of conforming items in the sample.}$$

As with the stable process, the selection of the sample from the lot creates a remainder composed of N − n items, such that x − r of the items are defective and (N − n) − (x − r) are conforming. Then

$$N - n = \text{Total number of items in the remainder,}$$

$$x - r = \text{Number of defective items in the remainder, and}$$

$$(N - n) - (x - r) = \text{Number of conforming items in the remainder.}$$

As before, Figure 13.2 represents the sequence of item groupings discussed in this section.

Here we can show that the number of defectives in the sample, r, is correlated with the number of defectives in the remainder, x − r. Thus, when p varies widely and unpredictably from lot to lot, the information from a sample provides insight into the remainder. Going back to the coin example, if lots of 1,000 biased coins were tossed repeatedly and samples of 50 of the 1,000 were drawn each time for inspection, the distribution of the fraction of heads in the samples and in the remainders wouldn't be distributed around 0.5; rather, they'd be related to the fraction of defectives in the lot and would be correlated. Consequently, the number of defectives in the sample and remainders would be correlated. Hence, acceptance sampling plans that determine the disposition of a remainder (accept, reject, or screen) based on the number of defectives in a sample are valid for a chaotic process.[12] Later in this chapter, we'll discuss the larger question of whether it's the most cost-effective plan given the chaotic nature of the process.

Note that as processes are stabilized as a result of quality efforts, acceptance plans that are valid for chaotic processes—albeit at high cost—will no longer be effective on the stable process.

Plan for Minimum Average Total Cost for Test of Incoming Materials and Final Product for Stable Processes

The kp Rule

Given a stable process and the knowledge that acceptance sampling plans aren't effective on such processes, we're left with only two of the inspection alternatives discussed at the beginning of this chapter: no inspection or 100 percent inspection. We discuss here Deming's kp rule, which specifies when to do no inspection and when to do 100 percent inspection so as to minimize the total cost of incoming and intermediate materials, final products, and repairing and testing those products that fail. The rule is derived in Appendix 13.2.

The assumptions for the use of the kp rule are listed next. These assumptions are not restrictive and are applicable to many common situations.[13]

1. All items are tested (inspected) before they move forward in the process. In other words, all nonconforming items are detected by a final inspection. When all items won't be subjected to a final inspection, the rule can be modified to reflect the possibility that a certain fraction of nonconforming parts, f, would be caught and the remaining fraction, $1 - f$, would continue on into production or into the hands of customers.[14]

2. Inspection is completely reliable. If an item is defective, it will fail inspection. As Deming wrote, "A defective part is one that by definition will cause the assembly to fail. If a part declared defective at the start will not cause trouble further down the line, or with the customer, then you have not yet defined what you mean by a defective part."[15]

3. The item vendor will give the buyer an extra supply of items, S, to replace any defective item found. The supplier adds the cost of these items onto her bill, either directly or indirectly. This cost is an overhead cost and would be present regardless of the inspection plan used. Hence, it need not be included in the cost function to be minimized.

The following notation is necessary to determine when to do 100 percent inspection and when to do no inspection. Let

p = The average incoming fraction of defective items in incoming lots of items. Recall that the process under study is stable and has a meaningful average incoming fraction of defective items, p, in incoming lots of items;

k_1 = The cost to initially inspect one item;

k_2 = The cost to dismantle, repair, reassemble, and test a good or service that fails because a defective item was used in its production.

If a process is stable around the fraction p, the kp rule states

1. *If k_1/k_2 is greater than p, then 0 percent inspection.* (No inspection minimizes the total cost.) This occurs if the fraction of incoming defective items, p, is very low, the cost of inspecting an incoming item is high, and the cost of the defective item getting into production is low; therefore, no inspection is needed. The rationale is that there's little risk or penalty associated with incoming defective items.

2. *If k_1/k_2 is less than p, then 100 percent inspection.* (One hundred percent inspection minimizes the total cost.) This occurs if the fraction of incoming defective items, p, is high, the cost of inspecting an incoming item is low, and the cost of the defective item getting into production is high; therefore 100 percent inspection is needed. The rationale here is that there's great risk and a penalty attached to incoming defective items.

3. *If k_1/k_2 equals p, then either 0 percent or 100 percent inspection.* A decision must be made as to whether 0 percent or 100 percent inspection should be done in this case. In general, if p isn't based on a substantial past history, 100 percent inspection is vital for safety's sake.

To summarize, the kp rule will minimize the total cost of incoming and intermediary materials and final product for a stable process by proper selection of a 0 percent or 100 percent inspection policy. If the process under study is stable, then whether item i is defective is independent of whether any other item is defective. Hence, item i should be inspected, or not inspected, according to whether p is greater than or less than k_1/k_2. Recall that item i is a randomly selected item, and policy set for item i applies to any item. We can thus extend the policy for item i to all items in the lot. And consequently, either all or no items in the lot should be inspected, depending on whether p is greater than or less than the breakeven point, k_1/k_2.

It's important to note that 0 percent inspection doesn't mean zero information. Small samples should always be drawn from every lot—or on a skip-lot basis—for information about the process under study. This information should be recorded on control charts to facilitate process improvement.[16] The cost of these small samples is assumed to be a cost of doing business, and consequently isn't considered in the cost function to be minimized.

The kp rule is appropriate between any two points in an organization's interdependent system of stakeholders, such as internally, in the vendor's processes, or between the firm and the vendor.

An Example of the kp Rule.   A car manufacturer is deciding whether to purchase $25 million worth of equipment that would test engines purchased from vendors. The vendor's process is stable. The following figures have been determined:

- The inspection cost to screen out incoming defective engines is $50 per engine ($k_1 = \$50$).
- The cost for corrective action if a defective engine gets into production is $500 per defective engine ($k_2 = \$500$).
- On average, 1 in 150 incoming engines is defective ($p = 1/150 = 0.0067$).

Consequently, $k_1/k_2 = 50/500 = 0.1$. Note that 0.1 is greater than 0.0067. Therefore, k_1/k_2 is greater than p, and the correct course of action would be to do no initial inspection on incoming engines to achieve minimum total cost.

If no engines are inspected, the auto company would expect to incur the $500 cost in 1 out of 150 engines. This translates into an average corrective action cost of $3.33 per engine ($500 × 1/150). By eliminating initial inspection, the company would save $46.67 per engine ($50 − $3.33) on average. As the company purchases 4,000 engines per day, this translates into a daily savings of $186,680 (4,000 × $46.67), not including the savings of $25 million for testing equipment, interest on that money, and time freed up to work on improving quality! The next step in the pursuit of quality is for the auto company to work with its engine vendor to reduce the fraction of defective engines.[17]

Exceptions to the kp Rule

Destructive Testing.   The kp rule doesn't apply to destructive testing in which an item is destroyed in conducting the test. The only solution in destructive testing is to achieve statistical control such that $p < k_1/k_2$, so that no inspection (other than routine small samples of the process) is the minimum cost policy. Note that achieving statistical control with $p < k_1/k_2$ is the best solution regardless of whether the test is destructive or nondestructive.[18]

Homogeneous Mixtures.   The kp rule doesn't apply to homogeneous mixtures— for example, "a jigger of gin or whiskey. We accept the fact that it matters little

---

FIGURE 13.3   Component Costs of k_1 and k_2

Some Inspection Costs k_1
Capital equipment
   Initial cost
   Depreciation (also considers residual value)
   Planned production volumes
   Cost of capital
Operating costs
   Labor
   Rent, utilities, maintenance
   Piece cost (outside vendor quote)

Possible Detrimental Costs k_2
Added costs of processing the nonconforming item further
Cost of sorting lots later to find a nonconforming item
Cost of repairing batches of assemblies later
Cost of lost production later if lots of parts or batches of assemblies must be
   guaranteed pending sorting and repair
Warranty costs
Cost of recalls
Law suits ($k_2 \rightarrow \infty$ for safety items)
Customer loyalty impinging upon future sales

---

whether we draw off a jigger from the top of the bottle or from the middle or from the bottom.''[19] In this case, the sample is identical in composition to the remainder; hence, we can make judgments about the remainder from the sample.

Component Costs of k_1 and k_2

Some costs to consider when calculating k_1 and k_2 are shown in Figure 13.3.[20] The costs required to compute k_1 are usually known and can be calculated. A firm's financial personnel should be helpful in computing k_1. However, the costs required to compute k_2 are generally unknown and frequently hard to compute (for example, the cost of customer dissatisfaction from recalls or lawsuits). A reasonable policy is to estimate k_2 without the more subjective costs. If $p > k_1/k_2$, there's no need to estimate the other components of k_2. On the other hand, if $p < k_1/k_2$ without including the subjective costs, an estimate of these missing costs may have to be made to determine more accurately the relationship between p and k_1/k_2.

Achieving Substantial Savings over 100 Percent Inspection in the Cost of Incoming Material and Final Product for Chaotic Processes

Given a chaotic process and the knowledge that acceptance sampling plans are appropriate for such processes, we have the three inspection alternatives discussed earlier: (1) no inspection, (2) 100 percent inspection, and (3) some form of acceptance sampling. Our choice between alternatives (1) and (3) to achieve savings over 100 percent inspection (alternative (2)) depends on the nature of the chaos in the process.

Mild Chaos

If the fraction defective in the process under study wanders in an unpredictable manner so that the fractions for the worst lots are below k_1/k_2 (Figure 13.4), then no inspection (except for routine sampling of the process) should be performed, as Figure 13.4 shows. Significant effort, however, should be directed toward stabilizing the process from the information in the routine samples.

If the fraction defective in the process under study wanders in an unpredictable manner so that the fractions for the best lots are above k_1/k_2, then 100 percent inspection should be performed (Figure 13.5). Again, significant effort should be directed toward stabilizing the process by using the information from 100 percent inspection.

These two cases are considered mild chaos because the distribution of the fraction defective is chaotic within bounds. Of course, these bounds can disappear at any moment. These aren't situations where we should be lulled into thinking that the chaos will always stay within bounds. Chaos is a wild beast that can run anywhere at anytime, including beyond any earlier boundary.

FIGURE 13.4   Mild Chaos with Low Fraction Defective

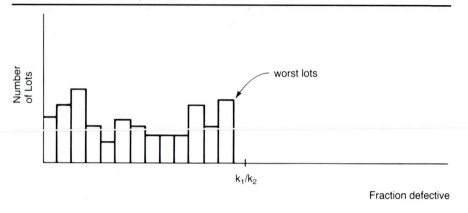

FIGURE 13.5   Mild Chaos with High Fraction Defective

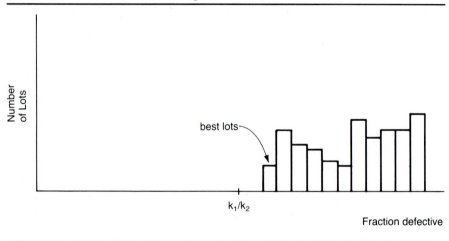

Severe Chaos

If the fraction defective in the process under study wanders in an unpredictable manner within a narrow range around k_1/k_2, the most practical plan is 100 percent inspection of all lots.[21] No effort at acceptance sampling can justify the cost of administering the plan.[22]

If the fraction defective in the process under study wanders in an unpredictable manner within a wide range[23] around k_1/k_2, Orsini[24] has devised a rule that yields substantial savings over 100 percent inspection:

If k_1/k_2 is less than 1/1,000, then inspect 100 percent of the incoming lots.

If k_1/k_2 is between 1/1,000 and 1/100, then test a sample of 200. If there are no defectives, then accept the remainder. Inspect the entire remainder if at least one defective item is found in the sample.

If k_1/k_2 is greater than $1/100$, then do no inspection.

This rule is also helpful in working with a vendor to bring her process into control. A running record of the samples of 200 can be kept, and the number of defectives can be charted, sample by sample. Feedback to the vendor is extremely helpful in identifying problems.

Exceptions to Rules for Chaos

The rules for chaos are subject to the same exceptions as the kp rule for stable processes. Further, if items come into a firm from an unknown vendor, the optimal policy is to perform 100 percent inspection until enough information has been collected to construct a control chart(s) for the vendor's process/product. Then we can select the best inspection plan. One hundred percent inspection should also be carried out for critical parts and safety items. (For safety items, k_2 is infinite.)

Summary

This chapter discussed different types of acceptance sampling plans: lot-by-lot plans, continuous sampling plans, and special sampling plans. Lot-by-lot acceptance sampling plans are used to inspect goods or services whenever the goods or services can be conveniently grouped into lots. Continuous flow acceptance sampling plans are used to inspect goods or services whenever the goods or services can't be grouped into lots—for example, goods on a conveyor belt. Special sampling plans are used for lots found in series. These acceptance sampling plans don't, however, minimize the average total cost of inspection of incoming, intermediate, and final goods and services. A theoretical argument invalidating acceptance sampling was presented.

Plans for minimizing the average total cost of testing incoming materials and final product, called the kp rule, were analyzed. One plan is used for stable processes, and the other plan is used for chaotic processes. Examples and exceptions to these plans were discussed. Both plans require the collection and control charting of either inspection data or routine samples to achieve process stability and pursue continuous and never-ending improvement.

APPENDIX 13.1
PROOF THAT THE NUMBER OF DEFECTIVES IN A SAMPLE IS INDEPENDENT OF THE NUMBER OF DEFECTIVES IN THE REMAINDER FOR LOTS DRAWN FROM A STABLE PROCESS

To prove: Given a stable process with 100p percent defective, the number of defectives in a sample of n items drawn from a lot of size N is independent of the number of defectives in the remaining N − n items.

Proof: n = Number of items in the sample

N = Number of items in the lot

r = Number of defective items in the sample

x = Number of defective items in the lot

x − r = Number of defective items in remaining N − n items

f(x) = The probability of x defectives in a lot of N items drawn from a stable process with 100p percent defective

f(r) = The probability of r defectives in a sample of n items drawn from a stable process with 100p percent defective

f(r|x) = The conditional probability of r defectives in a sample of n items drawn from a lot of N items given that the lot contains x defectives

f(r ∩ (x − r)) = The joint probability that there are r defectives in the sample of n items and x − r defectives in the remaining N − n items

We must show that f(r ∩ (x − r)) = f(r)f(x − r).

The joint probability of x defectives in the lot and r defectives in the sample is given by

$$f(x \cap r) = f(x)f(r|x)$$

The number of defectives, x, in a lot of size N is binomially distributed:

$$f(x) = \binom{N}{x} p^x q^{N-x}$$

Similarly, the number of defectives, r, in a sample of size n is binomially distributed:

$$f(r) = \binom{n}{r} p^r q^{n-r}$$

and the number of defectives, x − r, in the remaining N − n items is also binomially distributed:

$$f(x - r) = \binom{N - n}{x - r} p^{x-r} q^{(N-n)-(x-r)}$$

The number of defectives, r, in a sample of size n, given a total of x defectives in a lot of size N, has a hypergeometric distribution, as given by

$$f(r|x) = \frac{\binom{n}{r}\binom{N-n}{x-r}}{\binom{N}{x}}$$

Then

$$f(x \cap r) = \binom{N}{x} p^x q^{N-x} \frac{\binom{n}{r}\binom{N-n}{x-r}}{\binom{N}{x}}$$

This can be written as:

$$f(x \cap r) = p^{x-r+r} q^{N-x-n+n-r+r} \binom{n}{r}\binom{N-n}{x-r}$$

or, rearranging

$$f(x \cap r) = \binom{N-n}{x-r} p^{x-r} q^{(N-n)-(x-r)} \binom{n}{r} p^r q^{n-r}$$

or

$$f(x \cap r) = f(x-r)f(r)$$

This is the probability of x defectives in a lot of size N and r defectives in a sample of size n. But this is the same as the probability of $x - r$ defectives in $N - n$ items and r defectives in n items. So

$$f(x \cap r) = f((x-r) \cap r)$$

$$\text{Then } f((x-r) \cap r) = f(r)f(x-r)$$

If the joint probability of two events equals the product of their unconditional probabilities, the events must be independent. Thus, the number of defectives in a sample of n items is independent of the number of defectives in the remaining $N - n$ items.

APPENDIX 13.2
DERIVATION OF THE KP RULE FOR STABLE PROCESSES

Let

p = The average fraction of defective items in incoming lots of items,

k_1 = The cost to initially inspect one item,

k_2 = The cost to dismantle, repair, reassemble, and test a good or service that fails because a defective item was used in its production,

k = The average cost to test one or more items to find a conforming item from the supply, S, to replace a defective item found
= $k_1/(1 - p)$,

C_1 = The cost to initially inspect one item,

C_2 = The cost to repair a failed good or service.

Further, let

$$x_i = 1 \text{ if item i is defective, and}$$

$$x_i = 0 \text{ if item i is conforming.}$$

Now, if one item, item i, is randomly drawn from a lot, then the probability that it's defective is p. The cost to initially inspect item i is

$$C_1 = k_1 + kx_i \quad \text{if we test item i, and}$$

$$C_1 = 0 \qquad\qquad \text{if we don't test item i.}$$

C_1 is composed of the cost to initially test one item plus the cost to replace the item if it's found to be defective. Hence,

$$C_1 = k_1 + k \quad \text{if item i is tested and found to be defective,}$$

$$C_1 = k_1 \qquad \text{if item i is tested and found to conform, or}$$

$$C_1 = 0 \qquad \text{if item i isn't tested.}$$

The cost to repair a failed good or service due to item i is

$$C_2 = (k_2 + k)x_i \quad \text{if we don't initially test item i, and}$$

$$C_2 = 0 \qquad\qquad \text{if we do initially test item i.}$$

C_2 is composed of the cost to repair a failed good or service if item i wasn't initially inspected. Hence,

$C_2 = k_2 + k$ if item i wasn't initially inspected and item i is defective,

$C_2 = 0$ if item i wasn't initially inspected and item i is conforming, and

$C_2 = 0$ if item i was initially inspected.

C_1 and C_2 are mutually exclusive as they can't occur simultaneously for a given item; if one is positive, the other is zero. The total cost for item i is

$$C = C_1 + C_2$$

Figure 13A.1 summarizes the cost structure for inspection versus no inspection for item i.[25]

FIGURE 13A.1   **Cost Structure for Inspection Decision for Item i**

Inspect the Item?	C_1	C_2	*Total Cost* $C = C_1 + C_2$
Yes	$k_1 + kx_i$	0	$k_1 + kx_i$
No	0	$(k_2 + k)x_i$	$(k_2 + k)x_i$

FIGURE 13A.2   Cost Structure for Inspection Decision for Average Item

---

Inspect the Item?	C_1	C_2	Total Cost $C = C_1 + C_2$
Yes	$k_1 + kp$	0	$k_1 + kp$
No	0	$(k_2 + k)p$	$k_2p + kp$

Figure 13A.2 extends the cost structure for inspection versus no inspection to the average cost per item over the lot. In this case x_i is replaced by p because

$$p = \sum_{i=1}^{N} [x_i/N]$$

None of the other elements in Figure 13A.2 is affected because they're constants.

Now, the breakeven point between inspection and no inspection can be determined by setting the total cost for inspection equal to the total cost for no inspection. That is, at the breakeven point, Cost (Inspect) = Cost (Do Not Inspect)

$$k_1 + kp = k_2p + kp$$

$$k_1 = k_2p$$

$$p = k_1/k_2$$

Exercises

13.1 Discuss the three possible alternatives for the inspection of goods or services.

13.2 a. Explain the purpose of acceptance sampling.
   b. Explain the purpose of lot-by-lot acceptance sampling plans. Describe the situations in which lot-by-lot acceptance sampling plans are used as a basis for action on a lot of goods.
   c. Explain the purpose of Military Standard 414.
   d. Explain the purpose of Military Standard 105D.
   e. Explain the purpose of continuous flow acceptance sampling plans.
   f. Briefly describe the operation of continuous flow acceptance sampling plans.

13.3 Explain why acceptance sampling plans are theoretically incorrect for stable processes and shouldn't be used as a basis for action. Mathematically defend your explanation.

13.4 Explain why acceptance sampling is theoretically correct but not economical for chaotic processes and consequently shouldn't be used as a basis for action.

13.5 a. Describe the kp rule for stable processes.
   b. Explain the assumptions required to use the kp rule.

c. List several examples of k_1 inspection costs.

d. List several examples of k_2 inspection costs.

13.6 Explain the term *mild chaos* and its significance to taking action on incoming or intermediary material or on final product.

13.7 A radio manufacturer has a policy of inspecting every incoming radio speaker to ensure it conforms to specifications. What information would you need to question the wisdom of this inspection policy?

13.8 The production manager of Exercise 13.7's radio company learned about the kp rule. He used past inspection data concerning the proportion of defective radio speakers purchased per day to construct a p chart with variable sample size. The p chart indicated that the incoming stream of radio speakers was stable with respect to the fraction of defective radio speakers purchased each day. The average fraction of defective radio speakers was found to be 0.002. Further study showed that it costs approximately $0.50 to inspect an incoming radio speaker and that it costs approximately $7.50 to repair a radio with a defective speaker before it leaves the factory.

a. Use the kp rule to determine if 0 percent of 100 percent inspection should be used for incoming radio speakers.

b. What should management do about the incoming radio speaker process, given your answer in part a?

13.9 Explain the term *severe chaos*. Describe an inspection rule that can be used when a process exhibits severe chaos.

Endnotes

1. An excellent reference for details on acceptance sampling is Acheson Duncan, *Quality Control and Industrial Statistics,* 5th ed. (Homewood, IL: Richard D. Irwin, 1986), pp. 161–414.
2. Ibid., pp. 291–305.
3. Ibid., pp. 217–48.
4. Ibid., pp. 406–13.
5. Ibid., pp. 536–40.
6. Ibid., pp. 177–79.
7. Ibid., pp. 252–53.
8. Ibid., pp. 340–65.
9. Ibid., pp. 271–72.
10. A cautionary note: The above proof doesn't mean that statistical inference doesn't work—that is, that a random sample from a population doesn't provide information about the sampling frame. Rather, it only means that samples provide no information about remainders from stable processes.

A proof that a random sample from a stable process provides information about the process p is

Recall: $f(x)f(r|x) = f(r \cap (x - r)) = f(r)f(x - r) = f(r \cap x)$

Hence: $\sum_{x=0}^{N} f(r \cap x) = f(r) = \binom{n}{r} p^r(1 - p)^{n-r}$

This indicates that f(r) is the rth term of the binomial $(p + q)^n$, just as if the sample of n items was drawn (with replacement) directly from the process without the introduction of the lot between the process and the sample. This means that a random sample of n items from the process provides information on the process average via the estimation p = r/n.

11. Alexander Mood, "On the Dependence of Sampling Inspection Plans under Population Distributions," *Annals of Mathematical Statistics* 14 (1943), pp. 415–25. Also see W Edwards Deming, *Some Theory of Sampling* (New York: John Wiley & Sons, 1950), p. 258.

12. This proof will not be shown.

13. W E Deming, *Quality, Productivity and Competitive Position* (Cambridge, MA: Massachusetts Institute of Technology, Center for Advanced Engineering Study, 1982, pp. 267–311. Joyce Orsini, "Simple Rule to Reduce Total Cost of Inspection and Correction of Product in State of Chaos," doctoral dissertation, Graduate School of Business Administration, New York University, 1982. G P Papadakis, "The Deming Inspection Criteria for Choosing Zero or 100 Percent Inspection," *Journal of Quality Technology* 17 (July 1985), pp. 121–27.

14. Papadakis, "The Deming Inspection Criteria for Choosing Zero or 100 Percent Inspection," p. 123.

15. Deming, *Quality, Productivity and Competitive Position*, p. 268.

16. Ibid., p. 273.

17. Ibid., pp. 276–77.

18. Ibid., pp. 274–75.

19. Ibid., p. 285.

20. Papadakis, "The Deming Inspection Criteria for Choosing Zero or 100 Percent Inspection," p. 124.

21. A narrow range is defined as a coefficient of variation of 0.3 or less:

$$\frac{nq}{p} \leq 0.3$$

22. Deming, *Quality, Productivity and Competitive Position*, p. 271.

23. A wide range is defined as a coefficient of variation of more than 0.3:

$$\frac{nq}{p} > 0.3$$

24. Orsini, "Simple Rule to Reduce Total Cost of Inspection and Correction of Product in a State of Chaos."

25. Deming, *Quality, Productivity and Competitive Position*, p. 302.

VII Foundations of Quality Revisited

This section presents the relationship between W Edwards Deming's system of profound knowledge, the 14 points for management, and the reduction of variation, as well as some current thoughts on statistical studies and statistical practice.

In Chapter 14, a teaching aid, based on an experiment utilizing a funnel, demonstrates the interconnectedness between management thought and statistical theory. Finally, a focal point of the system of profound knowledge referred to in Chapter 1 and the 14 points is shown to be the reduction of variation in an organization's interdependent system of stakeholders.

Chapter 15 analyzes current thinking about statistical studies and statistical practice. It explores differences in the theory of statistics for enumerative studies and analytic studies, and assesses the far-reaching implications of these differences. This chapter reproduces a paper written by Deming dealing with a code of professional conduct for the statistician. It includes responsibilities of the statistician—and the subject matter expert (the client)—in conducting a statistical study.

Deming's 14 Points and the Reduction of Variation

The Purpose of Statistics

The purpose of statistics is to study and understand process and product variation, interactions among product and process variables, and operational definitions—and ultimately to take action to reduce variation. As Chapter 2 said, there are two types of statistical studies: enumerative studies and analytic studies. The purpose of an enumerative study is to survey the characteristics of a frame—one characteristic being variation—and to take action on the disposition of the material in the frame. The purpose of an analytic study is to examine the causes of variation in a process and reduce that variation and/or change the process's average.

The Purpose of Management

The purpose of management is to lead an organization in the direction of never-ending and continuous improvement. Never-ending improvement of an organization is pursued by continual reduction of variation of the organization's processes and products. This pursuit has both analytic and enumerative aspects.

An Aid to Understanding the Relationship between Statistics and Management

W Edwards Deming stated, "If anyone adjusts a stable process to try to compensate for a result that is undesirable, or for a result that is extra good, the output that follows will be worse than if he had left the process alone."[1] This is called *overcontrol of the process* or *tampering*. Recall Chapter 5's discussion of overadjustment. If management tampers with a process without profound knowledge of how to improve the process through statistical thinking, they'll increase the process's variation and reduce their ability to manage that process.

As Deming illustrated this, "A common example is to take action on the basis of a defective item, or on complaint of a customer. The result of his efforts to improve future output (only doing his best) will be to double the variance of the output, or even cause the system to explode. What is required for improvement is a fundamental change in the system, not tampering."[2]

The Funnel Experiment

Loss to an organization results from overcontrol of its processes, which include safety, training, hiring, supervision, union-management relations, policy formation, production, maintenance, shipping, purchasing, administration, and customer relations. This loss can be demonstrated by an experiment utilizing a funnel.[3] We'll describe the apparatus and procedure for conducting the experiment and then demonstrate its relationship to management's pursuit of continual reduction of variation.

FIGURE 14.1   **Funnel Experiment Equipment**

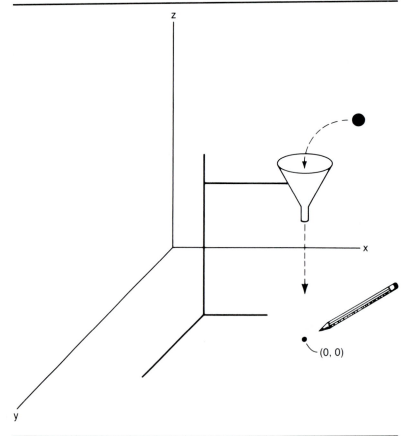

To conduct the experiment, shown in Figure 14.1, we need (1) a funnel, (2) a marble that will fall through the funnel, (3) a flat surface (e.g., a table top), (4) a pencil, and (5) a holder for the funnel.

The experiment involves five steps. (1) Designate a point on the flat surface as a target and consider this target to be the point of origin in a two-dimensional space, where x and y represent the axes of the surface; hence, at the target (x,y) = (0,0). (2) Drop a marble through the funnel. (3) Mark the spot where the marble comes to rest on the surface with a pencil. (4) Drop the marble through the funnel again and mark the spot where the marble comes to rest on the surface. (5) Repeat step (4) through 50 drops.

A rule for adjusting the funnel's position in relation to the target is needed to perform the fourth step. There are four possible rules, and the second rule can be handled in two ways.

Rule 1.   Set the funnel over the target at (0,0) and leave the funnel fixed through all 50 drops. This rule will produce a stable pattern of points on the surface; this pattern will approximate a circle, as in Figure 14.2.[4] Further, as we'll see, the variance of the diameters of all circles produced by repeated experimentation using Rule 1 will be smaller than the variance resulting from any other rule used in the fourth step of the experiment.

FIGURE 14.2   Rule 1

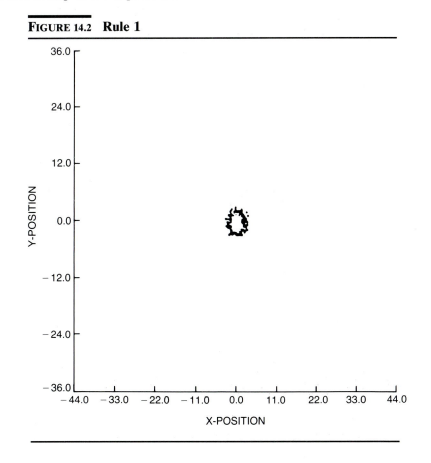

Management's use of the first rule demonstrates an understanding of the distinction between special and common variation, and the different types of managerial action required for each type of variation. Rule 1 implies that the process is being managed by people who know how to reduce variation.

Rule 2.   The funnel is set over the target at (0,0) prior to the initial drop. Let (x_k, y_k) represent the point where the kth marble dropped through the funnel comes to rest on the surface. Rule 2 states that the funnel should be moved the distance $(-x_k, -y_k)$ from its last resting point. In essence, this is an adjustment rule with a memory of the last resting point. This rule will produce a stable pattern of resting points on the surface, which will approximate a circle. However, the variance of the diameters of all circles produced by repeated experimentation using Rule 2 will have double the variance of the circular pattern produced using Rule 1, as Figure 14.3 shows.[5]

In terms of its application to management actions, Rule 2 implies that the process is being tampered with by people with inadequate knowledge of how to manage the process to reduce its variation. It implies acting on common variation as if it were special variation. Rule 2 is commonly used as a method of "attempting" to make things better in a process. Here are three examples.

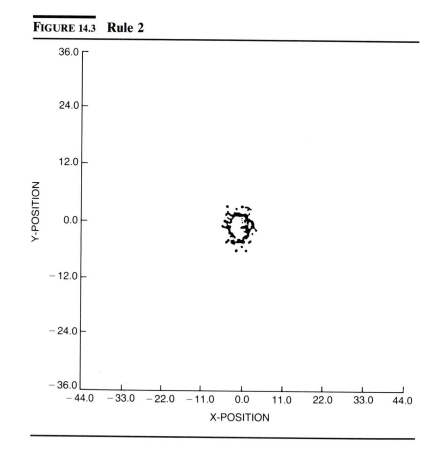

FIGURE 14.3   Rule 2

1. Automatic process control. The automatic adjustment of a process to hold output within specified tolerance limits is an example of Rule 2, assuming adjustments to the process are made from the last process measurement. This type of process adjustment procedure is frequently called *rule-based process control* (*RPC*). RPC is widely used in industry.

2. Operator adjustment. Operator adjustment to compensate for a unit of output's not being on target, or nominal, is an example of Rule 2, assuming adjustments to the process are made from the last process measurement.[6] This type of overcontrol frequently leads to a sawtooth-type pattern on an x-bar chart.

3. Stock market. The stock market's reaction to good or bad news is often an overreaction to phenomena, which follows Rule 2.

Rule 2a.   A variant on Rule 2 is often employed in industry. Rule 2a states if (x_k, y_k) is within a circle centered at $(0,0)$ with diameter d_{spec}, don't adjust the funnel. But if (x_k, y_k) is outside the circle centered at $(0,0)$ with diameter d_{spec}, use the adjustment rule specified in Rule 2. Rule 2a creates a "deadband" in which no process adjustment takes place. Research results obtained by Gitlow, Kang, and Kellogg demonstrate that Rule 2a, with any size deadband, yields the same result of doubling the variation of the process as does Rule 2, when compared with Rule 1.[7]

An example of Rule 2a can be seen in variance analysis in cost accounting. One method of monitoring performance in an organization is the use of efficiency and spending variances for the areas of direct labor, direct materials, and overhead. This is called variance analysis. Traditionally, manufacturers in the United States have relied on variance analysis to evaluate performance. If a particular variance is favorable, it's believed to indicate that excellent work is being done. If a particular variance is unfavorable, the converse is assumed to be true.

Variance analysis causes employees to react inappropriately to accounting variances. That is, variance analysis forces employees to react to variances as if they're only due to special causes as opposed to system causes. In the long run, variance analysis will double the variation of a stable processes being managed in accordance with cost accounting principles.

Rule 3.   The funnel is set over the target at $(0,0)$ prior to the initial drop. Let (x_k, y_k) represent the point where the kth marble dropped through the funnel comes to rest on the surface. Rule 3 states that the funnel should be moved a distance $(-x_k, -y_k)$ from the target $(0,0)$. In essence, this is an adjustment rule with no memory of the last resting point. This rule will produce an unstable, explosive pattern of resting points on the surface; as k increases without bound, the pattern will move farther and farther away from the target in some symmetrical pattern, such as the bow-tie–shaped pattern in Figure 14.4.[8]

Rule 3, like Rule 2, is commonly used as a method of "attempting" to improve the process. Rule 3 implies that the process is being tampered with by people with inadequate knowledge of how to manage the process to reduce its

FIGURE 14.4   Rule 3

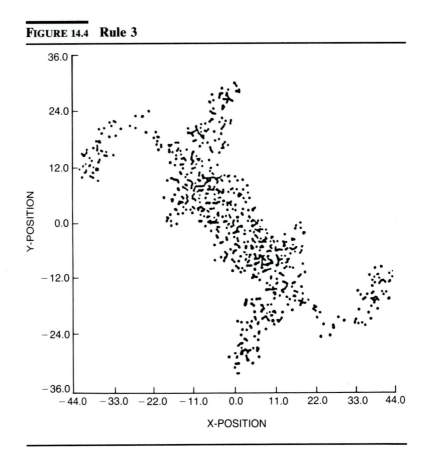

variation. It implies acting on common variation as if it were special variation. Here are five examples of Rule 3.

1. Automatic process control. The automatic adjustment of a process to hold output within specified tolerance limits is an example of Rule 3, assuming adjustments to the process are made from the target and not the last process measurement.

2. Operator adjustment. Operator adjustment to compensate for a unit of output's not being on target, or nominal, is an example of Rule 3, assuming adjustments to the process are made from the target and not the last process measurement.

3. Setting the current period's goal based on last period's overage or underage. A sales quota policy that states that if you're short of this month's goal by $25,000, you must increase next month's goal by $25,000 is an example of Rule 3.

4. Setting the inspection policy for the kth batch based on the k − 1st batch's record. An AQL inspection policy that recommends tightened or loosened inspection for a batch of material based on the history of the

prior batch is an example of Rule 3. A better policy would be to use the kp rule, as discussed in Chapter 13, in conjunction with a variables control chart to estimate process capability and p.

5. Making up the previous period's shortage during the current period. A production policy that requires production personnel to make up any shortages from last month's production run in this month's production run is an example of Rule 3.

Rule 4. The funnel is set over the target at (0,0) prior to the initial drop. Let (x_k, y_k) represent the point where the kth marble dropped through the funnel comes to rest on the surface. Rule 4 states that the funnel should be moved to the resting point, (x_k, y_k). In essence, this is an adjustment rule with no memory of either the last resting point or the position of the target at (0,0). This rule will produce an unstable, explosive pattern of resting points on the surface as k increases without bound, and it will eventually move farther and farther away from the target at (0,0) in one direction, as shown in Figure 14.5.[9]

Rule 4 is commonly used as a method of "attempting" to make things better in a process. Rule 4 implies that the process is being tampered with by people with

---

FIGURE 14.5   Rule 4

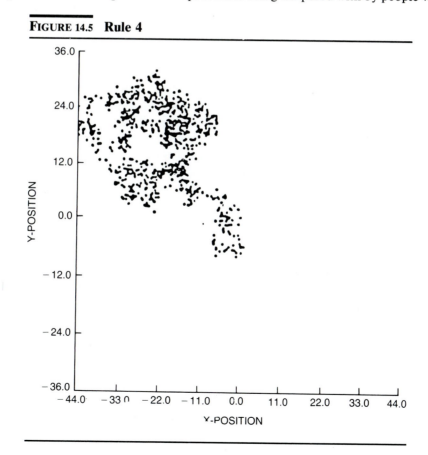

inadequate knowledge of how to manage the process to reduce its variation. It implies acting on common variation as if it were special variation. Many of the following 10 examples were discussed by Deming in his management seminars.

1. Make it like the last one. Using the last unit of output as the standard for the next unit of output will eventually produce material bearing no resemblance to the original piece—an example of Rule 4. A possible solution to this problem is to use a master piece as a point of comparison. A common complaint is that "we sold the master (model) piece" so a standard is no longer available for comparison purposes.

   Another example of this phenomenon is "a man who matches color from batch to batch for acceptance of material, without reference to the original swatch."[10] Two possible solutions to this problem are to use a standard color chip or to use the original color sample, assuming the color chip or the original color sample doesn't fade over time.

2. On-the-job training. Deming cites "A frightening example of rule 4 . . . where people on a job train a new worker. This worker is then ready in a few days to help to train a new worker. The methods taught deteriorate without limit. Who would know?"[11] Possible solutions to this problem are to formalize training with a video presentation or to utilize a master in the subject matter to do the training, to get a consistent and desired message to the trainees.

3. Budgeting. Setting the next period's budget as a percentage of the last period's budget is an example of Rule 4.

4. Policy setting. If they lack a mission statement to guide them and a theory of management, then executives meeting to establish policy for an organization will set policies that become increasingly less consistent and more confusing, so that eventually their policies will damage the organization.

5. The telephone game. The telephone game small children play is an example of Rule 4. The message gets continually more confusing, and after a time the current message bears no resemblance to the original message.

6. The grapevine. People who take action in their personal and professional lives based on information from the "grapevine" or "rumor mill" are using Rule 4 as a basis for action. Their next action is a function of only the most recent past action.

7. Engineering changes. Engineering changes to a product or process based on the latest version of a design without regard to the original design are made in accordance with Rule 4. Eventually, the current design will bear no resemblance to the original design.

8. Policy surveys. An executive who changes policy based on results of the latest employee survey, in a stream of employee surveys, is operating under Rule 4. Eventually the policy will have no bearing on its original intended purpose.

9. Adjusting work standards to reflect current performance. An organization that adjusts work standards to reflect current conditions is using Rule 4. Work standards should be replaced with control charts that allow management to understand the capability of a process and take action to improve the process by reducing the variation in the process.

10. Collective bargaining. Union–management negotiations in which successive contracts are a reaction to current conditions is an example of Rule 4.

Process Improvement

The experiment using the funnel illustrates how a system is improved not by overcontrol but by reducing variation. In the experiment, this means reducing the diameter of the circle created under Rule 1 either by moving the funnel closer to the surface[12] or by straightening and lengthening the tube portion of the funnel to reduce the dispersion among the resting points. Note that both methods for improvement are system changes. In terms of an organization, the corresponding reduction of variation also involves system changes. As management is responsible for the system, only management can make the necessary changes to reduce this variation in the system.

We next discuss how the concept of reduction of variation relates to the 14 points discussed in Chapter 1.

Deming's 14 Points

Deming's 14 points provide management with a guide for modifying its processes (the organizational counterparts of the five items and four rules required to perform the experiment) to create improvement and reduce variation (the organizational counterpart of reducing the diameter of the circle created with Rule 1 in the experiment using the funnel). The 14 points are a road map for management to pursue their responsibilities for process improvement and the need to end management by the organizational counterparts of Rules 2, 3, or 4.

Point 1: Create constancy of purpose toward improvement of product and service with a plan to become competitive, stay in business, and provide jobs.
Establishment of a mission statement is synonymous with setting a process's nominal or target level; it prevents management from operating according to Rule 4 of the funnel experiment. Getting all employees (management, salaried, and hourly), members of the board of directors, and shareholders to interpret the mission statement uniformly and to behave in accordance to the common interpretation of the mission statement is a problem of reducing variation.[13]

The mission statement should inspire employees, board members, and shareholders to think about innovation: what the consumer will want tomorrow. This will create uniformity in their concern for the customer and their commitment that there will be a tomorrow.

The mission statement should foster employee efforts for reducing variation. As employees focus on the reduction of variation to achieve customer satisfaction, a greater consistency in their actions will result.

An interesting example of this arose during a seminar at a Fortune 500 company. The company had a mission statement, and all employees received one half-day of training in the mission statement. The mission had four goals, including "create a great place to work" and "maximize profits." One new manager said he was concerned only with profit maximization, not creating a great place to work; his managerial actions reflected his view. His interpretation of the mission statement conflicted with other managers and created variability in management style. This created confusion, frustration, and a decrease in quality.

Point 2: Adopt the new philosophy. We are in a new economic age. We can no longer live with commonly accepted levels of delays, mistakes, defective material, and defective workmanship.
All people in an organization should embrace the system of profound knowledge as the focus of all action. As everyone uniformly embraces the system of profound knowledge, variation in how people view the organization—and in how they interpret their job responsibilities—will fall.

The new philosophy requires that conformance to specifications be replaced by continuous and never-ending improvement of the organization's processes to reduce variation and reap the rewards, such as decrease in rework, increase in productivity, increase in quality, decrease in unit cost, increase in price flexibility, increase in competitive position, increase in market share, increase in profit, increase in job security, the creation of more jobs, and the increased availability of funds to fuel further process improvements. A corollary of the new philosophy is that tightening specifications doesn't improve quality and lower cost—it only creates more out-of-specification material.

Point 3: Cease dependence on mass inspection. Require, instead, statistical evidence that quality is built in to eliminate the need for inspection on a mass basis.
Defect detection is dependent on mass inspection to sort conforming from defective material. Dependence on mass inspection does nothing to decrease variation. Moreover, inspection doesn't create a uniform product within specification limits—rather, product is bunched around specification limits; or at best, product is distributed within specification limits with large variance and tails truncated at the specification limits.

Some organizations that operate in a defect detection mode use automatic process control to hold processes within deadband limits. (See the previous discussion of Rule 2a). In fact, these systems may double the variation of the process, at best—when compared to organizations operating under Rule 1—rather than reduce the process's variation.

Defect prevention can only reduce variation such that products are within specification. Unfortunately, as a result of entropy, a process that operates just within specification limits will eventually go out of specification limits. Further, defect prevention leaves employees with the impression that they're successfully accomplishing their jobs—with respect to the reduction of variation—if they achieve "Zero Defects." This creates an atmosphere of laxness that will ultimately take its toll.

Never-ending improvement works at continuously reducing variation within specification limits. Traditional loss functions, where loss is zero until a specification limit is reached and then becomes positive and constant, are inadequate for today's marketplace. Rather, a Taguchi quadratic loss function (as discussed in Chapter 12) should be used as it demonstrates the economic wisdom of continuous reduction of variation.

The use of inspection should be governed by the kp rule. The kp rule requires a feedback mechanism from a stable process. This feedback mechanism must help in the continuous reduction of p under Rule 1 of the funnel experiment (in the Pareto sense). The kp rule is an operational tool that can help move a process from an attributes mentality (p chart) to a variables mentality (x-bar and R chart).

Point 4: End the practice of awarding business on the basis of price tag. Instead, depend on meaningful measures of quality, along with price. Move toward a single supplier for any one item on a long-term relationship of loyalty and trust. Purchasing is a process that can be continuously improved through the never-ending reduction of variation in purchasing procedures, incoming materials, number of suppliers for any given item, and number of purchasing agents for any given item.

A one-time purchase from a supplier is an enumerative decision in which we're interested only in the frame; the decision can be made without regard to the supplier's process. However, continuing purchases over time is an analytic decision and can't be made without regard to the supplier's process. For the analytic decision, continuous reduction of variation in the supplier's process is critical. Long-term relationships with a supplier make sense if the supplier consistently meets the organization's needs and will continue to improve its ability to do so in the long run.

Multiple supplier processes, each of which has small variations, combine to create a process with large variation. This means an increase in the variability of inputs to the organization, which is counter to the reduction of variation. Consequently, reducing the supply base from many suppliers to one supplier is a rational action. This idea applies to both external and internal suppliers.

The number of purchasing agents who work with any given supplier should be reduced from several purchasing agents to one purchasing agent to decrease variation in purchasing practices.[14]

An organization has a relationship with each of its many stakeholders. These relationships define a "distribution of relationships" for an organization. For example, purchasing agents are responsible for the location and shape of the distribution of relationships between an organization and its suppliers. Purchasing agents' role is to move the distribution in a desirable direction while decreasing its variance.

Point 5: Improve constantly and forever the system of production and service, to improve quality and productivity, and thus constantly decrease costs.

Continual improvement of a system around its aim is tied directly to reducing variation within and between the functions and activities of the system. The Taguchi loss function explains the need for continuous reduction of variation around the aim of the system.

Management must understand a system's capability. They must realize that only when a system is stable and exhibits only system variation, can management predict (with process knowledge) the system's future condition. This allows management to plan the future state of the system.

As we said in Chapter 1, we use the PDSA cycle to decrease the difference, or variation, between customer needs and process performance. The PDSA cycle is a useful analytic procedure for improving process and reducing variation. In Chapter 3, we discussed using operational definitions to create uniformity in the interpretation of product, process, or job characteristics. Operational definitions allow for the reduction of an important source of variation, variation caused by multiple views of a product's, process's, or job's definition.

Never-ending improvement strives to continuously reduce variation within specification limits for operationally defined process and product characteristics. An example of an area that can benefit from the never-ending improvement resulting from application of the PDSA cycle is union–management relations. Union–management relations are an analytic concern, not a one-time "sit down at the contract table" enumerative concern.

Point 6: Institute modern methods of training.

Training changes the skill distribution for each job skill. Management must understand the capability of the training process and the capability of the trainee job skill distribution to improve the trainee job skill distribution. Thus management must understand variation.

Statistical methods should be used to determine when training is finished. In chaos, more training of the same type is effective. In stability, more training of the same type isn't effective; management may have to find the trainee a new job for which he is trainable.

Everyone in an organization should similarly be trained in basic statistical methods and the organization should foster everyone's ability to understand variation. Statistical methods and statistical thinking will create a uniform way for employees to view their organization's processes. This will focus attention on causes of variation and point to methods to remove those causes, reduce variation in a process, and improve the organization. Statistical methods and statistical thinking allow management to separate systems problems (for which only they are accountable) from special problems that can be dealt with by all employees.

Point 7: Institute modern methods of supervision.

Managers should know how to lead their organizations, including employees and suppliers, toward continuous and never-ending improvement. To do this, each manager needs to know when people have special problems and how to continuously improve the system. Both functions require an understanding of variation, its two causes, and the appropriate managerial action for each type of variation.

A leader must understand that variation in a system can come from the individual, the system, or the interaction between the system and the individual. A leader must not rank the people who perform within the limits of a system's capability.

Point 8: Drive out fear, so that everyone may work effectively for the company.

Managers who don't understand variation rank individuals within a system; that is, they hold individuals accountable for system problems. This causes fear, which stifles the desire to change and improve a process. Fear creates variability between an individual's or team's actions and the actions required to surpass customer needs and wants.

In a fear-filled environment, a system of statistically based management won't work. This is because people in the system will view statistics as vehicles for policing and judging rather than as vehicles for providing opportunities for improvement.

Point 9: Break down organizational barriers—everyone must work as a team to foresee and solve problems.

Barriers between departments result in multiple interpretations of a given message. This increases variability in the actions taken with respect to a given message. Operational definitions create a common language for communication between departments and consequently reduce variability in the actions taken on processes and products by different departments. "Operationally defining the ultimate customer's needs and expectations so that everyone understands how he contributes to the success of the organization is a solid step to breaking down barriers between departments."[15]

Point 10: Eliminate arbitrary numerical goals, posters, and slogans for the work force that seek new levels of productivity without providing methods.

The system and its variation are the responsibility of management. Slogans and posters try to shift that responsibility to the worker. For example, a sign in a factory reading SAFETY IS BETTER THAN COMPENSATION! attempts to shift the burden for safety from factory management to the worker.

Point 11: Eliminate work standards and numerical quotas.

Work standards are negotiated values that have no bearing on a process or its capability. They create variability in performance by obscuring an employee's understanding of his job. Work standards create fears that undermine the smooth operation of the workplace and create an undesirable atmosphere, which is contrary to the Deming philosophy.

Work standards also place a cap on improvement. After the standard has been reached, employees stop working for fear of a new and higher standard. Thus, work standards create variability between actual performance and desired performance.

If a work standard is between a system's upper natural limit (UNL) and lower natural limit (LNL), there's a possibility that the standard can be met, but meeting the standard this way is simply a random lottery. If a work standard is above the system's UNL, then there's little chance that the standard will be met unless management changes the system. Rather than focusing on the standard as a means to productivity, management should focus on stabilizing and improving the process to increase productivity.

The fallacy inherent in quotas and management by objectives (MBO) is illustrated when considering Point 11 in terms of the funnel experiment. A quota is equivalent to declaring that an employee must achieve a given number of hits per time period within a circle centered at (0,0) with diameter d_{spec}. This is equivalent to saying that an employee will be evaluated on performance that's solely a function of random chance. Try to imagine the frustration this can create as the results are really beyond the employee's control. MBO confounds common variation (like the random variation in the funnel experiment) and individual performance.

It's unfair and counterproductive to hold employees accountable for circumstances beyond their control. This kind of accountability leads to fear and frustration, often resulting in overcontrol, or the use of Rules 2, 3, or 4. As we've seen, this will only increase variation, thereby making it even harder to achieve the quota or MBO.

Point 12: Remove barriers that rob employees of their pride of workmanship.

Barrier 1: Managerial Ignorance Concerning Variation and Statistical Thinking. Management must be educated so they understand variation and its impact on decision making. By understanding the importance of not acting on common causes of variation as if they were special causes of variation, management can change an organization's systems to focus on continuous and never-ending improvement.

Barrier 2: Performance Appraisal Systems (PAS). The *performance appraisal system* (PAS) destroys teamwork by encouraging every person and every department to focus on individual goals—rather than on the organization's goals—to obtain a positive rating in their performance appraisal. This results in serious suboptimization with respect to the organization's goals. Simultaneous seeking of different—possibly conflicting—goals creates variability in management's behavior, leading to confusion and fear as to exactly what everyone's job is. Eliminating the PAS makes it feasible to focus everyone's attention on the organization's goals.

PAS also reduce initiative or risk taking because once an objective has been reached, effort stops. This is counter to the notion of continuous and never-ending improvement.

PAS can cause employees to lower their planned performance levels to increase the chances of meeting their objectives. Again, this creates variability between the desired objective and the negotiated objective, which invariably lowers productivity.

Further, PAS can foster banking of performance to create a cushion for the next PAS time period. This banking distorts the information upon which management makes decisions and subsequently creates variability between the desired objective and the actual situation.

PAS can increase variability in employee performance, resulting from actions such as rewarding everyone who's above average and penalizing everyone who's below average. In such a situation, below-average employees try to emulate above-average employees. However, as the employees who are above average and those who are below average are part of the same system (only common variation is present), the below-average ones are adjusting their behavior based on common variation. We know from the experiment using the funnel that this doubles performance variation at best, but explodes performance variation at worst.

The PAS assume that people are directly and solely responsible for their output. This assumption fails to distinguish between the individual's effect and the system's effect on output. In other words, it fails to appreciate the causes of variation and that most variation in performance can be attributed to the system, not the individual. The system is the responsibility of management.

PAS focuses on the short term. The variability between where an organization wants to be after five years and where it will actually be after 20 quarters of short-term focus can be substantial.[16] Further, adjusting an organization's direction based on short-term results is akin to Rule 4 in the experiment using the funnel: what you do next depends solely on where you are now. This type of action will explode the variance and create disastrous results.

We can use the funnel experiment to demonstrate the absurdity of PAS, as Figure 14.6 shows.

1. Establish two concentric circles centered on the target of (0,0).
2. Name the area within the circle of smallest diameter (the innermost circle) "excellent performance." Name the area between the smallest circle and the next-larger circle "above average" and so on until the last area is to be named. Name this last area (the area between the second largest circle and the largest circle) "poor performance."
3. Allow each experimenter one drop of the marble. Record the area in which the marble comes to rest on the surface.
4. Label the experimenter in accordance with the name of the area in which his marble comes to rest on the surface.
5. Reward the experimenter accordingly. "Excellent performance" gets high merit pay, while "poor performance" gets chastisement and no cost-of-living pay.
6. Observe what happens over time to the points where the marble comes to rest on the surface as a result of this rating system.

The experimenters whose marbles fell in the outer circles will move to Rules 2, 3, or 4 out of frustration and fear in their attempts to emulate the experimenters whose marbles came to rest in the innermost circle. Of course, we know that this is overcontrol based on common variation and will at best double the process variation (Rule 2) and at worst explode the variation in the system (Rules 3 or 4).

FIGURE 14.6   Performance Appraisal Systems

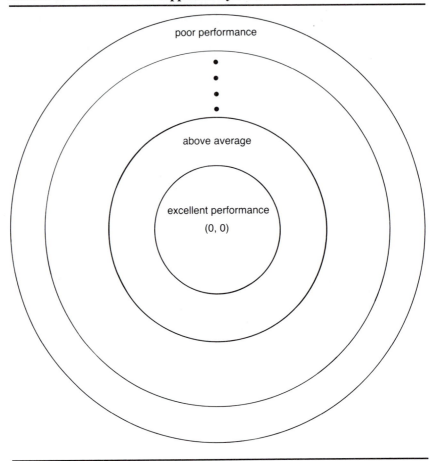

A Proposal That Fosters Consistency.   Management must understand variation to recognize whether an employee is operating in the system. An employee can exist in only one of three categories: in the system, out of the system on the negative side, and out of the system on the positive side. The latter two categories signify that the employee needs special attention. Further, if an employee shows improving or deteriorating performance for more than eight periods in a row (see Rule 5 in Chapter 5), that can be used as evidence that the employee needs special attention; this is an illustration of the institution of modern methods of supervision (Deming's point seven).

If all employees are within the system, then to reward or punish employees on the basis of performance (or merit) would be destructive. This is because punished employees will try to emulate rewarded employees, when in fact there's no difference between the two types of employees; they're both in the system and any difference only results from common variation. This type of reward/punish management will at best double the variation between employees, as illustrated by Rule 2 of the experiment using the funnel. In this situation, any improvement must result from improvements in the system, which will benefit all employees.

If one or more employees aren't within the system (special employees), then management must: (1) determine if the special employees form their own special system and (2) determine if the rest of the employees are within the old revised system. Next, management must determine the sources of variation for the special employees and use this information to reduce variation, and create consistency and improvement.[17]

Barrier 3: Daily Production Reports.   Daily production reports focus attention on yesterday's production, relative to the long-term production daily average, as a basis for action, without any acknowledgment of variation. This illustrates Rules 2 and 3 of the funnel experiment. At best, this will double the variation in production (Rule 2) and, at worst, will explode the variation in the system of production into cycles of ever-increasing variability (Rule 3).

Daily production reports can create banking of production to meet tomorrow's quota. This reduces the accuracy of the information on which management bases its actions, and, hence, increases the variability of its decisions.

Daily production reports can create pressure to explain any variances from target production, when in fact variances may result from common variation. This leads to overcontrol and increases production variation.

A better method is to distribute production reports less frequently than every day—say, every week—and to include with the latest production information a control chart relative to the process's past history. In this way, management can see whether variation was caused by special or common causes.

Barrier 4: Cost of Quality.   Two types of figures are important to decision making: visible figures (for example, break-even figures) and invisible figures (for example, the cost of an unhappy customer). If an organization is managed only on visible figures, decisions will be based on misleading information. This will result in greater variability in decision making, depending on what visible figures are used.

Point 13: Institute a vigorous program of education and training.
As an organization improves, it will free up resources that can be used to educate and train its employees. These improvements will lower variability in processes, products, and jobs, continuing the never-ending cycle of improvement.[18]

Management must use only attrition to reduce the number of people in an organization. This will create a body of highly trained employees who are the resources that the organization should invest in through education and training. If management sporadically fires and hires employees depending on market conditions, it will create a body of employees whose performance is extremely variable as a result of a lack of training, a lack of common purpose, and a lack of commitment to the organization.

Point 14: Create a structure that will push the prior 13 points every day.

The current paradigm of Western management is shaped by reactive forces. Therefore, it has an explosive and high degree of variation in its application (Rule 4 of the funnel experiment). The transformation must emanate out of a new paradigm shaped by the system of profound knowledge, not reactive forces. This new paradigm will have a stable, reducible degree of variation in its application. (Improve the funnel under Rule 1 of the funnel experiment.)

Variation tells us that all individuals have different reasons for wanting to, or not wanting to, accomplish the transformation. Variation also tells us that different individuals will have different interpretations of what's involved in the transformation.

The statistical leader in an organization must work to reduce variability with respect to management's ability to use statistical thinking in their decision making, and in the application and teaching of statistical thinking and statistical methods in the organization. The statistical leader should strive to reduce variation between managers with respect to their views on the organization's mission and quality.

Summary

This chapter explains the significance of reducing variation to each of Deming's 14 points. It begins by discussing statistics and management, showing that both are concerned with the never-ending reduction of variation. An experiment shows that many types of management actions increase variation, rather than decrease it. Thus we see that management must begin to act based on knowledge of variation.

The main portion of the chapter explains how each of Deming's 14 points helps an organization to continuously reduce variation in all of its processes.

Exercises

14.1 Explain the purpose of the experiment using the funnel.

14.2 Explain Rule 1, giving an example of how it's used in an organization.

14.3 Explain Rule 2, giving an example of how it's used in an organization.

14.4 Explain Rule 2a, giving an example of how it's used in an organization.

14.5 Explain Rule 3, giving an example of how it's used in an organization.

14.6 Explain Rule 4, giving an example of how it's used in an organization.

Endnotes

1. W E Deming, *Out of the Crisis* (Cambridge, MA: Massachusetts Institute of Technology Center for Advanced Engineering Study, 1986), p. 327.

2. Ibid.

3. W E Deming, *The New Economics for Industry, Government, Education* (Cambridge, MA: MIT Center for Advanced Engineering Study, 1993), pp. 194–209.

4. T J Boardman and H Iyer, *The Funnel* (Fort Collins, CO: Colorado State University Press, 1986), p. 1.

5. Ibid.

6. Deming, *Out of the Crisis.* See example 5, pp. 359–60.

7. H Gitlow, K Kang, and S Kellogg, "Process Tampering: An Analysis of On/Off Deadband Process Controlling," *Quality Engineering*, 5 no. 2, 1992–93, pp. 239–310.

8. Boardman and Iyer, *The Funnel,* p. 1.

9. Ibid.

10. Deming, *Out of the Crisis,* p. 329.

11. Ibid., p. 330.

12. Private conversation with William Latzko, North Bergen, NJ.

13. William Scherkenbach, *The Deming Route to Quality and Productivity: Road Maps and Roadblocks* (Washington, DC: CeePress Books, 1986), pp. 133–34.

14. Ibid., pp. 133–36.

15. Ibid., p. 82.

16. Ibid., pp. 55–57.

17. Ibid., pp. 62–69.

18. Ibid., p. 126.

CHAPTER 15 Some Current Thinking about Statistical Studies and Practice

Introduction

This chapter relates some current views on enumerative and analytic studies, as espoused most recently by W Edwards Deming[1] and by Thomas Nolan, Ron Moen, and Lloyd Provost.[2] We also discuss principles of professional conduct for the statistician, closing with a reprinted article by Deming.

Enumerative and Analytic Studies

In Chapter 2 we distinguished between enumerative studies and analytic studies. We saw that (1) the purpose of an enumerative study is to determine the disposition of material in a frame and (2) that a statistical inference can be made only with respect to a frame. Consequently, an enumerative study conducted in Miami provides no basis for a statistical inference in Los Angeles or Spokane.

Also in Chapter 2, we saw that the purpose of an analytic study is to take action on a process to improve its performance in the future. As a statistical inference with a quantifiable degree of belief can be made only with respect to a frame—and it's impossible to get a frame of future events—statistical inferences with a quantifiable degree of belief can't be made in analytic studies. Figure 15.1 summarizes differences between enumerative and analytic studies. We see that a frame is necessary in an enumerative study to take action on the disposition of material in the frame, at a given point in time; this is a static situation. A frame makes possible random sampling and quantification of a major source of uncertainty in an enumerative study; that is, sampling error. Consequently, the theory of estimation, including sampling distributions, standard errors, and confidence intervals, is completely applicable.

The major source of uncertainty in an analytic study is prediction. Unfortunately, current statistical theory provides no aid for quantifying future effects of variables that will affect the process. There's no statistical theory for quantifying

---

FIGURE 15.1   Important Aspects of Enumerative and Analytic Studies

	Enumerative	*Analytic*
Aim	Descriptive/ inference	Prediction
Focus	Description of material in frame	Cause-and-effect system
Methods of access	Frame	Models of the process such as flowcharts and cause-and-effect diagrams (no frame)
Major sources of uncertainty	Sampling error	Extrapolation to the future (How well can I predict?)
Is the major source of uncertainty quantifiable?	Yes	No
Environment of the study	Static	Dynamic
Type of sample	Random	Judgment

SOURCE: R Moen, T Nolan, and L Provost, *Improving Quality through Planned Experimentation* (New York: McGraw-Hill, 1991), p. 54.

this type of uncertainty. Further, in Figure 15.1 we see that there's no frame for an analytic study; hence, there can be no random sampling and quantification of sampling error. The theory of estimation isn't applicable. Thus, we face a dilemma: how can we predict future behavior of a dynamic process without a quantifiable degree of belief? The best answer, given current knowledge, is that we must rely on a combination of expertise concerning the process under study plus statistical thinking and methods.

Errors in Analytic Studies

Two types of errors can occur in any study. A *type one error* occurs when action is taken on a process when it should have been left alone. A *type two error* occurs when we fail to take action on a process when action is appropriate. It's impossible to calculate the probability of either type of error in an analytic study because we can't know how things would have turned out if an alternative action, other than the one chosen, had been selected. Without the benefit of a quantifiable degree of belief of making type one and type two errors, we need a methodology for combining expertise with statistical notions.

Design of Analytic Studies

We can increase our degree of belief in a prediction from an analytic study by considering (1) sequential building of knowledge, (2) testing over a wide range of conditions, and (3) selection of units for the study.

Sequential Building of Knowledge.   The Deming philosophy requires that an organizational (team) focus be given to quality improvement efforts. In other words, improvement efforts should be based on an organizational belief that these efforts will decrease the difference between customer needs and process performance. Given the selection of a process for improvement efforts, the process must be described by documenting and defining it as discussed in Chapter 3. The PDSA cycle can then be used to decrease the difference between customer needs and process performance. Experiments performed in the iterations of the PDSA cycle, in combination with theory about the process from subject matter experts, may increase the degree of belief in predictions about future behavior of the process under study. The degree of belief in predictions about the process is increased as sequential predictions about the process's future behavior come closer to the actual performance of the process.

Note that there can be no significant prediction about the future behavior of a process without knowledge of that process. Further, knowledge of the process is continually improved and increased through successive iterations of the PDSA cycle. It's the predictive ability of knowledge that's the ultimate measure of its value.

Testing over a Wide Range of Conditions.   The degree of belief in the predictive value of the knowledge gained from an analytic study is increased if the analytic study yields the same results over a wide range of conditions. For example, one contribution of Taguchi's off-line quality control methodology is that it facilitates the testing of various product configurations over a wide range of conditions. The conditions enter the experiment in the form of the noise matrix we discussed in Chapter 12. Another example of testing over a wide range of conditions is being able to predict the outcome of using a particular supplier's product—regardless of the batch from which it comes, or the time of year it's delivered, or how it's used.

We can summarize by saying that the degree of belief in the predictive power of the knowledge gained from an analytic study is increased as the same results are obtained over repeated trials under a wide range of conditions. Only an expert in the subject matter under study can answer questions such as how wide a range of conditions is adequate to have a degree of belief sufficient to make a prediction, or how close to actual conditions must the experimental conditions be to have a degree of belief high enough to make predictions. These questions can't be answered by statistical theory—still, statistical methods can provide guidance for increasing the degree of belief in the knowledge gained from analytic studies.

Selection of the Units for the Study.   As there's no frame in an analytic study, there can be no random sample and quantifiable degree of uncertainty in a prediction about the future. Judgment samples are used to conduct analytic studies.[3] The judgment of a subject matter expert determines both the conditions under which a process will be studied and the measurements that will be taken for each set of conditions. Moreover, it's the expert who judges whether the results of an analytic study provide a sufficient degree of belief to take action on a process.

In analytic studies, judgment samples are almost always superior to random samples. For example, consider an analytic study to determine which of two machines is less sensitive to worker-to-worker variation. We need this information to purchase whichever machine yields more uniform output, regardless of operator. Suppose we have funds to include only 10 of 50 operators in the study. In this case, a random sample of the 10 operators would yield a quantifiable degree of uncertainty in the measurement variation, but the conditions under which the operators were studied will never be seen again; in the future, there will be new operators, new materials, and different training. For this study, a judgment sample is more appropriate. For example, a judgment sample might include the five most experienced operators and the five least experienced operators.

Suppose both machines perform best when used by an experienced operator and worst when used by an inexperienced operator. The degree of belief that the machine with the lower worker-to-worker variation is the machine to purchase will be greater if the variation is estimated from a judgment sample rather than from a sample of 10 randomly selected workers.

Analysis of Data from Analytic Studies

The purpose of an analytic study is to improve knowledge so that predictions can be made about the future behavior of a process. Yet Deming stated of analytic studies, "Analysis of variance, t-tests, confidence intervals, and other statistical techniques taught in the books, however interesting, are inappropriate because they provide no basis for prediction and because they bury the information contained in the order of production."[4] Thus, most techniques that are useful in enumerative studies are inappropriate for analytic studies.

The standard deviation of a stable process can be used to distinguish past common variation from past special variation. But calculation of the standard deviation based on the assumption of a stable process doesn't consider the most important source of uncertainty in an analytic study—predicting the process's future behavior.

A Stable Process.   When a process is stable, it's easier to determine the effect that changes to the process have on the process's future behavior. Hence, a stable process provides a forum for a subject matter expert to conduct experiments to gain knowledge to predict the future behavior of the process.

Unfortunately, stability in the past doesn't guarantee stability in the future. Past conditions that created the stable process may never be seen again. As a result, responsibility for prediction still rests with the subject matter expert.

Graphical Analysis.   Because the distribution of a process that existed in the past can't be used to predict the distribution of a future process, we need an approach to prediction that doesn't rely on past distribution statistics.

The general approach for conducting analytic studies relies on graphical techniques, such as control charts, which utilize both statistical knowledge and subject matter knowledge to learn about the process to predict its future behavior.

Knowledge of the process should give meaning to control chart patterns and help distinguish common variation from special variation. For example, knowledge of a process might lead researchers to expect that when eight or more points in a row increase in value, tool wear is the likely special source of variation. This prediction about the process's behavior is a combination of subject matter knowledge and statistical knowledge.

Interactions between process variables can dramatically alter process behavior depending on the process variables that are operational at any given time. For example, one machine might be better when used by an experienced operator, while another machine might be better when used by an inexperienced operator. Graphical techniques will allow a process and its interactions to be studied in a dynamic fashion over a wide range of conditions.

Principles of Professional Practice: An Article by Deming

Every statistician and every subject matter expert working with a statistician should be familiar with the areas of responsibility for statisticians and their clients. The rest of this chapter is a reprint of W E Deming's article "Principles of Professional Statistical Practice."[5]

Purpose and Scope

1. The Statistician's Job. The statistician's responsibility, whether as a consultant on a part-time basis, or on regular salary, is to find problems that other people could not be expected to perceive. The statistician's tool is statistical theory (theory of probability). It is the statistician's use of statistical theory that distinguishes him from other experts. Statistical practice is a collaborative venture between statistician and experts in subject matter. Experts who work with a statistician need to understand the principles of professional statistical practice as much as the statistician needs them.

Challenges face statisticians today as never before. The whole world is talking about safety in mechanical and electrical devices (in automobiles, for example); safety in drugs; reliability; due care; pollution; poverty; nutrition; improvement of medical practice; improvement of agricultural practice; improvement in quality of product; breakdown of service; breakdown of equipment; tardy buses, trains and mail; need for greater output in industry and in agriculture; enrichment of jobs.

These problems cannot be understood and cannot even be stated, nor can the effect of any alleged solution be evaluated, without the aid of statistical theory and methods. One cannot even define operationally such adjectives as *reliable, safe, polluted, unemployed, on time* (arrivals), *equal* (in size), *round, random, tired, red, green,* or any other adjective, for use in business or in government, except in statistical terms. To have meaning for business or legal purposes, a standard (as of safety, or of performance or capability) must be defined in statistical terms.

2. Professional Practice. The main contribution of a statistician in any project or in any organization is to find the important problems, the problems that other people cannot be expected to perceive. An example is the 14 points that top management learned in Japan in 1950. The problems that walk in are important, but are not usually *the* important problems [4].

The statistician also has other obligations to the people with whom he works, examples being to help design studies, to construct an audit by which to discover faults in procedure before it is too late to make alterations, and at the end, to evaluate the statistical reliability of the results.

The statistician, by virtue of experience in studies of various kinds, will offer advice on procedures for experiments and surveys, even to forms that are to be filled out, and certainly in the supervision of the study. He knows by experience the dangers of (1) forms that are not clear, (2) procedures that are not clear, (3) contamination, (4) carelessness, and (5) failure of supervision.

Professional practice stems from an expanding body of theory and from principles of application. A professional man aims at recognition and respect for his practice, not for himself alone, but for his colleagues as well. A professional man takes orders, in technical matters, from standards set by his professional colleagues as unseen judges, never from an administrative superior. His usefulness and his profession will suffer impairment if he yields to convenience or to opportunity. A professional man feels an obligation to provide services that his client may never comprehend or appreciate.

A professional statistician will not follow methods that are indefensible, merely to please someone, or support inferences based on such methods. He ranks his reputation and profession as more important than convenient assent to interpretations not warranted by statistical theory. Statisticians can be trusted and respected public servants.

Their careers as expert witnesses will be shattered if they indicate concern over which side of the case the results seem to favor. "As a statistician, I couldn't care less" is the right attitude in a legal case or in any other matter.

Logical Basis for Division of Responsibilities

3. Some Limitations in Application of Statistical Theory. Knowledge of statistical theory is necessary but not sufficient for successful operation. Statistical theory does not provide a road map toward effective use of itself. The purpose of this article is to propose some principles of practice, and to explain their meaning in some of the situations that the statistician encounters.

Statisticians have no magic touch by which they may come in at the stage of tabulation and make something of nothing. Neither will their advice, however wise in the early stages of a study, ensure successful execution and conclusion. Many a study launched on the ways of elegant statistical design, later boggled in execution, ends up with results to which the theory of probability can contribute little.

Even though carried off with reasonable conformance to specifications, a study may fail through structural deficiencies in the method of investigation (questionnaire, type of test, technique of interviewing) to provide the information needed. Statisticians may reduce the risk of this kind of failure by pointing out to their clients in the early stages of the study the nature of the contributions that they themselves must put into it. (The word *client* will denote the expert or group of experts in a substantive field.) The limitations of statistical theory serve as signposts to guide a logical division of responsibilities between statistician and client. We accordingly digress for a brief review of the power and the limitations of statistical theory.

We note first that statistical inferences (probabilities, estimates, confidence limits, fiducial limits, etc.), calculated by statistical theory from the results of a study, will relate only to the material, product, people, business establishments, and so on, that the frame was in effect drawn from, and only to the environment of the study, such as the method of investigation, the date, weather, rate of flow and levels of concentration of the components used in tests of a production process, range of voltage, or of other stress specified for the tests (as of electrical equipment).

Empirical investigation consists of observations on material of some kind. The material may be people; it may be pigs, insects, physical material, industrial product, or records of transactions. The aim may be enumerative, which leads to the theory for the sampling of finite populations. The aim may be analytic, the study of the causes that make the material what it is, and which will produce other material in the future. A definable lot of material may be divisible into identifiable sampling units. A list of these sampling units, with identification of each unit, constitutes a *frame*. In some physical and astronomical investigations, the sampling unit is a unit of time. We need not detour here to describe nests of frames for multistage sampling. The important point is that without a frame there can be neither a complete coverage of a designated lot of material nor a sample of any designated part thereof. Stephan introduced the concept of the frame, but without giving it a name [7].

Objective statistical inferences in respect to the frame are the speciality of the statistician. In contrast, generalization to cover material not included in the frame, nor to ranges, conditions, and methods outside the scope of the experiment, however essential for application of the results of a study, are a matter of judgment and depend on knowledge of the subject matter [5].

For example, the universe in a study of consumer behavior might be all the female homemakers in a certain region. The frame might therefore be census blocks, tracts or other small districts, and the ultimate sampling unit might be a segment of area containing households. The study itself will, of course, reach only the people that can be found at home in the segments selected for the sample. The client, however, must reach generalizations and take action on the product or system of marketing with respect to all female homemakers, whether they be the kind that are almost always at home and easy to reach, or almost never at home and therefore in part omitted from the study. Moreover, the female homemakers on whom the client must take action belong to the future, next year, and the next. The frame only permits study of the past.

For another example, the universe might be the production process that will turn out next week's product. The frame for study might be part of last week's product. The universe might be traffic in future years, as in studies needed as a basis for estimating possible losses or gains as a result of a merger. The frame for this study might be records of last year's shipments.

Statistical theory alone could not decide, in a study of traffic that a railway is making, whether it would be important to show the movement of (for example) potatoes, separately from other agricultural products in the northwestern part of the United States; only he who must use the data can decide.

The frame for a comparison of two medical treatments might be patients or other people in Chicago with a specific ailment. A pathologist might, on his own judgment, without further studies, generalize the results to people that have this ailment anywhere in the world. Statistical theory provides no basis for such generalization.

No knowledge of statistical theory, however profound, provides by itself a basis for deciding whether a proposed frame would be satisfactory for a study. For example, statistical theory would not tell us, in a study of consumer research, whether to include in the frame people that live in trailer parks. The statistician places the responsibility for the answer to this question where it belongs, with the client: Are trailer parks in your problem? Would they be in it if this were a complete census? If yes, would the cost of including them be worthwhile?

Statistical theory will not of itself originate a substantive problem. Statistical theory cannot generate ideas for a questionnaire or for a test of hardness, nor can it specify what would be acceptable colors of dishes or of a carpet; nor does statistical theory originate ways to teach field workers or inspectors how to do their work properly or what characteristics of workmanship constitute a meaningful measure of performance. This is so in spite of the fact that statistical theory is essential for reliable comparisons of questionnaires and tests.

4. Contributions of Statistical Theory to Subject Matter. It is necessary, for statistical reliability of results, that the design of a survey or experiment fit into a theoretical model. Part of the statistician's job is to find a suitable model that he can manage, once the client has presented his case and explained why statistical information is needed and how the results of a study might be used. Statistical practice goes further than merely to try to find a suitable model (theory). Part of the job is to adjust the physical conditions to meet the model selected. Randomness, for example, in sampling a frame, is not just a word to assume; it must be created by use of random numbers for the selection of sampling units. Recalls to find people at home, or tracing illegible or missing information back to source documents, are specified so as to approach the prescribed probability of selection.

Statistical theory has in this way profoundly influenced the theory of knowledge. It has given form and direction to quantitative studies by molding them into the requirements of the theory of estimation, and other techniques of inference. The aim is, of course, to yield results that have meaning in terms that people can understand.

The statistician is often effective in assisting the substantive expert to improve accepted methods of interviewing, testing, coding, and other operations. Tests properly designed will show how alternative test or field procedures really perform in practice, so that rational changes or choices may be possible. It sometimes happens, for example, that a time-honored or committee-honored method of investigation, when put to statistically designed tests, shows alarming inherent variances between instruments, between investigators, or even between days for the same investigator. It may show a trend, or heavy intraclass correlation between units, introduced by the investigators. Once detected, such sources of variation may then be corrected or diminished by new rules, which are of course the responsibility of the substantive expert.

Statistical techniques provide a safe supervisory tool to help to reduce variability in the performance of worker and machine. The effectiveness of statistical controls, in all stages of a survey, for improving supervision, to achieve uniformity, to reduce errors, to gain speed, reduce costs, and to improve quality of performance in other ways is well known. A host of references could be given, but two will suffice [2, 3].

5. Statistical Theory as a Basis for Division of Responsibilities. We may now, with this background, see where the statistician's responsibilities lie.

In the first place, his specialized knowledge of statistical theory enables him to see which parts of a problem belong to substantive knowledge (sociology, transportation, chemistry of a manufacturing process, law), and which parts are statistical. The responsibility as logician in a study falls to him by default, and he must accept it. As logician, he will do well to designate, in the planning stages, which decisions will belong to the statistician and which to the substantive expert.

This matter of defining specifically the areas of responsibility is a unique problem faced by the statistician. Business managers and lawyers who engage an expert on corporate finance, or an expert on steam power plants, know pretty well what such experts are able to do and have an idea about how to work with them. It is different with statisticians. Many people are confused between the role of theoretical statisticians (those who use theory to guide their practice) and the popular idea that statisticians are skillful in compiling tables about people or trade, or who prophesy the economic outlook for the coming year and which way the stock market will drift. Others may know little about the contributions that statisticians can make to a study or how to work with them.

Allocation of responsibilities does not mean impervious compartments in which you do this and I'll do that. It means that there is a logical basis for allocation of responsibilities and that it is necessary for everyone involved in a study to know in advance what he or she will be accountable for.

A clear statement of responsibilities will be a joy to the client's lawyer in a legal case, especially at the time of cross-examination. It will show the kind of question that the statistician is answerable for, and what belongs to the substantive experts. Statisticians have something to learn about professional practice from law and medicine.

6. Assistance to the Client to Understand the Relationship. The statistician must direct the client's thoughts toward possible uses of the results of a statistical study of the entire frame without consideration of whether the entire frame will be studied or only a small sample of the sampling units in the frame. Once these matters are cleared, the statistician may outline one or more statistical plans and explain to the client in his own language what they mean in respect to possible choices of frame and environmental conditions, choices of sampling unit, skills, facilities and supervision required. The statistician will urge the client to foresee possible difficulties with definitions or with the proposed method of investigation. He may illustrate with rough calculations possible levels of precision to be expected from one or more statistical plans that appear to be feasible, along with rudimentary examples of the kinds of tables and inferences that might be forthcoming. These early stages are often the hardest part of the job.

The aim in statistical design is to hold accuracy and precision to sensible levels, with an economic balance between all the possible uncertainties that afflict data—built-in deficiencies of definition, errors of response, nonresponse, sampling variation, difficulties in coding, errors in processing, difficulties of interpretation.

Professional statistical practice requires experience, maturity, fortitude, and patience, to protect the client against himself or against his duly appointed experts in subject matter who may have in mind needless but costly tabulations in fine classes (five-year age groups, small areas, fine mileage brackets, fine gradations in voltage, etc.), or unattainable precision in differences between treatments, beyond the capacity of the skills and facilities available, or beyond the meaning inherent in the definitions.

These first steps, which depend heavily on guidance from the statistician, may lead to important modifications of the problem. Advance considerations of cost, and of the limitations of the inferences that appear to be possible, may even lead clients to abandon a study, at least until they feel most confident of the requirements. Protection of a client's bank account, and deliverance from more serious losses from decisions based on inconclusive or misleading results, or from misinterpretation, is one of the statistician's greatest services.

Joint effort does not imply joint responsibility. Divided responsibility for a decision in a statistical survey is as bad as divided responsibility in any venture—it means that no one is responsible.

Although they acquire superficial knowledge of the subject matter, the one thing that statisticians contribute to a problem, and which distinguishes them from other experts, is knowledge and ability in statistical theory.

Summary Statement of Reciprocal Obligations and Responsibilities

7. Responsibilities of the Client. The client will assume responsibility for those aspects of the problem that are substantive. Specifically, he will stand the ultimate responsibility for:

a. The type of statistical information to be obtained.

b. The methods of test, examination, questionnaire, or interview, by which to elicit the information from any unit selected from the frame.

c. The decision on whether a proposed frame is satisfactory.

d. Approval of the probability model proposed by the statistician (statistical procedures, scope, and limitations of the statistical inferences that may be possible from the results).

e. The decision on the classes and areas of tabulation (as these depend on the uses that the client intends to make of the data); the approximate level of statistical precision or protection that would be desirable in view of the purpose of the investigation, skills and time available, and costs.

The client will make proper arrangements for:

f. The actual work of preparing the frame for sampling, such as serializing and identifying sampling units at the various stages.

g. The selection of the sample according to procedures that the statistician will prescribe, and the preparations of these units for investigation.

h. The actual investigation; the training for this work, and the supervision thereof.

i. The rules for coding; the coding itself.

j. The processing, tabulations and computations, following procedures of estimation that the sampling plans prescribe.

The client or his representative has the obligation to report at once any departure from instructions, to permit the statistician to make a decision between a fresh start or an unbiased adjustment. The client will keep a record of the actual performance.

8. Responsibilities of the Statistician. The statistician owes an obligation to his own practice to forestall disappointment on the part of the client, who if he fails to understand at the start that he must exercise his own responsibilities in the planning stages, and in execution, may not realize in the end the fullest possibility of the investigation, or may discover too late that certain information that he had expected to get out of the study was not built into the design. The statistician's responsibility may be summarized as follows:

a. To formulate the client's problem in statistical terms (probability model), subject to approval of the client, so that a proposed statistical investigation may be helpful to the purpose.

b. To lay out a logical division of responsibilities for the client, and for the statistician, suitable to the investigation proposed.

c. To explain to the client the advantages and disadvantages of various frames and possible choices of sampling units, and of one or more feasible statistical plans of sampling or experimentation that seem to be feasible.

d. To explain to the client, in connection with the frame, that any objective inferences that one may draw by statistical theory from the results of an investigation can only refer to the material or system that the frame was drawn from, and only to the methods, levels, types, and ranges of stress presented for study. It is essential that the client understand the limitations of a proposed study, and of the statistical inferences to be drawn therefrom, so that he may have a chance to modify the content before it is too late.

e. To furnish statistical procedures for the investigation—selection, computation of estimates and standard errors, tests, audits, and controls as seem warranted for detection and evaluation of important possible departures from specifications, variances between investigators, nonresponse, and other persistent uncertainties not contained in the standard error; to follow the work and to know when to step in.

f. To assist the client (on request) with statistical methods of supervision, to help him to improve uniformity of performance of investigators, gain speed, reduce errors, reduce costs, and to produce a better record of just what took place.

g. To prepare a report on the statistical reliability of the results.

9. The Statistician's Report or Testimony. The statistician's report or testimony will deal with the statistical reliability of the results. The usual content will cover the following points:

a. A statement to explain what aspects of the study his responsibility included, and what it excluded. It will delimit the scope of the report.

b. A description of the frame, the sampling unit, how defined and identified, the material covered, and a statement in respect to conditions of the survey or experiment that might throw light on the usefulness of the results.

c. A statement concerning the effect of any gap between the frame and the universe for important conclusions likely to be drawn from the results. (A good rule is that the statistician should have before him a rough draft of the client's proposed conclusions.)

d. Evaluation (in an enumerative study) of the margin of uncertainty, for a specified probability level, attributable to random errors of various kinds, including the uncertainty introduced by sampling, and by small independent random variations in judgment, instruments, coding, transcription, and other processing.

e. Evaluation of the possible effects of other relevant sources of variation, examples being differences between investigators, between instruments, between days, between areas.

f. Effect of persistent drift and conditioning of instruments and of investigators; changes in technique.

g. Nonresponsive and illegible or missing entries.

h. Failure to select sampling units according to the procedure prescribed.

i. Failure to reach and to cover sampling units that were designated in the sampling table.

j. Inclusion of sampling units not designated for the sample but nevertheless covered and included in the results.

k. Any other important slips and departures from the procedure described.

l. Comparisons with other studies, if any are relevant.

In summary, a statement of statistical reliability attempts to present to readers all information that might help them to form their own opinions concerning the validity of conclusions likely to be drawn from the results.

The aim of a statistical report is to protect clients from seeing merely what they would like to see; to protect them from losses that could come from misuse of the results. A further aim is to forestall unwarranted claims of accuracy that a client's public might otherwise accept.

Any printed description of the statistical procedures that refer to this participation, or any evaluation of the statistical reliability of the results, must be prepared by the statistician as part of the engagement. If a client prints the statistician's report, it will be printed in full.

The statistician has an obligation to institute audits and controls as a part of the statistical procedures for a survey or experiment, to discover any departure from the procedures prescribed.

The statistician's full disclosure and discussion of all the blemishes and blunders that took place in the survey or experiment will build for him a reputation of trust. In a legal case, this disclosure of blunders and blemishes and their possible effects on conclusions to be drawn from the data are a joy to the lawyer.

A statistician does not recommend to a client any specific administrative action or policy. Use of the results that come from a survey or experiment are entirely up to the client. Statisticians, if they were to make recommendations for decision, would cease to be statisticians.

Actually, ways in which the results may throw light on foreseeable problems will be settled in advance, in the design, and there should be little need for a client or for anyone else to reopen a question. However, problems sometimes change character with time (as when a competitor of the client suddenly comes out with a new model), and open up considerations of statistical precision and significance of estimates that were not initially in view.

The statistician may describe in a professional or scientific meeting the statistical methods that he develops in an engagement. He will not publish actual data or substantive results or other information about a client's business without permission. In other words, the statistical methods belong to the statistician: the data to the client.

A statistician may at times perform a useful function by examining and reporting on a study in which he did not participate. A professional statistician will not write an opinion on another's procedures or inferences without adequate time for study and evaluation.

Supplemental Remarks

10. Necessity for the Statistician to Keep in Touch. A statistician, when he enters into a relationship to participate in a study, accepts certain responsibilities. He does not merely draft instructions and wait to be called. The people whom he serves are not statisticians, and thus cannot always know when they are in trouble. A statistician asks questions and will probe on his own account with the help of proper statistical design, to discover for himself whether the work is proceeding according to the intent of the instructions. He must expect to revise the instructions a number of times in the early stages. He will be accessible by mail, telephone, telegraph, or in person, to answer questions and to listen to suggestions.

He may, of course, arrange consultations with other statisticians on questions of theory or procedure. He may engage another statistician to take over certain duties. He may employ other skills at suitable levels to carry out independent tests or reinvestigation of certain units, to detect difficulties and departures from the prescribed procedure, or errors in transcription or calculation.

It must be firmly understood, however, that consultation or assistance is in no sense a partitioning of responsibility. The obligations of a statistician to his client, once entered into, may not be shared.

11. What Is an Engagement? Dangers of Informal Advice. It may seem at first thought that statisticians ought to be willing to give the world informally and impromptu the benefit of their knowledge and experience, without discussion or agreement concerning participation and relationships. Anyone who has received aid from a doctor of medicine who did his best without a chance to make a more thorough examination can appreciate how important the skills of a professional man can be, even under handicap.

On second thought, most statisticians can recall instances in which informal advice backfired. It is the same in any professional line. A statistician who tries to be helpful and give advice under adverse circumstances is in practice and has a client, whether he intended it so or not; and he will later find himself accountable for the advice. It is important to take special precaution under these circumstances to state the basis of understanding for any statements or recommendations, and to make clear that other conditions and circumstances could well lead to different statements.

12. When Do the Statistician's Obligations Come to a Close? A statistician should make it clear that his name may be identified with a study only as long as he is active in it and accountable for it. A statistical procedure, contrary to popular parlance, is not installed. One may install new furniture, a new carpet, or a new dean, but not a statistical procedure. Experience shows that a statistical procedure deteriorates rapidly when left completely to nonprofessional administration.

A statistician may draw up plans for a continuing study, such as for the annual inventory of materials in process in a group of manufacturing plants, or for a continuing national survey of consumers. He may nurse the job into running order, and conduct it through several performances. Experience shows, however, that if he steps out and leaves the work in nonstatistical hands, he will shortly find it to be unrecognizable. New people come on the job. They may think that they know better than their predecessor how to do the work; or they may not be aware that there ever were any rules or instructions, and make up their own.

What is even worse, perhaps, is that people that have been on a job a number of years think that they know it so well that they cannot go wrong. This type of fault will be observed, for example, in a national monthly sample in which households are to be revisited a number of times; when left entirely to nonstatistical administration, it will develop faults. Some interviewers will put down their best guesses about the family, on the basis of the preceding month, without an actual interview. They will forget the exact wording of the question, or may think that they have something better. They will become lax about calling back on people not at home. Some of them may suppose that they are following literally the intent of the instructions, when in fact (as shown by a control), through a misunderstanding, they are doing something wrong. Or they may depart wilfully, thinking that they are thereby improving the design, on the supposition that the statistician did not really understand the circumstances. A common example is to substitute an average-looking sampling unit when the sampling unit designated is obviously unusual in some respect [1].

In the weighing and testing of physical product, people will in all sincerity substitute their judgment for the use of random numbers. Administration at the top will fail to rotate areas in the manner specified. Such deterioration may be predicted with confidence unless the statistician specifies statistical controls that provide detective devices and feedback.

13. The Single Consultation. It is wise to avoid a single consultation with a commercial concern unless satisfactory agenda are prepared in advance and satisfactory arrangements made for absorbing advice offered. This requirement, along with an understanding that there will be a fee for a consultation, acts as a shield against a hapless conference which somebody calls in the hope that something may turn up. It also shows that statisticians, as professional people, although eager to teach and explain statistical methods, are not on the lookout for chances to demonstrate what they might be able to accomplish.

Moreover, what may be intended as a single consultation often ends up with a request for a memorandum. This may be very difficult to write, especially in the absence of adequate time to study the problem. The precautions of informal advice apply here.

14. The Statistician's Obligation in a Muddle. Suppose that fieldwork is under way. Then the statistician discovers that the client or duly appointed representatives have disregarded the instructions for the preparation of the frame, or for the selection of the sample, or that the fieldwork seems to be falling apart. "We went ahead and did so and so before your letter came, confident that you would approve," is a violation of relationship. That the statistician may approve the deed done does not mitigate the violation.

If it appears to the statistician that there is no chance that the study will yield results that he could take professional responsibility for, this fact must be made clear to the client, at a sufficiently high management level. It is a good idea for the statistician to explain at the outset that such situations, while extreme, have been known.

Statisticians should do all in their power to help clients to avoid such a catastrophe. There is always the possibility that a statistician may be partly to blame for not being sufficiently clear nor firm at the outset concerning his principles of participation, or for not being on hand at the right time to ask questions and to keep apprised of what is happening. Unfortunate circumstances may teach the statistician a lesson on participation.

15. Assistance in Interpretation of Nonprobability Samples. It may be humiliating, but statisticians must face the fact that many accepted laws of science have come from theory and experimentation without benefit of formal statistical design. Vaccination for prevention of smallpox is one; John Snow's discovery of the source of cholera in London is another [6]. So is the law $F = ma$ in physics; also Hooke's law, Boyle's law, Mendel's findings, Keppler's laws, Darwin's theory of evolution, the Stefan–Boltzmann law of radiation (first empirical, later established by physical theory). All this only means, as everyone knows, that there may well be a wealth of information in a nonprobability sample.

Perhaps the main contribution that the statistician can make in a nonprobability sample is to advise the experimenter against conclusions based on meaningless statistical calculations. The expert in the subject matter must take the responsibility for the effects of selectivity and confounding of treatments. The statistician may make a positive contribution by encouraging the expert in the subject matter to draw whatever conclusions he believes to be warranted, but to do so over his own signature, and not to attribute conclusions to statistical calculations or to the kind of help provided by a statistician.

16. A statistician will not agree to use of his name as advisor to a study, nor as a member of an advisory committee, unless this service carries with it explicit responsibilities for certain prescribed phases of the study.

17. A statistician may accept engagements from competitive firms. His aim is not to concentrate on the welfare of a particular client, but to raise the level of service of his profession.

18. A statistician will prescribe in every engagement whatever methods known to him seem to be most efficient and feasible under the circumstances. Thus, he may prescribe for firms that are competitive methods that are similar or even identical word for word in part or in entirety. Put another way, no client has a proprietary right in any procedures or techniques that a statistician prescribes.

19. A statistician will, to the best of his ability, at his request, lend technical assistance to another statistician. In rendering this assistance, he may provide copies of procedures that he has used, along with whatever modification seems advisable. He will not, in doing this, use confidential data.

References

[1] Bureau of the Census (1954). Measurement of Employment and Unemployment. Report of Special Advisory Committee on Employment Statistics, Washington, D.C.

[2] Bureau of the Census (1964). Evaluation and research program of the censuses of population and housing. 1960 series ER60. No. 1.

[3] Bureau of the Census (1963). The Current Population Survey and Re-interview Program. *Technical Paper No. 6.*

[4] Center for Advanced Engineering Study (1983). *Management for Quality, Productivity, and Competitive Position.* Massachusetts Institute of Technology, Cambridge, Mass.

[5] Deming, W. E. (1960). *Sample Design in Business Research.* Wiley, New York, Chap. 3.

[6] Hill, A. B. (1953). Observation and experiment. *N. Eng. J. Med.* **248,** 995–1001.

[7] Stephan, F. F. (1936). Practical problems of sampling procedure. *Amer. Sociol. Rev.,* **1,** 569–80.

Bibliography

See the following works, as well as the references just given, for more information on the topic of principles of professional statistical practice.

Brown, T. H. (1952). The statistician and his conscience. *Amer. Statist.,* **6**(1), 14–18.

Burgess, R. W. (1947). Do we need a "Bureau of Standards" for statistics? *J. Marketing,* **11,** 281–282.

Chambers, S. P. (1965). Statistics and intellectual integrity. *J. R. Statist. Soc. A,* **128,** 1–16.

Court, A. T. (1952). Standards of statistical conduct in business and in government. *Amer. Statist.,* **6,** 6–14.

Deming, W. E. (1954). On the presentation of the results of samples as legal evidence. *J. Amer. Statist. Ass.,* **49,** 814–825.

Deming, W. E. (1954). On the Contributions of Standards of Sampling to Legal Evidence and Accounting. Current Business Studies, Society of Business Advisory Professions, Graduate School of Business Administration, New York University, New York.

Eisenhart, C. (1947). The role of a statistical consultant in a research organization. *Proc. Int. Statist. Conf.,* **3,** 309–313.

Freeman, W. W. K. (1952). Discussion of Theodore Brown's paper [see Brown, 1952]. *Amer. Statist.,* **6**(1), 18–20.

Freeman, W. W. K. (1963). Training of statisticians in diplomacy to maintain their integrity. *Amer. Statist.,* **17**(5), 16–20.

Gordon, R. A. et al. (1962). Measuring Employment and Unemployment Statistics. President's Committee to Appraise Employment and Unemployment Statistics. Superintendent of Documents, Washington, D.C.

Hansen, M. H. (1952). Statistical standards and the Census. *Amer. Statist.*, **6**(1), 7–10.

Hotelling, H. et al. (1948). The teaching of statistics (a report of the Institute of Mathematical Statistics Committee on the teaching of statistics). *Ann. Math. Statist.*, **19**, 95–115.

Jensen, A. et al. (1948). The life and works of A. K. Erlang. *Trans. Danish Acad. Tech. Sci.*, No. 2 (Copenhagen).

Molina, E. C. (1913). Computation formula for the probability of an event happening at least c times in n trials. *Amer. Math. Monthly*, **20**, 190–192.

Molina, E. C. (1925). The theory of probability and some applications to engineering problems. *J. Amer. Inst. Electr. Eng.*, **44**, 1–6.

Morton, J. E. (1952). Standards of statistical conduct in business and government. *Amer. Statist.*, **6**(1), 6–7.

Shewhart, W. A. (1931). *Economic Control of Quality of Manufactured Product.* D. Van Nostrand, New York.

Shewhart, W. A. (1939). *Statistical Method from the Viewpoint of Quality Control.* The Graduate School, Department of Agriculture, Washington, D.C.

Shewhart, W. A. (1958). Nature and origin of standards of quality. *Bell Syst. Tech. J.*, **37**, 1–2.

Zirkle, C. (1954). Citation of fraudulent data. *Science*, **120**, 189–190.

Summary

This chapter presented some current thoughts on statistical studies and principles for statistical practice. These thoughts were related to the quantification of uncertainty in enumerative and analytic studies, and, in particular, to the impossibility of using enumerative methods to calculate the probabilities of the two types of errors that can occur in analytic studies. The discussion of the design of an analytic study included the importance of sequential building of knowledge to the development of an ability to predict the future with a comfortable degree of belief, the importance of testing a plan of action over a wide range of conditions to improve the degree of belief in a prediction based on the plan, and the importance of a judgment sample to increasing the degree of belief in a prediction from an analytic study. Some issues concerning the analysis of data from analytic studies include the importance of a stable process in gaining new knowledge about a process and the significance of graphical techniques for gaining knowledge from a process.

Finally, principles of professional practice were presented. The discussion included the statistician's job, what constitutes professional practice for a statistician, some limitations in application of statistical theory, contributions of statistical theory to subject matter, using statistical theory as a basis for the division of responsibilities between a statistician and client, the statistician's responsibility to assist the client in understanding the results of a statistical study, other responsibilities of the statistician and client, and the statistician's report or testimony.

Endnotes

1. This material was presented at a seminar entitled "Deming Seminar for Statisticians," March 28–30, 1988, New York University, New York.

2. R Moen, T Nolan, and L Provost, *Improving Quality through Planned Experimentation* (New York: McGraw-Hill, 1991), pp. 53–61.

3. W E Deming, "On Probability as a Basis for Action," *The American Statistician 29*, no. 4 (November 1975), pp. 146–52.

4. W E Deming, *Out of the Crisis* (Cambridge, MA: Massachusetts Institute of Technology, Center for Advanced Engineering Study, 1986), p. 132.

5. W E Deming, "Principles of Professional Statistical Practice." Reprinted with permission from Kotz-Johnson, *Encyclopedia of Statistical Science 7* (New York: John Wiley & Sons, 1986), pp. 184–93.

Appendix: Statistical Tables

TABLE 1 Control Chart Constants

Number of Observations in Subgroup, n	A_2	A_3	A_6	B_3	B_4	c_4	d_2	d_3	d_4	D_3	D_4	D_5	D_6	E_2
2	1.880	2.659		0.000	3.267	0.7979	1.128	0.853	0.954	0.000	3.267	0.000	3.865	2.660
3	1.023	1.954	1.187	0.000	2.568	0.8862	1.693	0.888	1.588	0.000	2.574	0.000	2.745	1.772
4	0.729	1.628		0.000	2.266	0.9213	2.059	0.880	1.978	0.000	2.282	0.000	2.375	1.457
5	0.577	1.427	0.691	0.000	2.089	0.9400	2.326	0.864	2.257	0.000	2.114	0.000	2.179	1.290
6	0.483	1.287		0.030	1.970	0.9515	2.534	0.848	2.472	0.000	2.004	0.000	2.055	1.184
7	0.419	1.182	0.509	0.118	1.882	0.9594	2.704	0.833	2.645	0.076	1.924	0.078	1.967	1.109
8	0.373	1.099		0.185	1.815	0.9650	2.847	0.820	2.791	0.136	1.864	0.139	1.901	1.054
9	0.337	1.032	0.412	0.239	1.761	0.9693	2.970	0.808	2.915	0.184	1.816	0.187	1.850	1.010
10	0.308	0.975		0.284	1.716	0.9727	3.078	0.797	3.024	0.223	1.777	0.227	1.809	0.975
11	0.285	0.927	0.350	0.321	1.679	0.9754	3.173	0.787	3.121	0.256	1.744			
12	0.266	0.886		0.354	1.646	0.9776	3.258	0.778	3.207	0.283	1.717			
13	0.249	0.850		0.382	1.618	0.9794	3.336	0.770	3.285	0.307	1.693			
14	0.235	0.817		0.406	1.594	0.9810	3.407	0.762	3.356	0.328	1.672			
15	0.223	0.789		0.428	1.572	0.9823	3.472	0.755	3.422	0.347	1.653			
16	0.212	0.763		0.448	1.552	0.9835	3.532	0.749	3.482	0.363	1.637			
17	0.203	0.739		0.466	1.534	0.9845	3.588	0.743	3.538	0.378	1.622			
18	0.194	0.718		0.482	1.518	0.9854	3.640	0.738	3.591	0.391	1.608			
19	0.187	0.698		0.497	1.503	0.9862	3.689	0.733	3.640	0.403	1.597			
20	0.180	0.680		0.510	1.490	0.9869	3.735	0.729	3.686	0.415	1.585			
21	0.173	0.663		0.523	1.477	0.9876	3.778	0.724	3.730	0.425	1.575			
22	0.167	0.647		0.534	1.466	0.9882	3.819	0.720	3.771	0.434	1.566			
23	0.162	0.633		0.545	1.455	0.9887	3.858	0.716	3.811	0.443	1.557			
24	0.157	0.619		0.555	1.445	0.9892	3.895	0.712	3.847	0.451	1.548			
25	0.153	0.606		0.565	1.435	0.9896	3.931	0.709	3.883	0.459	1.541			
More than 25	$3/\sqrt{n}$			$1 - 3/\sqrt{2n}$	$1 + 3/\sqrt{2n}$									

SOURCE: A_2, A_3, B_3, B_4, c_4, d_2, d_3, D_3, D_4, E_2 reprinted with permission from ASTM Manual on the Presentation of Data and Control Chart Analysis (Philadelphia, Penn.: ASTM, 1976), pp. 134–36. Copyright ASTM.

A_6, d_4, D_5, D_6 reprinted with permission from D. J. Wheeler and D. S. Chambers, Understanding Statistical Process Control (Knoxville: Statistical Process Controls, Inc., 1986), pp. 307, 309, 312.

TABLE 2   2,500 Four-Digit Random Numbers

5347	8111	9803	1221	5952	4023	4057	3935	4321	6925
9734	7032	5811	9196	2624	4464	8328	9739	9282	7757
6602	3827	7452	7111	8489	1395	9889	9231	6578	5964
9977	7572	0317	4311	8308	8198	1453	2616	2489	2055
3017	4897	9215	3841	4243	2663	8390	4472	6921	6911
8187	8333	1498	9993	1321	3017	4796	9379	8669	9885
1983	9063	7186	9505	5553	6090	8410	5534	4847	6379
0933	3343	5386	5276	1880	2582	9619	6651	7831	9701
3115	5829	4082	4133	2109	9388	4919	4487	4718	8142
6761	5251	0303	8169	1710	6498	6083	8531	4781	0807
6194	4879	1160	8304	2225	1183	0434	9554	2036	5593
0481	6489	9634	7906	2699	4396	6348	9357	8075	9658
0576	3960	5614	2551	8615	7865	0218	2971	0433	1567
7326	5687	4079	1394	9628	9018	4711	6680	6184	4468
5490	0997	7658	0264	3579	4453	6442	3544	2831	9900
4258	3633	6006	0404	2967	1634	4859	2554	6317	7522
2726	2740	9752	2333	3645	3369	2367	4588	4151	0475
4984	1144	6668	3605	3200	7860	3692	5996	6819	6258
2931	4046	2707	6923	5142	5851	4992	0390	2659	3306
3046	2785	6779	1683	7427	0579	0290	6349	0078	3509
2870	8408	6553	4425	3386	8253	9839	2638	0283	3683
1318	5065	9487	2825	7854	5528	3359	6196	5172	1421
6079	7663	3015	4029	9947	2833	1536	4248	6031	4277
1348	4691	6468	0741	7784	0190	4779	6579	4423	7723
3491	9450	3937	3418	5750	2251	0406	9451	4461	1048
2810	0481	8517	8649	3569	0348	5731	6317	7190	7118
5923	4502	0117	0884	8192	7149	9540	3404	0485	6591
8743	8275	7109	3683	5358	2598	4600	4284	8168	2145
2904	0130	5534	6573	7871	4364	4624	5320	9486	4871
6203	7188	9450	1526	6143	1036	4205	6825	1438	7943
3885	8004	5997	7336	5287	4767	4102	8229	2643	8737
4066	4332	8737	8641	9584	2559	5413	9418	4230	0736
4058	9008	3772	0866	3725	2031	5331	5098	3290	3209
7823	8655	5027	2043	0024	0230	7102	4993	2324	0086
9824	6747	7145	6954	0116	0332	6701	9254	9797	5272
6997	7855	6543	3262	2831	6181	1459	7972	5569	9134
3984	2307	4081	0371	2189	9635	9680	2459	2620	2600
6288	8727	9989	9996	3437	4255	1167	9960	9801	4886
5613	6492	2945	5296	8662	6242	3106	7618	9531	3926
9080	5602	4899	6456	6746	6018	1297	0384	6258	9385
0966	4467	7476	3335	6730	8054	9765	1134	7877	4501
3475	5040	7663	1276	3222	3454	1810	5351	1452	7212
1215	7332	7419	2666	7808	5363	5230	0000	0570	6353
6938	0773	9445	7642	1612	0930	6741	6858	8793	3884
9335	6456	4376	4504	4493	6997	1696	0827	6775	6029
3887	3554	9956	8540	0491	6254	7840	0101	8618	2207
5831	6029	7239	6966	1247	9305	0205	2980	6364	1279
8356	1022	9947	7472	2207	1023	2157	2032	2131	5712
2806	9115	4056	3370	6451	0706	6437	2633	7965	3114
0573	7555	9316	8092	5587	5410	3480	8315	0453	8136

SOURCE: Reprinted from *A Million Random Digits with 100,000 Normal Deviates* by the RAND Corporation (New York: The Free Press, 1955). Copyright 1955 and 1983 by The RAND Corporation.

TABLE 2 **2,500 Four-Digit Random Numbers** (*continued*)

2668	7422	4354	4569	9446	8212	3737	2396	6892	3766
6067	7516	2451	1510	0201	1437	6518	1063	6442	6674
4541	9863	8312	9855	0995	6025	4207	4093	9799	9308
6987	4802	8975	2847	4413	5997	9106	2876	8596	7717
0376	8636	9953	4418	2388	8997	1196	5158	1803	5623
8468	5763	3232	1986	7134	4200	9699	8437	2799	2145
9151	4967	3255	8518	2802	8815	6289	9549	2942	3813
1073	4930	1830	2224	2246	1000	9315	6698	4491	3046
5487	1967	5836	2090	3832	0002	9844	3742	2289	3763
4896	4957	6536	7430	6208	3929	1030	2317	7421	3227
9143	7911	0368	0541	2302	5473	9155	0625	1870	1890
9256	2956	4747	6280	7342	0453	8639	1216	5964	9772
4173	1219	7744	9241	6354	4211	8497	1245	3313	4846
2525	7811	5417	7824	0922	8752	3537	9069	5417	0856
9165	1156	6603	2852	8370	0995	7661	8811	7835	5087
0014	8474	6322	5053	5015	6043	0482	4957	8904	1616
5325	7320	8406	5962	6100	3854	0575	0617	8019	2646
2558	1748	5671	4974	7073	3273	6036	1410	5257	3939
0117	1218	0688	2756	7545	5426	3856	8905	9691	8890
8353	1554	4083	2029	8857	4781	9654	7946	7866	2535
1990	9886	3280	6109	9158	3034	8490	6404	6775	8763
9651	7870	2555	3518	2906	4900	2984	6894	5050	4586
9941	5617	1984	2435	5184	0379	7212	5795	0836	4319
7769	5785	9321	2734	2890	3105	6581	2163	4938	7540
3224	8379	9952	0515	2724	4826	6215	6246	9704	1651
1287	7275	6646	1378	6433	0005	7332	0392	1319	1946
6389	4191	4548	5546	6651	8248	7469	0786	0972	7649
1625	4327	2654	4129	3509	3217	7062	6640	0105	4422
7555	3020	4181	7498	4022	9122	6423	7301	8310	9204
4177	1844	3468	1389	3884	6900	1036	8412	0881	6678
0927	0124	8176	0680	1056	1008	1748	0547	8227	0690
8505	1781	7155	3635	9751	5414	5113	8316	2737	6860
8022	8757	6275	1485	3635	2330	7045	2106	6381	2986
8390	8802	5674	2559	7934	4788	7791	5202	8430	0289
3630	5783	7762	0223	5328	7731	4010	3845	9221	5427
9154	6388	6053	9633	2080	7269	0894	0287	7489	2259
1441	3381	7823	8767	9647	4445	2509	2929	5067	0779
8246	0778	0993	6687	7212	9968	8432	1453	0841	4595
2730	3984	0563	9636	7202	0127	9283	4009	3177	4182
9196	8276	0233	0879	3385	2184	1739	5375	5807	4849
5928	9610	9161	0748	3794	9683	1544	1209	3669	5831
1042	9600	7122	2135	7868	5596	3551	9480	2342	0449
6552	4103	7957	0510	5958	0211	3344	5678	1840	3627
5968	4307	9327	3197	0876	8480	5066	1852	8323	5060
4445	1018	4356	4653	9302	0761	1291	6093	5340	1840
8727	8201	5980	7859	6055	1403	1209	9547	4273	0857
9415	9311	4996	2775	8509	7767	6930	6632	7781	2279
2648	7639	9128	0341	6875	8957	6646	9783	6668	0317
3707	3454	8829	6863	1297	5089	1002	2722	0578	7753
8383	8957	5595	9395	3036	4767	8300	3505	0710	6307

TABLE 2 **2,500 Four-Digit Random Numbers** (*continued*)

5503	8121	9056	8194	1124	8451	1228	8986	0076	7615
2552	9953	4323	4878	4922	0696	3156	2145	8819	0631
8542	7274	9724	6638	0013	0566	9644	3738	5767	2791
6121	4839	4734	3041	3939	9136	5620	7920	0533	3119
2023	0314	5885	1165	2841	1282	5893	3050	6598	2667
9577	8320	5614	5595	8978	6442	0844	4570	8036	6026
0760	1734	0114	8330	9695	6502	3171	8901	7955	4975
0064	1745	7874	3900	3602	9880	7266	5448	6826	3882
6295	8316	6150	3155	8059	4789	7236	7272	0839	3367
7935	1027	8193	2634	0806	6781	0665	8791	7416	8551
4833	6983	5904	8217	9201	5844	6959	5620	9570	8621
0584	0843	7983	5095	3205	3291	1584	1391	4136	8011
2585	0220	0730	5994	7138	7615	1126	3878	6154	2260
2527	1615	8232	7071	9808	3863	9195	4990	7625	3397
7300	2905	1760	4929	4767	9044	6891	0567	2382	8489
8131	9443	2266	0658	3814	0014	1749	5111	6145	6579
1002	4471	5983	8072	6371	6788	2510	4534	5574	6761
8467	5280	8912	3769	2089	8233	2262	0614	0577	0354
2929	5816	2185	3373	9405	8880	5460	0038	6634	6923
5177	9407	7063	4128	9058	8768	1396	5562	2367	3510
4216	5625	6077	5167	3603	7727	8521	1481	9075	2267
7835	6704	2249	5152	3116	3045	2760	4442	9638	2677
0955	5134	3386	8901	7341	8153	7739	3044	9774	1815
1577	6312	3484	0566	0615	4897	5569	6181	9176	2082
1323	9905	9375	3673	4428	4432	1572	3750	4726	1333
5058	0357	3847	7323	6761	7278	7817	1871	9909	6411
9948	5733	1063	7490	9067	1964	6990	6095	1796	3721
5467	3952	7378	4886	6983	6279	6520	6918	0557	7474
9934	7154	1024	7603	3170	7686	8890	6957	2764	0033
3549	4023	3486	5535	1284	6809	5264	3273	6701	4678
9817	2538	0384	2392	4795	1035	7011	1117	6329	9990
0267	8615	5686	0259	0164	4220	7995	3776	8234	7195
3693	4287	8163	7995	0706	4162	9680	9238	8886	6858
5685	1277	2430	7366	8426	2466	1668	0223	6602	6413
0546	2889	1427	2377	8859	1708	3388	8878	3901	5711
1502	2023	6338	7112	0662	0741	9498	3232	7942	7038
9561	0803	8146	9106	8885	5658	0122	2809	1972	7146
0902	4037	0573	5512	7429	4919	3166	4260	3036	9642
8143	9995	5246	6766	9732	6980	2124	6592	1262	9289
2143	5933	5862	9482	6548	0964	4101	8510	1611	3207
9583	7614	1163	8028	1778	9793	1282	7389	6600	2752
9981	4463	4374	9979	8682	1211	3170	0502	2815	0420
7721	3114	5054	1160	5093	0249	0918	9587	8584	7195
1326	0260	7983	6605	8027	0853	2867	3753	7053	8235
4428	7173	2662	5469	1490	5213	8111	7454	7885	3199
7052	4595	7963	5737	0505	3196	3337	1323	8566	8661
8838	1122	2508	7146	0981	4600	1906	6898	1831	7417
8316	7399	1720	7944	6409	4979	1193	4486	8697	3453
5021	7172	3385	4514	0569	2993	1282	0159	0845	5282
9768	2934	6774	8064	1362	2394	4939	8368	3730	9535

TABLE 2   2,500 Four-Digit Random Numbers (*continued*)

1236	2389	3150	9072	1871	8914	5859	9942	2284	0826
3889	3023	3423	2257	7442	2273	2693	4060	1078	8012
8078	5541	3977	9331	1827	2114	5208	7809	8563	8114
0239	7758	0885	2356	3354	4579	1097	4472	2478	0969
7372	7018	6911	7188	8014	7287	3898	2340	6395	4475
6138	1722	5523	1896	3900	9350	1827	4981	5280	6967
3916	4428	1497	9749	2597	3360	6014	3003	7767	4929
8090	7448	3988	1988	3731	0420	4967	3959	0105	4399
0905	6567	6366	3403	0657	8783	2812	4888	5048	5573
3342	2422	3204	6008	2041	8504	5357	3255	6409	5232
7265	6947	7364	7153	5545	1957	1555	2057	1212	5003
0414	3209	8358	6182	3548	3273	6340	9149	3719	0276
8522	1419	5221	6074	2441	5785	3188	5126	8229	7355
5488	0357	9167	5950	0861	3379	2901	8519	6226	2868
3325	5151	8203	4523	3935	3322	5946	6554	7680	1698
7597	1595	3240	8208	0221	5714	3352	4719	9452	7325
9063	7531	3538	3445	4924	1146	2510	7148	8988	9970
6506	1549	9334	3356	1942	6682	0304	9736	0815	4748
6442	0742	8223	9781	3957	0776	6584	2998	1553	9011
2717	1738	7696	7511	4558	9990	4716	5536	2566	2540
3221	3009	8727	5689	1562	3259	8066	0808	1942	8071
5420	5804	7235	8982	0270	1681	8998	3738	4403	5936
5928	6696	8484	7154	6755	3386	8301	6621	6937	2390
8387	5816	0122	9555	2219	6590	3878	0135	4748	2817
8331	5708	0336	8001	3960	4069	5643	6405	0249	5088
6454	2950	1335	7864	9262	1935	6047	5733	5213	0711
3926	0007	5548	0152	7656	2257	2032	8462	3018	4390
2976	0567	2819	6551	1195	7859	6390	2134	1921	9028
0631	0299	0146	2773	9028	1769	6451	3955	3469	0321
9754	4760	5765	5910	2185	4444	0797	5429	8467	7875
8296	8571	1161	9772	5351	5378	9894	3840	7093	1131
7687	3472	1252	9064	1692	1366	1742	8448	6830	8524
8739	7888	8723	9208	9563	6684	2290	6498	8695	5470
7404	1273	5961	3369	1259	4489	6798	7297	8979	1058
4789	4141	6643	7004	5079	4592	9656	6795	5636	4472
8777	7169	6414	5436	9211	3403	5906	6205	6204	3352
9697	6314	7221	8004	1199	4769	9562	7299	2904	8589
4382	1328	7781	8169	2993	7075	0202	3237	0055	8668
5720	8396	4009	3923	6595	5991	9141	5557	8842	4557
4906	7217	8093	0601	9032	6368	0793	9958	4901	2645
9425	8427	9579	1347	8013	2633	5516	7341	4076	4517
6814	8138	8238	1867	4045	9282	3004	3741	4342	4513
1220	9780	3361	2886	4164	1673	8886	3263	4198	8461
8831	8970	2611	1241	1943	6566	6098	5976	1141	1825
5672	8035	2961	6305	1525	4468	6468	4235	5102	7768
0713	1232	0107	1930	8704	5892	2845	8106	9397	6665
2118	6455	5561	3608	2433	8439	1602	1220	7755	7566
0215	1225	8873	4391	0365	2109	6080	6324	2684	3581
9095	8523	3277	0730	3618	4742	1968	3318	4138	0324
8010	9130	1285	4129	0032	1501	1957	9113	1272	9260

TABLE 2 2,500 Four-Digit Random Numbers (*concluded*)

9263	7824	1926	9545	5349	2389	3770	7986	7647	6641
7944	7873	7154	4484	2610	6731	0070	3498	6675	9972
5965	7196	2738	5000	0535	9403	2928	1854	5242	0608
3152	4958	7661	3978	1353	4808	5948	6068	8467	5301
0634	7693	9037	5139	5588	7101	0920	7915	2444	3024
2870	5170	9445	4839	7378	0643	8664	6923	5766	8018
6810	8926	9473	9576	7502	4846	6554	9658	1891	1639
9993	9070	9362	6633	3339	9526	9534	5176	9161	3323
9154	7319	3444	6351	8383	9941	5882	4045	6926	4856
4210	0278	7392	5629	7267	1224	2527	3667	2131	7576
1713	2758	2529	2838	5135	6166	3789	0536	4414	4267
2829	1428	5452	2161	9532	3817	6057	0808	9499	7846
0933	5671	5133	0628	7534	0881	8271	5739	2525	3033
3129	0420	9371	5128	0575	7939	8739	5177	3307	9706
3614	1556	2759	4208	9928	5964	1522	9607	0996	0537
2955	1843	1363	0552	0279	8101	4902	7903	5091	0939
2350	2264	6308	0819	8942	6780	5513	5470	3294	6452
5788	8584	6796	0783	1131	0154	4853	1714	0855	6745
5533	7126	8847	0433	6391	3639	1119	9247	7054	2977
1008	1007	5598	6468	6823	2046	8938	9380	0079	9594
3410	8127	6609	8887	3781	7214	6714	5078	2138	1670
5336	4494	6043	2283	1413	9659	2329	5620	9267	1592
8297	6615	8473	1943	5579	6922	2866	1367	9931	7687
5482	8467	2289	0809	1432	8703	4289	2112	3071	4848
2546	5909	2743	8942	8075	8992	1909	6773	8036	0879
6760	6021	4147	8495	4013	0254	0957	4568	5016	1560
4492	7092	6129	5113	4759	8673	3556	7664	1821	6344
3317	3097	9813	9582	4978	1330	3608	8076	3398	6862
8468	8544	0620	1765	5133	0287	3501	6757	6157	2074
7188	5645	3656	0939	9695	3550	1755	3521	6910	0167
0047	0222	7472	1472	4021	2135	0859	4562	8398	6374
2599	3888	6836	5956	4127	6974	4070	3799	0343	1887
9288	5317	9919	9380	5698	5308	1530	5052	5590	4302
2513	2681	0709	1567	6068	0441	2450	3789	6718	6282
8463	7188	1299	8302	8248	9033	9195	7457	0353	9012
3400	9232	1279	6145	4812	7427	2836	6656	7522	3590
5377	4574	0573	8616	4276	7017	9731	7389	8860	1999
5931	9788	7280	5496	6085	1193	3526	7160	5557	6771
2047	6655	5070	2699	0985	5259	1406	3021	1989	1929
8618	8493	2545	2604	0222	5201	2182	5059	5167	6541
2145	6800	7271	4026	6128	1317	6381	4897	5173	5411
9806	6837	8008	2413	7235	9542	1180	2974	8164	8661
0178	6442	1443	9457	7515	9457	6139	9619	0322	3225
6246	0484	4327	6870	0127	0543	2295	1894	9905	4169
9432	3108	8415	9293	9998	8950	9158	0280	6947	6827
0579	4398	2157	0990	7022	1979	5157	3643	3349	7988
1039	1428	5218	0972	2578	3856	5479	0489	5901	8925
3517	5698	2554	5973	6471	5263	3110	6238	4948	1140
2563	8961	7588	9825	0212	7209	5718	5588	0932	7346
1646	4828	9425	4577	4515	6886	1138	1178	2269	4198

References

Akao, Y. *Quality Function Deployment: Integrating Customer Requirements into Product Design*. Cambridge, Mass.: Productivity Press, 1990.

AT&T. *Statistical Quality Control Handbook*. Indianapolis, Ind.: AT&T, 1956. 10th printing, May 1984.

J. F. Beardsley & Associates, International, Inc. *Quality Circles: Member Manual*. San Jose, Calif.: J. F. Beardsley, 1977.

Boardman, T. J., and H. Iyer. *The Funnel*. Fort Collins, Colo.: Colorado State University Press, 1986.

Deming, W. E. *Some Theory of Sampling*. New York: John Wiley & Sons, 1950.

_____. "On the Distinction between Enumerative and Analytic Surveys." *Journal of the American Statistical Association* 48, 1953, pp. 244–55.

_____. "On Some Statistical Aids toward Economic Production." *Interfaces* 5, August 1975, pp. 1–15.

_____. "On Probability as a Basis for Action." *The American Statistician* 29(4), November 1975, pp. 146–52.

_____. "On the Use of Judgment-Samples." *Reports of Statistical Applications* 23, March 1976, pp. 25–31.

_____. *Quality, Productivity, and Competitive Position*. Cambridge, Mass.: MIT Center for Advanced Engineering Study, 1982.

_____. *Out of the Crisis*. Cambridge, Mass.: MIT Center for Advanced Engineering Study, 1986.

_____. "Principles of Professional Statistical Practice." *Encyclopedia of Statistical Sciences* 7. Edited by Kotz-Johnson. New York: John Wiley & Sons, 1986.

_____. *The New Economics for Industry, Government, Education*. Cambridge, Mass.: MIT Center for Advanced Engineering Study, 1993.

Duncan, A. *Quality Control and Industrial Statistics*. 5th ed. Homewood, Ill.: Richard D. Irwin, 1986.

Eureka, W., and N. Ryan. *The Customer Driven Company: Managerial Perspectives on QFD.* Dearborn, Mich.: SAI Press, 1988.

Feigenbaum, A. V. "Total Quality Control." *Harvard Business Review,* November 1956.

Fitzgerald, J. M., and A. F. Fitzgerald. *Fundamentals of Systems Analysis.* New York: John Wiley & Sons, 1973.

Florida Power and Light. *FPL's Total Quality Management.* Miami, FL.

Ford Motor Company. "Statistical Process Control Case Study." *Introduction to Ford's Operating Philosophy and Principles and Statistical Management Methods—Participant Notebook,* September 1983, pp 7.E.9.–7.E.18.

————. *Continuing Process Control and Process Capability Improvement,* February 1984.

Gitlow, H. "Definition of Quality." *Proceedings—Case Study Seminar—Dr. Deming's Management Methods: How They Are Being Implemented in the U.S. and Abroad.* Andover, Mass.: G.O.A.L., November 1984, pp. 4–18.

———— and S. Gitlow. *The Deming Guide to Quality and Competitive Position.* Englewood Cliffs, N.J.: Prentice-Hall, 1987.

———— and S. Gitlow. *Total Quality Management in Action.* Englewood Cliffs, NJ: Prentice-Hall, 1994.

———— and P. Hertz. "Product Defects and Productivity." *Harvard Business Review,* September–October 1983, pp. 131–41.

————; K. Kang; and S. Kellogg. "Process Tampering: An Analysis of On/Off Deadband Process Controlling." *Quality Engineering* 5, no. 2, 1992–93, pp. 239–310.

———— and R. Oppenheim. *Stat City: Understanding Statistics Through Realistic Applications.* 2nd ed. Homewood, Ill.: Richard D. Irwin, 1986.

Golomski, W. A. "Quality Control—History in the Making." *Quality Progress,* July 1976, pp. 16–18.

Grant, E. L., and R. S. Leavenworth. *Statistical Quality Control.* 5th ed. New York: McGraw-Hill, 1980.

Harrington, H. J. "Quality's Footprints in Time." IBM Technical Report TR 02.1064. San Jose, Calif., September 20, 1983.

Hauser, J., and D. Clausing. "House of Quality." *Harvard Business Review,* May–June 1988, pp. 63–73.

Imai, M. *KAIZEN—The Keys to Japan's Competitive Success.* New York: Random House, 1986.

Ishikawa, K. *Guide to Quality Control.* Tokyo: Asian Productivity Organization, 1976. 11th printing, 1983. (Available through UNIPUB, Box 433, Murray Hill Station, New York, N.Y., 10157.)

————and D. Lu. *What Is Total Quality Control? The Japanese Way.* Englewood Cliffs, N.J.: Prentice Hall, 1985.

Joiner, B. *Fourth Generation Management.* New York: McGraw-Hill, 1994.

Juran, J. *Quality Control Handbook.* 3rd ed. New York: McGraw-Hill, 1979.

Kackar, R. "Taguchi's Quality Philosophy: Analysis and Commentary." *Quality Progress,* December 1986, pp. 21–29.

Kane, E. J. "IBM's Quality Focus on the Business Process." *Quality Progress,* April 1986, pp. 26–33.

Kane, V. "Process Capability Indices." *Journal of Quality Technology* 18, January 1986, pp. 41–52.

Kano, N., F. Seraku, F. Takahashi, and S. Tsuji. "Attractive Quality versus Must Be Quality." *Hinshitsu* (Quality) 14, no. 2, 1984, pp. 39–48.

King, B. *Better Design in Half the Time.* Metheun, Mass.: GOAL/QPC, 1987.

Melan, E. H. "Process Management in Service and Administrative Operations." *Quality Progress,* June 1985, pp. 52–59.

Mendenhall, W., L. Ott, and R. Schaeffer. *Elementary Survey Sampling.* Belmont, Calif.: Duxbury, 1993.

Moen, R., T. Nolan, and L. Provost, *Improving Quality Through Planned Experimentation.* New York: McGraw-Hill, Inc., 1991.

Mood, A. M. "On the Dependence of Sampling Inspection Plans under Population Distributions." *Annals of Mathematical Statistics* 14, 1943, pp. 415–25.

Neave, H. *The Deming Dimension.* Knoxville, Tenn.: SPC Press, 1990.

Nolan, T. W. "Analytic Studies." Working paper, Associates in Process Management, 1988.

Orsini, J. "Simple Rule to Reduce Total Cost of Inspection and Correction of Product in State of Chaos." Ph.D. dissertation, Graduate School of Business Administration, New York University, 1982.

Papadakis, G. P. "The Deming Inspection Criteria for Choosing Zero or 100 Percent Inspection." *Journal of Quality Technology* 17(3), July 1985, pp. 121–27.

Rehg, V. *Quality Circle Manual for Coordinators and Leaders.* Wright Patterson AFB, Ohio.

Rice, W. B. *Control Charts in Factory Management.* New York: John Wiley & Sons, 1947.

Scherkenbach, W. *The Deming Route to Quality and Productivity: Road Maps and Roadblocks.* Washington, D.C.: Ceepress Books, 1986.

————. *Deming's Road to Continual Improvement.* Knoxville, Tenn.: SPC Press, 1991.

Scholtes, P. *The Deming User's Manual—Parts 1 and 2.* Madison, Wis.: Joiner Associates, 1988.

Shewhart, W. A. *Economic Control of Quality of Manufactured Product.* New York: D. Van Nostrand Co., Inc., 1931.

Silver, G. A., and J. B. Silver. *Introduction to Systems Analysis.* Englewood Cliffs, N.J.: Prentice-Hall, 1976.

Taguchi, G., and Y. Wu. *Introduction to Off-Line Quality Control.* Nagoya, Japan: Central Japan Quality Control Association, 1980.

Wheeler, D. J., and D. S. Chambers. *Understanding Statistical Process Control.* Knoxville, Tenn.: Statistical Process Controls, Inc., 1986.

————. *Understanding Statistical Process Control,* 2nd Edition. Knoxville, Tenn.: Statistical Process Controls, Inc., 1992.

Index